A New Unifying Biparametric Nomenclature that Spans all of Chemistry

The science of incorporating daily over 2,000 new names to a base of over 42 million compounds while still maintaining order

A New Unifying Biparametric Nomenclature that Spans all of Chemistry

The science of incorporating daily over 2,000 new names to a base of over 42 million compounds while still maintaining order

Seymour B. Elk

Elk Technical Associates
New Milford, New Jersey
U.S.A.

2004

ELSEVIER

**Amsterdam - Boston - Heidelberg - London - New York - Oxford
Paris - San Diego - San Francisco - Singapore - Sydney - Tokyo**

ELSEVIER B.V.	ELSEVIER Inc.	ELSEVIER Ltd	ELSEVIER Ltd
Sara Burgerhartstraat 25	525 B Street	The Boulevard	84 Theobalds Road
P.O. Box 211, 1000 AE	Suite 1900, San Diego	Langford Lane, Kidlington,	London WC1X 8RR
Amsterdam, The Netherlands	CA 92101-4495, USA	Oxford OX5 1GB, UK	UK

First edition 2004

Library of Congress Cataloging in Publication Data
A catalog record is available from the Library of Congress.

British Library Cataloguing in Publication Data
A catalogue record is available from the British Library.

ISBN: 0-444-51685-9

∞ The paper used in this publication meets the requirements of ANSI/NISO Z39.48-1992 (Permanence of Paper).
Printed in The Netherlands.

Preface

As a byproduct of historical development, there are different, unrelated systems of nomenclature for "inorganic chemistry", vs. "organic chemistry", vs. "polymer chemistry", vs. "natural products chemistry", vs. etc. With each new discovery in the laboratory, as well as each new theoretical proposal for a chemical, the lines that traditionally have separated these "distinct" subsets of matter continually grow more blurred. This lack of uniformity in characterizing and naming chemicals increases the communication difficulties between differently trained chemists, as well as other scientists, and greatly impedes progress. With the set of known chemicals numbering over 42,000,000 (in Chemical Abstracts' data base) and continually growing (about 2,000 new additions every day), the desirability for a unified system for naming all chemicals simultaneously grows. Moreover, in order to meet the requirements of disparate groups of scientists, and of society in general, the name assigned to a given chemical should not only uniquely describe that substance, but also should be a part of a readily recognizable order for the entire field. For these purposes, a topology-based "bi-parametric" system of nomenclature is herein proposed. Individual bonds between each pair of adjacent atoms are integrated directly into the nomenclature in a systematic manner, in contradistinction to the present collage of mostly add-on prefixes and suffixes. The foundation upon which this system is built is the synergy that exists between the name assigned and the geometrical structure of the relevant "entity" (molecule, ion, or monomer). Major advantages of the proposed nomenclature include:

(1) Treating chemistry as a unified science, for which there is a comprehensive system of "canonical" names that encompasses each of the historically distinct "fiefdoms" which had evolved their own, often incompatible, rules for taxonomy and nomenclature;

(2) Recognizing the obsolescence of a two-dimensional world view of chemistry, and of integrating the influence of the third dimension directly into the nomenclature;

(3) Providing a framework in which newly formulated compositions of matter can be canonically named within the system, as well as providing a means for expanding the system when new, unanticipated forms are discovered in the laboratory or are proposed in the literature;

(4) Eliminating non-equivalent meanings and symbols for what should be identical terms in the historically evolved, but illogically separated, subsystems of nomenclature that are endemic today;
(5) Correcting inconsistencies, such as prescribing the wrong bond order between atoms in some molecules, as well as assigning ambiguous names in others;
(6) Eliminating the reliance on historically evolved tables and arcane rules for encoding and decoding these tables;
(7) Discontinuing the unwarranted allocation of precision to empirical concepts;
(8) Segregating various topological concepts from metric ones that have been illogically merged;
(9) Assigning a single unambiguous canonical name to both forms of a tautomer. This is notwithstanding that distinct, isolatable entities do not exist.

At this time it should be noted that in the process of creating such a unified nomenclature, there is the need to re-examine and occasionally to reformulate the geometrical foundations upon which the present understanding of chemistry is based. This sometimes means viewing from different perspectives some of the "elementary" physics that underlie chemical taxonomy.

The underlying principle behind most of modern chemical nomenclature lies in the naming of a presumed geometrical arrangement of relevant chemical moieties (atoms and bonds). The more accurate the geometrical description, the more useful the nomenclature will be. Consequently, as new advances in understanding both the geometry and the chemistry of molecules, ions, crystals, polymers, etc. evolve, simultaneously so should the means of naming them. In other words, there is the need for the nomenclature to be continuously updated so that it reflects the current state of knowledge.

Unlike the disjoint sets of approaches to taxonomy and nomenclature for "organic chemistry" vs. "inorganic chemistry" vs. "polymer chemistry", etc., which form the cornerstone of all of the various nomenclature systems in common usage today, a common graph theory based, bi-parametric, alternating code of atoms and bonds that is equally applicable to each of these individual domains is proposed. In this system the detailed formula will be **all** of the name that is needed. Advantages to such an approach include:

(1) A more precise correlation between the various bonding types which historically gave rise to different nomenclature schemes in the

"fiefdoms" of inorganic vs. organic chemistry. By focusing on the mathematical similarities in contrast to the chemical differences, the different perspectives that arose to describe related concepts are finessed. For example, by viewing the "inorganic" concept of chelation in terms of graph theory cycles, one can produce a fusion of the taxonomy of multi-dentate "inorganic" structures with "organic" ring structures; thereby allowing for the postulation of a common nomenclature;

(2) Replacement of the tedious system of morphemic suffixes in use in IUPAC organic nomenclature (-ane, -ene, -yne for the various bond unsaturations vs. the unrelated, but "seemingly parallel" set of suffixes that are assigned to selected functional groups: -one, -al, oic acid, etc.) by a system that has complete dichotomy between bond order and other functionality, as well as obviation of the collage of affixes endemic in IUPAC inorganic nomenclature (μ, η, κ, λ, etc.). Furthermore, in both domains, the various prefixes (bi, bis, di, etc,) that denote the number of a given kind of substituent group in a molecule are replaced by single, unambiguous numbers;

(3) Creation of a single, unified, systemic formulations for addending modules at specified locations to an evolving skeletal base; thereby replacing the tedious process of needing to consult long lists of tabulated data — much of which is based on uncoordinated self-contained systems of organization or logic which vary from one table to the next;

(4) Elimination of the dependence on the antiquated, admittedly empirical, concept of "oxidation number" in inorganic chemistry, as well as reliance on the (not admitted) topologically inappropriate concept of smallest set of smallest rings in organic chemistry — whose mathematical raison d'etre is a two dimensional world view;

(5) Creation of a new perspective for understanding molecular rearrangements, especially tautomerism. Based on the needs that arose when trying to assign canonical names to the different tautomers, a new insight has been gained that is extendable to other such phenomena.

One of the most significant changes over existing systems is the introduction of a selective use of non-integer bonds directly into the nomenclature. Not only does such an introduction subsume the underlying concepts sometimes expressed as "half-bond" (3 center 2 electron bond) structures in the boranes, as well as "bond and a half" (Robinson) ring structures in aromatic compounds, etc., but also this approach points the way to a logical system in which use of both integer and non-integer bonds become the norm, rather than the exception, for assigning a canonical name to compounds of any genre in any of the historical fiefdoms. In addition to this being a unifying factor for these hitherto disjoint domains, other benefits of this approach are the formulation of more appropriate descriptions of the bonding in:

(1) multi-atom anions, without having to resort to, what we believe is, an ill-conceived extension of Lewis structure;
(2) molecules that have an extended aromaticity, but for which the traditional single vs. double bond alternation is not evident, such as in many ring compounds containing nitrogen atoms;
(3) compounds in which selected bonds are unambiguously "fixed" to be either always single or always double, while others "resonate" between single and double bonds;
(4) tautomers, by creating an "alpha" bonded ring and assigning a name that simultaneously encompasses both relevant forms, such as: keto-enol, imine-enamine, oxime-nitroso;
(5) compounds which may be described by fractional bonds that are not half-integer, but which bear a chemical similarity to the more familiar half-integer bonds.

Similarly, various of the more esoteric organic compounds, such as the cyclophanes, as well as the many compounds that exist primarily as labile ring dimers formed by hydrogen bonds, etc. are better described by the use of non-integer bonds. Moreover, despite the nearly century and a half recognition of the major dichotomy in the chemistry of compounds that have been categorized as "aliphatic" vs. "aromatic" and the shorter time span in which chemists have been aware of aromaticity vs. anti-aromaticity, before our proposed introduction of the "beta" bond, there has been no convenient way in which these fundamental chemical differences could be finessed. In other words, by making the nomenclature more efficient, problems in the description of chemical properties that had been previously ignored were shown to have a simple solution.

Furthermore, precisely because the perspective chosen in assigning canonical names is everywhere global, in contradistinction to the nearly universal present usage of a local perspective, some other important observations are:

(1) Use of any type of Euler-polynomial based system, such as smallest set of smallest rings, is inappropriate for most fisular compounds — especially for that class of compounds which subsumes overlap compounds, paddlanes, propellanes, etc., as well as for the analogous,

but differently cataloged, inorganic compounds, such as the cryptands. Because the proposed nomenclature does not have the inherent defects endemic to such an approach, organic and inorganic compounds may be treated similarly;

(2) Much of the anticipated similarity between geometric isomers is not fulfilled. To the contrary, intra-molecular bonding is a sufficiently important attribute that various cis compounds may be viewed in the context of there existing additional "pseudo" rings that have been formed by hydrogen bonding. This is in contradistinction to the "corresponding" trans isomer, for which such bonding is not geometrically attainable. Because these isomers often exhibit vastly different chemical properties, downplaying their differences in the nomenclature is disingenuous;

(3) Inadequacies in the presently accepted geometrical vs. topological description of the boranes abound. Although the assignment of "better" canonical names to such boron compounds will not compensate for errors in their description, nevertheless, by the attempts to maintain consistency in assigning such names, the limitations of the present and the need for a new taxonomy scheme are highlighted. Note that the proposed nomenclature is sufficiently malleable to be able to assign a canonical name to whatever geometry is acceptable at the moment, based on what is observed in the laboratory. Since one is nomenclating the geometry of a model, whenever such further knowledge allows for the postulation of a better model, the nomenclature may then be modified in order to correct any deficiencies;

(4) A deeper appreciation of the field of macro-molecules, especially in the domain of polymers, is creating by examining the mathematics of an unending concatenation of congruent modules. The field commonly referred to as "polymers" is divided into those aggregations that lack the regularity to meet this mathematical ideal (herein designated as "multimers") for which a consistent descriptive nomenclature is unattainable and those that do, which retain the name "polymer". For this latter category a consistent extension of the nomenclature for finite molecules is promulgated;

(5) For the above limited field of polymers, as well as the shift in focus from source-based to structure-based, further elucidation is achieved when one is compelled to assign a canonical name that is capable of differentiating between "similar" polymers. One of the fall-outs of this is the establishment of a canonical ordering of the atoms in the

polymer that designates where that aggregation called a "monomer" should begin and end. In this manner, a consistent cataloging of polymers is achievable. A second one is the elimination of the category of syndiotacticity, replacing it with an isotacticity having a monomer of twice the former length;

(6) The evolving domain of radial, as well as linear, addition of modules to form an expanding moiety, in a manner akin to the development of polymers, referred to as "dendrimers", is examined and nomenclated;

(7) The direct inclusion of topology in the description of isomers, once a very insignificant part of chemical nomenclature, is now a factor to be reckoned with, not only for the small class traditionally referred to as "topological" (including catenanes, rotaxanes, and knots), but also as new compositions of matter, such as the endothelial fullerenes, are formulated.

TABLE OF CONTENTS

Chapter 1	INTRODUCTION	1
Chapter 2	NON-INTEGER BONDS	49
Chapter 3	OTHER SIGNIFICANT DIFFERENCES FROM EXISTING SYSTEMS	115
Chapter 4	OXIDATION NUMBERS	167
Chapter 5	THE BORANES AND RELATED ALUMINUM COMPOUNDS	180
Chapter 6	SPIRO AND RELATED COMPOUNDS	206
Chapter 7	TOPOLOGICALLY RESTRAINED COMPOUNDS	258
Chapter 8	POLYMERS	269
Chapter 9	MOLECULAR REARRANGEMENT	293
Index		i1

Chapter 1

Introduction

CHAPTER ABSTRACT:
Chemical nomenclature today lacks uniformity! In each of the historically evolved subdivisions of chemistry there are different, unrelated algorithms, which assign names to molecules, ions, and monomers. These protocols are not only independent of one another, they are, also, often incompatible. A unified system of nomenclature, which spans these subdivisions, is needed in order to be able to maintain consistency in naming diverse compositions of matter. The historical evolution of these separate, uncoordinated systems of taxonomy and nomenclature, along with the rapid growth in both the number and the variety of new chemicals that fail to fit neatly into one of these domains, has made research much more difficult. In order to remedy this situation, a re-examination and clarification of many of the terms used to describe chemical structure has been undertaken. This produces an expanded world-view that emphasizes the three dimensionality of chemical moieties, with special attention to the mathematical foundations that underlie all of chemical structure. Diverse historical perspectives that have, at times, stressed these differences, while masking the similarities among chemical has produced mutually exclusive subsets of chemistry. In place of this historical mindset comes a new perspective on the place of nomenclature in chemical thought. No longer is it just a "necessary evil" in order to be able to distinguish one chemical from another for indexing and cataloging. Instead, when closely examined, the consistency that has to be built into a system that has the capacity to describe, as well as to differentiate between, "similar" chemicals often suggests new lines of research to pursue, as well as novel formulations of matter that have not yet been discovered. Special features of this system include:

(1) An alternating "bi-parametric" listing of atoms and bonds, rather than merely naming atoms and then "addending" bonds (as an afterthought).

(2) An expanded set of standardized bonds that, as well as being applicable to all subdivisions of chemistry, produces a more accurate description of the connectivity between pairs of atoms.

(3) A complete dichotomy between bond saturation and functional groups. The practice of affixing suffixes to a "parent" stem for both of these purposes when assigning names to organic molecules is eliminated. Multi-atom functional groups in both organic and inorganic chemistry are described by listing the sequence of atoms and the connecting bonds that describes the

"constitution" of that functional group. All measures of bond saturation are described using the expanded set of bond descriptors, which includes some new standardized intermediate values and some "pseudo-integers" as well as the traditional set of small integers.

(4) A "global", rather than the presently used "local", perspective is used to assign canonical names to all chemical moieties.

(5) Recognizing the empirical nature of oxidation numbers in inorganic chemistry nomenclature, and ending the use of this antiquated concept.

(6) Replacing the different words to describe numeric prefixes by single, unambiguous integers.

Progress in chemistry has been greatly hindered because the various domains (inorganic, organic, polymer, natural products, etc.) do not use a common language. The lines of demarcation between divisions have, especially in recent years, become so blurred that new discoveries and developments are often slowed down, rather than assisted, by this compartmentalized thinking and organization. Part of the reason for this fragmentation is historical. A major feature of its predecessor, alchemy, was that names were given to the various potions for proprietary purposes. Although the main purpose in naming such a potion was advertising its magical powers, a secondary intent, almost as important in many cases, was to hide the composition of this potion from other would-be practitioners (i.e., sorcerers) [1]. Thankfully, as chemistry became less a study of the occult, and more a science, this practice was abandoned.

The first important development in forming a systematic chemical nomenclature can be traced to attempts to standardize the symbols used. Lavosier, in the last two decades of the eighteenth century, developed a system of chemical symbolism that was closely related to an algebraic language [2-3]. Simultaneously, chemists divided the set of all known chemical compounds into those that could be obtained from living organisms (henceforth called "organic") and those that could not ("inorganic"). The assumption that organic compounds originated because of some "vital force" led to a whole different set of rules (and names) for these compounds. Moreover, this partitioning of compounds into "organic" vs. "inorganic" fit conveniently with the next major development in chemical nomenclature: division, by Berzelius, of a chemical name into an electropositive and an electronegative part [4]. This binary division was well-suited for that part of chemical nomenclature referred to as "inorganic" (and is still in use today); however, it had little value in the then newly-emergent "organic" domain. Development of a "modern" organic nomenclature was not undertaken until the end of the nineteenth century, when competing national interests forced such an endeavor. The scope of this reform, however, was limited only to the sub-discipline of organic compounds.

Meanwhile, despite the objections raised by some to the observation that "organic" compounds could be prepared from "inorganic" materials, all that these critics could do was to raise the question:

> 'Does it make sense to draw such a line separating this part of chemistry from the rest?'

Then, when confronted with the unabashed answer "yes", to raise the second question:

> 'Can it be done in a logical, consistent manner?'

Unfortunately, logical consistency is seldom able to compete successfully against tradition; consequently, such objections were considered unimportant. To the contrary, the idea of subdividing compounds into "organic" vs. "inorganic" was regarded as an ***intuitively obvious*** choice. However, with the evolution of scientific thought in the late nineteenth and early twentieth centuries, especially in geometry and physics (two subjects which greatly impact the place of chemistry in the modern world), just what is "intuitively obvious" took on a new meaning. After over two millennia of unquestioning belief in the staid, old subject of geometry, the entire foundation developed by Euclid was re-examined and his "truths" downgraded from "self-evident" to only one of many ways to view the world. This renaissance, which resulted in creating first, "projective geometry", then the geometry of higher dimensional spaces, then "non-euclidean geometry" and finally much of what is now classified as "topology" including "graph theory", has had a tremendous impact on chemical taxonomy, and, consequently, on chemical nomenclature. Simultaneously, in physics, the extension of classical mechanics into the realm of the very small (quantum theory), the very large (astronomy), and the very fast (relativity) lead to the realization that chemistry is merely that branch of science associated with matter, rather than being a separate discipline unto itself. Moreover, just as the lines of demarcation between one subdivision of science and another became recognized as a matter of convenience, similarly, the boundaries that separated the historical subdivisions of chemistry can now be viewed as artificial ones, without physical significance. One of the consequences of this evolved perspective is the desirability for the formulation of an all-encompassing, systematic, standardized naming system that spans all of chemistry, rather than the present collage of unrelated nomenclatures that can be interconverted only with extreme difficulty.

Meanwhile, returning to the historical roots of chemistry, one notes that had there been serious attempts to develop such a unified nomenclature in earlier times, these would have been considered, if not absolutely impossible, then certainly highly impractical. Due to competing national interests and egos, the far less daunting task of establishing a generally accepted basis for naming just the very small set of "organic" compounds was a major undertaking. Nevertheless, despite personal animosities, there was a generally recognized need for such a system. Consequently, the belligerents first convened an international convention in 1889 in Geneva, Switzerland with the intent of internationalizing and standardizing a common nomenclature for "organic chemistry". During the next three years various proposals were floated by correspondence between the participants who again met in 1892. At this meeting, after much rancor, an agreed upon set of "nomenclature for organic chemistry" rules was adopted. Meanwhile, to the chemistry mainstream of that era, these results were dismissed as being irrelevant, inasmuch as they applied for only a small subset of the known chemicals. It was not until 1922 that a commission, the International Union of Pure and Applied Chemistry (IUPAC) was established "to improve and standardize chemical nomenclature" [5].

During the twentieth century, not only did organic chemistry grow to become the largest subdivision of chemistry, but also other newly emergent subsets of chemistry* independently developed their own set of nomenclature rules. This development may be viewed as following closely upon the mentality of alchemy, and the resulting partitioning of chemistry into its present sub-divisions as creating "fiefdoms": Within each individual fiefdom is a different perspective as to what is important for characterizing and nomenclating molecules. For example, when three or more atoms are connected (in pairs) to form a circle (what mathematicians call a "cycle"), organic chemists usually see a "ring"; that is, they view the various atoms that form this ring as being of equal importance. Inorganic chemists, on the other hand, normally focus on individual atoms and consider this same arrangement as one atom (usually a metal) grasping both ends of a "chain" of other atoms (usually all non-metals) to form a "chelation". Because of these distinct "world views", different terms are used to describe the same, or a nearly similar, idea [6]. The fall-out from this choice of terminology is that communication between differently trained chemists (as well as with mathematicians and scientists in other fields) is made much more difficult. This is precisely one of the areas that the proposed nomenclature is intended to address.

Instead of only focusing on the problems created by the process of devising a system of canonical names that will be applicable for all of the

* There has emerged as sub-disciplines of organic chemistry: "polymer chemistry", "natural products chemistry", "biochemistry", etc. Also, "inorganic chemistry" belatedly also developed sub-disciplines (bioinorganic chemistry, inorganic polymer chemistry, etc.), as well as a special sub-discipline "boron chemistry" which, in many respects is closer to "traditional" organic than it is to inorganic chemistry. Chapter 5 of this book is devoted to examining both the nomenclature and the science of boron compounds.

varied chemical moieties, this process can also be viewed as creating an opportunity to expand the store of knowledge by examining synergies with other important parameters. In particular, one parameter of primary [7] importance is the geometrical structure of the moiety. Nowadays, chemical nomenclature is based primarily on naming a presumed geometrical arrangement of atoms. The more accurate the geometric description is, the more useful the nomenclature will be. In formulating the optimal nomenclature a basic question is: 'How are the individual atoms connected?' Consequently, with each gain in understanding of the connectivity of the atoms to form these molecules, crystals, polymers, etc., it is desirable that the nomenclature be constantly up-graded in order to reflect these improvements. In other words, exception is taken to the philosophy expressed by Cahn and Dermer [8], which permeates the practice of nomenclature today, that:

"no system of nomenclature can start afresh..."

To the contrary, unless one is willing to adapt to new ideas, when the accumulated store of knowledge dictates, all progress will be stifled.*

With such a perspective in mind, attention is directed to the observation that all nomenclature systems commonly used today rely upon a uniparametric approach [9]. A mathematical model of relevant moieties is formulated followed by the assignment of a canonical name to each member of the set described by the math model. If one could assume that the model was absolutely accurate, the descriptor set created would have all of the properties of importance that the actual chemical moiety has. However, this is a gross assumption! Instead of the traditional linear† representation augmented with morphemic suffixes to identify bonding, as is the practice by IUPAC [18] in "organic chemistry" and is also carried over into nodal nomenclature [19-20], a biparametric alternating code of atoms and bonds, in which the detailed formula will be all of the name that is needed, has been formulated.‡ Note that the complexity§ of the system being nomenclated, both with regard to presently

---

* A major advance in understanding mathematics occurred when the Hindu-Arabic system of positional notation replaced the then prominent additive system, exemplified by Roman numerals.

† In contradistinction to a more esoteric basis system, such as a comprehensive system based on prime numbers (of which the Matula-Elk system [10-15] is an example, as well as limited application systems that use either prime numbers or, for special subsets of compounds, different base number systems rather than the traditional base 10 system [16-17].

‡ Unlike nodal nomenclature, the proposed system has no need to introduce a word stem, such as -nodane.

§ A mathematical model for molecular complexity, such as the one developed by Bertz [21], has merit in helping to organize those properties which contribute to a heuristic idea of this subject. Using the language of graph theory [22] such a model will include a hierarchy of types of nodes (atoms) and of types of edges (bonds), as well as various other concepts such

used nomenclatures and to the new system proposed in this treatise, has made it desirable to defer development of various metric considerations until a later time.

At this point, it is important to remember that acceptance of new ideas in science is a very slow process. Many generations were to pass before there was even a small amount of recognition, and then only among a few members of the chemistry community, that the same scientific principles are applicable in each of the historical subdivisions; i.e., that there is nothing living, "organic", vs. not living, "inorganic", about a particular molecule. This is in stark contrast to the generally accepted perspective nowadays that life depends not only on the carbon atom, but also on many of the other elements. In the intervening time, however, the separate protocols for assigning names that were formulated in the individual sub-disciplines became so deeply rooted in their own domain, that these fiefdoms now resist all attempts at reform or standardization.

This opposition to change is especially counterproductive today because it is occurring at a time when the overlap between these historical domains is growing rapidly. Consequently, as more mathematics enters the university chemistry curriculum, many chemists now concur that the "traditional" boundaries which separate the various historical subdivisions are not only obsolete, but also act as a major obstacle to progress. Additionally, there has not been any consensus regarding how canonical names for new discoveries, as well as for older known chemicals that were originally named in a different fiefdom, are to be assigned.

Moreover, 'precisely what makes for a good nomenclature?' is a question with probably as many answers as there are persons answering the question. One historically important answer given by Read and Milner [24] and amplified by Goodson [25] is the "wish-list" of the following nine properties:

1. The names should be linear character strings, to permit lexicographic ordering.
2. A structural formula should give rise to a unique name.
3. The name should permit the retrieval of the structural formula.
4. The coding process should be simple, and preferably it should be possible for a chemist to code a formula without recourse to a computer.
5. The decoding process should also be simple.

---

as branches, cycles, etc. Unfortunately, because of the different values that a particular researcher places on these respective contributors, there shall exist inherent heuristic elements in ***every*** such system. Consequently, ambiguity is unavoidable and any system devised will be only as good as the insight that the formulator has built into it. Furthermore, any system so devised will eventually lead to a Goedelian impasse [23].

6. The coding process should not depend upon chemical intuition; that is, there should exist an efficient algorithm for coding, and computer implementation of this algorithm should be feasible.
7. Names should be brief
8. Names should be pronounceable.
9. Names should be easily comprehenisble to chemists.

To this list Goodson added his (and Chemical Abstracts's) specialization:

10. Names should be capable of being divided into convenient components, i.e., heading parent, substituents, stereochemistry, and other descriptive terms.

Most, but not all, of these items were taken into consideration in the formulation of the system described in the following pages of this text. In particular, there is ambiguity in item #8. If one assumes that pronounceability refers to distinctions in the oral, rather than the written, language, this attribute is regarded as inconsequential. By such a requirement, all homonyms would have to be avoided. Such a proviso is violated often in IUPAC nomenclature; e.g., the names "fluorine" for element with atomic number 9 vs. "fluorene" (a three ring hydrocarbon described in Table 1 of Chapter 2), etc. Conversely, there is no disaffection for a sequence of locant descriptors that use the letters of the alphabet, even though no pronounceable acronym has been formed.

Meanwhile, in formulating any system of nomenclature, an item of great concern to both the expert and the beginning student is the vocabulary. The more complicated the system the greater the need for precision in the choice of terminology.

(1) In science, "term" is used when a precise definition is being emphasized vs. "word" when more ambiguity is allowed [26]. When dealing with "terms", communication is greatly improved when the "denotation" (precisely what has been spelled out in a given definition) as well as the "connotation" (what the user of the word might infer) intended by their use is recognized. For example, to talk about a "flu shot", rather than an "influenza inoculation" **may** be acceptable in the language of the layman, but it is incorrect in a chemistry or biology journal.*

* There is a set of two (or possibly three) strains of virus associated with the disease influenza. It is only these particular virus strains that the "so-called" **flu shot** is intended to be effective against. Note that the science journal that described this inoculation [27] was very meticulous in its choice of terminology. It deliberately never used the word "flu". To the lay public, the word "flu" refers to any bad rhinovirus, of which there are over 300 known such viruses today, as well as thousands of the larger class of all viruses. Moreover, these numbers are rapidly growing with new discoveries. To the scientists who designed, synthesized and analyzed these chemicals, the inoculation that they developed is intended to ward off only two viruses, not hundreds or thousands. The misinterpretation of these results by writers in the popular press, and even some, who should have known better, in science

(2) The term "orismology" has been resurrected from being an arcane synonym of "terminolgy" to denote a study of the entire evolution of the ideas inherent in a term, rather than merely being limited to the specific connotation in the present usage of that term — which is the usual meaning associated with "terminology" [28]. The evolution of chemistry may be viewed as a history of change in both the denotation and the connotation of important words. This is described in an on-going series of articles on the orismology of these terms in chemistry [29-36]. Seemingly minor word changes have often reflected major advances that are occurring in our understanding of the science. A familiar example of this is that the term "aromatic" has come a long way from its original association with aroma. In Part 8 of this series [36] what one may call "The General Rule of Orismology" is introduced:

> "In any evolving body of knowledge, such as science, there are no terms that remain synonymous for long"

With increased knowledge of a specific chemical or process, the difference between two molecules that previously had been viewed as being only a minor variation of a general idea, and thus that could be subsumed by a single term, is now recognized as being of sufficient importance that distinct terms are **required** in order to be able to adequately describe this difference. Consequently, it is not rare that terms once considered to be absolutely synonymous are reinterpreted so that one of these terms describes the new variation, while the other retains the old denotation.

(3) The noun "moiety" incorporates into a single, general class distinct chemical structures that are the basic units into which matter is subdivided for purposes of taxonomy.* Note that for two of these structures (molecules and ions) there is an isolatable aggregation of atoms forming this basic unit and that all such aggregations are congruent. For polymers, on the other hand, the concept of isolatability is a mathematically inspired extension which is achieved by partitioning the polymer into congruent units, called "monomers" (See Chapter 8), that are bounded by unpaired electrons; i.e., "dangling" bonds.

---

magazines has created a highly undesirable fall-out. Because of this incorrect interpretation of a single word, unrealistic expectations in the general populace is initially raised. When these expectations can not be fulfilled (having never been promised in the first place), there is created a marked decrease in trust for medicine in particular, and science in general.

*Although the common usage of the **word** moiety is "one of the portions into which something is divided" [37], it is herein regarded as a **term** with this specific meaning in chemistry. Throughout this book, the adjective chemical will be implied whenever "moiety" is used.

(4) The adjective "canonical" was adopted from theology to indicate **the** prescribed standard to be used; e.g., a canonical name is that combination of symbols (letters, numbers, marks of punctuation, etc.) which uniquely describe a geometrical arrangement of atoms. An individual letter, number, or mark of punctuation is often referred to as a "morpheme". For example, the **y** in alkyne.

(5) There is a distinction that should be made between the familiar verb "name" versus, what may seem pedantic to the casual user, "nomenclate". The connotation associated with "nomenclate" is one of an existent system in which one assigns some pre-determinable, precise name to a given form. For this reason, most of this book shall be concerned with "nomenclating", rather than "naming", all sorts of moieties. The presence, or absence, of a "canonical" system to be used in assigning this name is the line that separates "nomenclating" from "naming".

(6) The term "heuristic" is used as both an adjective and as a noun with the connotation of an intuitive idea. The American Heritage Dictionary [38] defines the adjective as:

> "of or relating to a usually speculative formulation serving as a guide in the investigation or solution of a problem"

and then gives a more detailed explanation when applied to the field of computer technology, while The Webster New World Dictionary of Computer Terms [39] describes the noun as:

> "A method of solving a problem by using rules of thumb acquired from experience. Unlike an algorithm, a heuristic cannot guarantee a solution, but it may provide the only way to approach a complex problem."

(7) The definition given to the term "algorithm" is: "a procedure for solving a mathematical problem in a finite number of steps that frequently involves repetition of an operation; *broadly*: a step by step procedure for solving a problem or accomplishing some end" [40]. The creation of a detailed algorithm that is capable of assigning canonical names to each of the known, as well as all of the as yet unknown but mathematically possible, combination of atoms is the goal of this book.

(8) The term "constitution" refers to the way that individual atoms are connected to one another in a moiety. One common way of describing constitution is by a connectivity table. See Chapter 9.

(9) The term "isomer", derived from the prefix "iso-" meaning equal and the suffix "-mer" meaning parts [41] has the added connotation that there is something different about the two moieties being compared. This scenario is analogous to comparing two figures in geometry. It is meaningless to assert that two triangles are equal inasmuch as there is not

one common parameter that is being compared. Were one to be interested only in the original meaning (and usage) of the word "geometry" (to measure the earth; for the purpose of redefining land boundaries after the Nile River has flooded and receded, and if all of the recovered land was considered to be of equal value for farming), then the only measurement of interest might have been the area, which can be expressed by a single number (with its appropriate units). With such an objective, the idea of equality vs. inequality of the triangles would be reduced to the unambiguous comparison of two numbers. However, shape is a more complicated idea that requires additional measurements, such as angles, as well as lengths. For the triangles, the heuristic of "equality of shape", a concept called "similarity", requires that each of the corresponding angles be equal. Although this is a necessary condition, it is not a sufficient one. (The angles of a rectangle are equal to the angles of a square, all angles in both figures are 90°.). Additionally, the concept of similarity requires that the corresponding sides be proportional. In a like manner, in chemistry there is the concept of "equality" when the moieties being compared have the same number of each type of atom; i.e., they have the same molecular formula. The correspondence to geometric shape may now be expressed in terms of the connectivity of the individual atoms and the resulting geometric pattern that this connectivity induces. Each of the different means of describing selected attributes of these connections give rise to a particular type of isomerism. These shall be quantified using two important distance measurements described below (see item 12).

(10) The distinction between the term "configuration", defined as "the stable structural makeup of a chemical compound especially with reference to the space relations of the constituent atoms" [42] and "conformation",

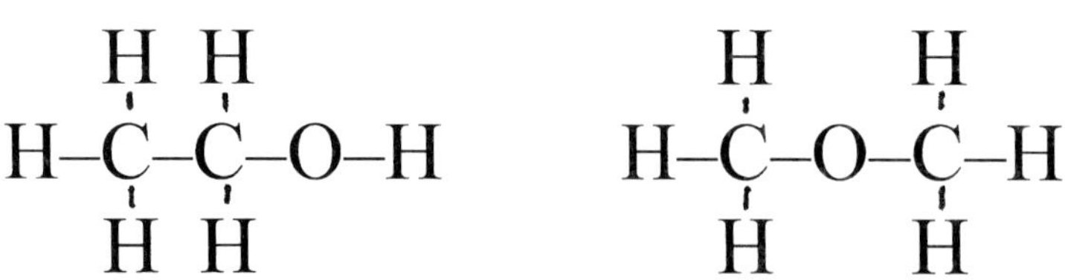

Ethyl alcohol (ethanol) Dimethyl ether

Fig. 1. An example of two structural isomers and their structural formulas

defined as "any of the spatial arrangements of a molecule that can be obtained by rotation of the atoms about a single bond" [43] shall be of importance in defining the measurements of distance that are relevant to chemical structure and nomenclature.

(11) Various ways of writing the formula to describe a moiety parallel the amount of information known about that moiety. For example, consider the two molecules illustrated in Figure 1. One can not tell from the "molecular formula", C_2H_6O, which isomer is being described. The "structural formula", on the other hand, advises precisely which atoms are attached to each other.

(12) Two distinct distance measurements useful in nomenclating moieties are the "graph theoretical distance" (GTD) and the "metric distance" (MD) [44]. GTD is defined as the length of the shortest path* between two selected vertices in the graph.† This dimensionless integer is a common descriptive feature of a configuration. Most traditional nomenclature has, without using the term, focused on this measure. MD, on the other hand, is a physical measurement that cuts across space, rather than following along a connecting path between two atoms and has a unit such as nanometers. When two moieties that do not have equal MDs between corresponding atoms in their most stable conformation (i.e., they are not congruent) have the same molecular formula, this is a convenient definition of "isomers". See item 9 above.

(13) Isomers that have equal GTDs between corresponding atoms are classified as "stereoisomers". Those with unequal GTDs between at least one pair of atoms are "structural isomers"; also referred to as "constitutional isomers". For example, in Figure 1, the GTD between the two carbon atoms in ethanol = 1, while in dimethyl ether GTD = 2.

---

*"Path" is both a heuristic **word** in common usage and a clearly delineated **term** in graph theory. In this latter capacity [45], one begins with a "walk", which is defined as 'an alternating sequence of points and lines (more accurately line segments) beginning and ending with a point, such that each line is incident with two (specified) points (preceding and following it)'. If the lines of a walk are all distinct it is called a "trail". If the points (and thus necessarily the lines) of a walk are all distinct, it is called a "path". If a path starts and ends at the same point, it is called a "cycle". Two important paths and cycles in chemistry are the "Eulerian" [46] and the "Hamiltonian" [47] ones. A **Eulerian path** is a path that goes through every edge exactly once. If this Eulerian path ends at its starting point it is an **Eulerian cycle**. Similarly, if path/cycle goes through every vertex exactly once, it is a **Hamiltonian path** or **cycle**. Note that an Eulerian path can pass through a given vertex more than once, and that a Hamiltonian path need not cover every edge. Both of these scenarios are the norm, rather than the exception.

†The atoms and bonds in a chemical moiety are respectively represented by the vertices and edges in a graph.

These two isomers, each of which has C_2H_6O as its empirical formula, are structural isomers, not stereoisomers. Care must be taken that the historically evolved choice of terms does not create confusion: Structural isomers have different structural formulas. It is the stereoisomers that have the same structural formula. The class of steroisomers will be further subdivided as the details of the nomenclature are developed. A flowchart characterizing the different types of isomers is given in [48].

(14) The term "locant" is described by Dyson [49] as follows: "A numerical subscript indicates the number of atoms involved and the figures at the end of each operation are *locants* to locate the position. This will be clear enough from the examples given." The inadequacy of this description as even a rough definition or an algorithm can be traced to its reliance on figures and examples. Nevertheless, the term "locant number" is useful for identifying which particular atom in a chain is attached to some designated atom or group of atoms. The single atom or group of atoms that is being attached is referred to as a "ligand". The atom to which it is being attached is usually part of the "parent" (see item 16 below). Modern geometry textbooks recognize that reliance on figures (and examples) readily leads to the creation of ridiculous proofs, such as "all triangles are isosceles", etc. [50]. Note that these descriptors can be used to support heuristic ideas, but NOT for formalizing mathematical (or scientific) results.

(15) When a group of contiguous atoms (along with their internal bonds) act as a unit, often with properties distinct from the individual atoms and bonds, this is referred to as a "functional group".* For pragmatic purposes, the simplest functional groups are two carbon atoms connected by either a single, double, or triple bond.† The presence of more than one of a single functional group may create a larger functional group with different properties than the individual smaller functional groups. For example, as well as the properties associated with an isolated double bond, two other distinct combinations of carbon atoms and double bonds

---

*This term is usually applied only to covalently bonded atoms, inasmuch as ionic compounds do not remain together to act as a unit. For example, when a sodium atom gives up an electron or when a chlorine atom accepts an electron, the resulting Na^+ and Cl^- ions act independently of their original neutral atom source. Every one of the ions formed are equally attracted by any other oppositely charged ion.

†Mathematically, there is no restraint on the formation of a quadruple bond between two carbon atoms; however, the energy constraints on such a combination, along with other considerations make such a molecule if not impossible, so highly unlikely that it need not be considered further. The existence of quadruple-bonds between pairs of atoms, other than carbon, are known and the nomenclature that is proposed must be able to canonically name such moieties.

give rise to chemically different molecules, which are referred to as "cumulenic" and "conjugated". The nomenclature associated with the cumulenic combination shall employ the same symbol as for isolated double bonds; however, the conjugated combination will be represented by a new symbol. This will be described in Chapter 2. Meanwhile, an attribute of importance to be noted is the inclusion of morphemes (see item #4) to designate functional groups in IUPAC nomenclature. The common nomenclature practice of explicitly naming functional groups was replaced over a century ago by a code containing these morphemes. In the proposed nomenclature, rather than having a list of affixes to be memorized, this information is encoded as a particular sequence of atoms and bonds, referred to as a "signature". For example, the partial signature* of an "alcohol" is the linear sequence of carbon atom, single bond, oxygen atom, single bond, hydrogen atom (see Figure 1 above). It is this sequence that one looks for, rather than the memorized suffix **–ol**, which is much more prone to typographical errors. Similarly, a different connectivity sequence will indicate an ether (carbon atom, single bond, oxygen atom, single bond, carbon atom), etc. Note that both IUPAC and the proposed nomenclatures have abandoned explicitly identifying the functional group, relying instead on implicitly naming it. The difference between these two algorithms is strictly a choice of coding (a memorized affix or a sequence of atoms and bonds).

(16) The term "parent compound" is a heuristic term that refers to a reference compound (whether physically existent or not) that has a minimum number of descriptors. This is an updated description ("definition") of the term presented in part 6 of the orismology series. See [33]. Although a frequent connotation of this term, as well as the historical implications of it, is that other compounds are created from it, in many instances there is ambiguity. For example, ethane is normally obtained starting from ethanol as the raw material, rather than vice versa. By such a process ethanol could be designated as the "parent" since ethane was derived from it. On the other hand, for organizational (and thus nomenclating) purposes, the heuristic parent is ethane and all compounds that have replaced one or more of the hydrogen atoms with other ligands are "daughter" compounds. This usage of the term gained prominence in

*This is only a partial signature in that two functional groups can have this same sequence as part of their signature – in this case both alcohols and phenols. In order to differentiate between these two groups one must examine a longer sequence, especially the bond immediately preceding the carbon atom. This idea will be described in more detail in Chapter 2.

1981 when Chemical Abstracts updated its Ring Index (originally issued in 1940, revised in 1960 and then supplemented three times [51])* with a new method of organization for compiling the set of known organic molecules which it called the Parent Compound Handbook [52]. Ambiguity arises when determining what is the parent for substituents containing carbon atoms. For example, it is generally agreed that a smaller alkyl group attached to a larger one is part of the parent compound (2-methylbutane is the parent compound, rather than just butane). However, when a cyanide (-CN) group is attached to an n-alkyl chain, is this additional carbon atom to be regarded in the same manner as though it were a non-carbon ligand, such as a chlorine and thus not a part of the parent compound, or should one consider that the ligand is the triply-bonded nitrogen atom with the carbon atom of this nitrile being a part of the parent compound? The decision as to what is the parent compound grows even more confusing when heteroatoms, such as oxygen and sulfur, are in the longest chain. In other nomenclatures besides IUPAC, the criterion for "parenting" may be different. In clarifying some of the attributes of nodal nomenclature {see [18] and [19]}, Gottlieb and Kaplan [53] try to evade the issue of parenting by allowing two alternate schemes for nomenclating heteroatoms. The problems introduced by this approach will be examined later in this chapter, after an examination of how nodal nomenclature assigns a canonical name to the simpler class of hydrocarbons.

Meanwhile, in the process of formulating the proposed, unifying, systemic nomenclature, there will be the need to introduce many other terms, which have limited, highly specialized meanings. The first time each such term is encountered it will be described in detail.

Having given the above definitions and clarifications, it should be noted that all of the conformers of a moiety will have the same GTDs between corresponding atoms, but will have at least one different MD. The simplest chemical structure for which there is interest in conformers is the hydrogen-suppressed picture of butane. (A model of "n-butane" including its hydrogen atoms is given in Figure 2.) For the hydrogen suppressed structure there is only a single parameter involved in both the GTD and the MD matrices. Because both of these distance matrices are symmetric, for convenience of presentation (but obviously not for any matrix mathematics, such as multiplication), one may combine these two matrices into a single square array with the names of the

---

*Each of the addenda were literally "**added at the end**", rather than being integrated into the totality of the original work; thus the verb "addend" connotes this afterthought. Observe that integration of addended material is only achieved with the formulation of a new work — in this case the Parent Compound Handbook.

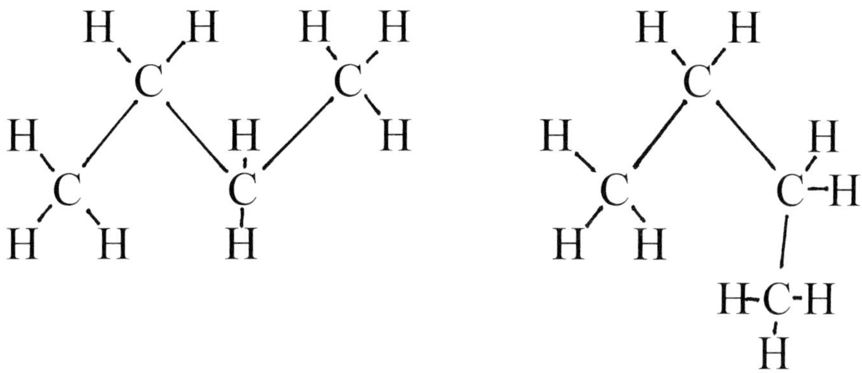

trans conformer cis conformer

Fig. 2: The extreme conformers of n-butane

atoms on the principal diagonal, GTD as an upper diagonal submatrix and MD as a lower diagonal submatrix. This is presented as Table 1. Note the above MDs were computed using simple trigonometry for these "boat" (minimum) and "chair" (maximum) conformers, with all lengths being 154 nm and all angles 109°28'. Although other intermediate conformers could be described, there is little scientific interest in them individually. However, in certain selected instances, there may be interest in an integration of all of these, using the calculus definition of "average":

$$\frac{1}{2\pi}\int_{0}^{2\pi} f(\theta)\, d\theta$$

A further observation is that in all standard nomenclatures, including many that have been devised independently of IUPAC, the entire purpose of the nomenclature is to assign names to "configurations". The identical name is assigned to all conformers, even the extreme ones. The question as to whether this is a desirable attribute, or a defect, in these traditional systems will be explored in Chapter 3.

At this point, observe that for many subject areas, the items being assigned names are all "familiar" (at least to the expert in the field) and there is a logical order, defined by a single "parameter"[54], which can be used to tabulate these items in a sequential manner. For example, the parameter used for a dictionary is the set of letters of the alphabet, and the sequence followed is the agreed-upon fiat of that alphabet. Now, even though new words are constantly being introduced into the language, their ordering in the dictionary has been pre-established. To create a nomenclature, which is analogous to

Table 1:
Combined GTD (upper)/MD (lower – in nanometers) array for n-Butane

C_1	1	2	3
154	C_2	1	2
251	154	C_3	1
a	251	154	C_3

where $336 \leq a \leq 388$

creating a language, it is necessary to formulate an algorithm to follow when assigning names to the members of some single parameter set.

In the evolution of language, the various sounds were designated by a symbol (a letter) and to this selected set of symbols an arbitrary order was promulgated (an alphabet). Although the arbitrariness of this order is evident by the various languages in use throughout history, within each language there is an agreed upon order — even if it is not based on any logical foundation. Moreover, for those languages which use the same alphabet, it is a simple matter to create a dictionary in which multiple languages are being listed simultaneously. For example, if the languages being collated are English, French and Spanish, there is a single parameter (the Latin alphabet); consequently, such a listing is viable, even if its usefulness is limited. On the other hand, if the languages are English, Greek and Hebrew, which use three different alphabets, there is the need to make a heuristic decision which of several possible orderings to follow in tabulating the words. There is no a priori way to select an ordering for **a** vs. **α** vs. א. Alternately, one could regard these three different letters as interchangeable and thus have a single ordering, etc. This, however, need not be a problem. In establishing a language, one can, by fiat, dictate an order. Having so decreed, the desired dictionary has been created, even though it may contain a large number of unused combinations, referred to as "nonsense words", such as qxqz, or worse aαאbβ.

Comparing this to chemistry, one might expect the problems to be much simpler inasmuch as there is a logical order to the various "letters" that form the chemistry "alphabet"; namely, the elements are ordered in increasing atomic number. What is different is that, unlike in a language, wherein the ordering is linear (each letter in a word can be both preceded and succeeded by at most one letter), in chemistry multiple ligands can be attached to a single coordinating atom. This added complexity may be described using graph theory as a sequence of "stars", where a star is defined as a complete bigraph* [55].

---

*A "complete graph" is a graph in which every vertex is connected to every other vertex. For example, a complete graph having five vertices (represented by the symbol K_5) is illustrated

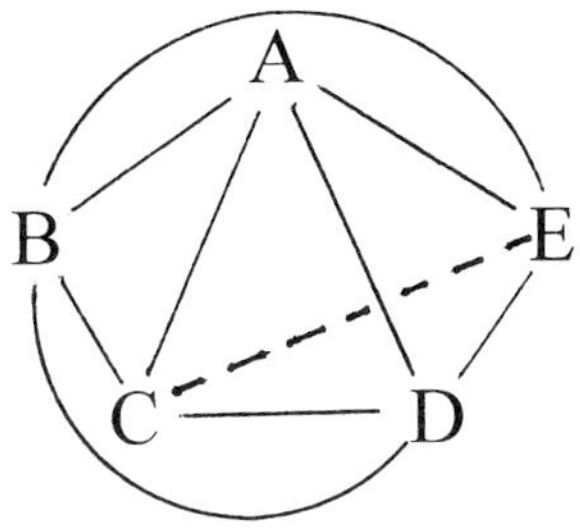

Fig.3: K_5

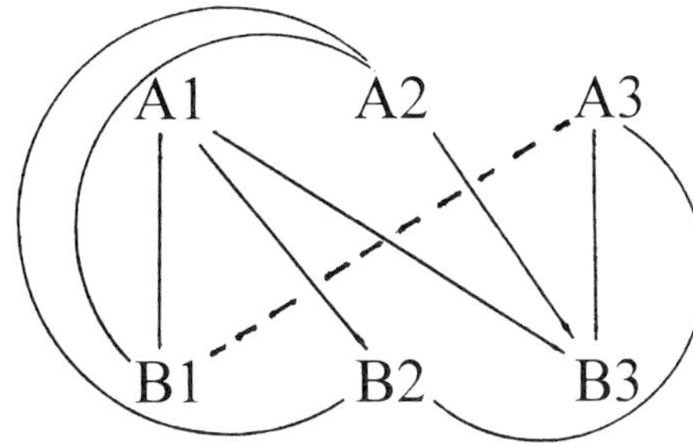

Fig. 4: $K_{3,3}$

The logical choice for a nomenclature now seems to be to begin the algorithm by focusing on the longest chain in the above-described sequence of stars. For Figure 2, using the protocol designated in IUPAC nomenclature of organic compounds [56], this sequence is the four carbon atoms; thus the name butane.* Furthermore, if one of the hydrogen atoms on one of the two end carbon atoms had been replaced by, say, a fluorine atom, or a group of atoms, such as a hydroxyl group, IUPAC would still consider the longest chain as the four carbon atoms and supplement the "stem name" (butane) with either a preceding name for the substituted hydrogen atom (e.g., 1-Fluorobutane, 1-Hydroxybutane, etc., where the 1- is a locant number — see definition # 14 above) or preferably, for selected common groups of atoms, a suffix, such as –ol to represent the hydroxyl group; namely 1-Butanol.†

---

here as Figure 3. A "bigraph" (also called a "bipartite graph" or a "bicolorable graph") is a graph in which the set of vertices are partitioned into two disjoint sets such that every vertex is adjacent only to members of the other set. A "complete bigraph" is a graph that is complete in the sense that every vertex in one set is connected to all of the vertices in the other set. When considering a star as an example of a complete bigraph the central atom is one of the sets and all of the other atoms (all of which have GTD = 1 from this central atom and GTD = ∞ from one another) form the other set. This is represented by the symbol $K_{1,n}$. A second important complete bigraph having three vertices in each set, represented by the symbol $K_{3,3}$, is illustrated as Figure 4.

*Or in the logical sequence of replacing common names, as suggested by Goodson [57] "tetrane".

†An alternate scheme, especially prominent in British publications, is to include the locant number immediately preceding the special suffix; viz., Butan-1-ol. This latter scheme becomes prevalent when there is more than one type of ligand attached to this backbone longest chain. Cahn and Dermer [58] advise: "Locants are placed as early in a name as does not cause confusion." They further indicate that this American practice applies to locants of a single type; however, when there is the need for more than one such descriptor, "the locant appearing first in the name is placed first on the left and the others directly precede their suffix, e.g., 3-hexen-5-yn-2-ol.... British custom is to place the locant always immediately in front of its suffix ... hex-3-en-5-yn-2-ol, but most chemists in other countries dislike this as splitting spoken words unnecessarily." In other words, there is neither a consistent logic nor

The problem of consistency in IUPAC nomenclature intensifies when one or more of the hydrogen atoms on a hydrocarbon molecule are replaced by metal atoms, rather than the more familiar non-metals. By such a substitution, the molecule is now deemed to be in a new and different domain referred to as "organometallic chemistry"*, instead of "organic chemistry", with its own set of nomenclature rules that borrow heavily from both organic and inorganic nomenclature, but is not consistently in either. In particular, focus is directed to an organo-metallic reagent that is important because in its two extreme conformers it can react in two different manners. This molecule, known originally by the old common name of "n-Butyllithium", but today mostly by "Butyllithium"† has as its canonical IUPAC organic name: 1-Lithiobutane vs. its canonical IUPAC inorganic names [61] of either Butan-1-ido-lithide (systemic addition name) or Butan-1-yl-lithide (systemic substitution name).

This difference between the perspectives of "organic" chemists (who view all ligands that have replaced a hydrogen atom in the parent compound as comparable and thus would nomenclate the lithium compound as "1-Lithio-", in exactly the same way as they had nomenclated the fluorine compounds, "1-Fluoro-") vs. "inorganic" chemists (who consider this molecule as a binary compound having an organic cation, C_4H_9, and an inorganic anion, Li — despite that it is only partially ionic), is another of the ambiguities that will be resolved by the proposed nomenclature. To the contrary, for purposes of cataloging and indexing, there will no longer be the question: should one list the metal first or last in the name?

Meanwhile, one should note that in chemistry, the process of establishing a "nomenclature" has, in addition to its analogy to the dictionary, a geometrical counterpart in terms of dimension. As well as the formulation, at the systems level, of topologically different models to be used for allocating moieties into distinct taxonomy classes [62], a corresponding development at the unit level is herewith included.

(0) One might envision the starting point, or 0-dimensional space, as merely naming the number and type of elements, i.e., the empirical formula. This was seen to be insufficient, due to the existence of isomerism.

---

is there uniformity in what should be a relatively simple, straight-forward IUPAC name assignment.

*The term "organometallic" refer to compounds that contain a carbon-metal bond.[59] The term spans compounds that are primarily ionic (such as when the metal is sodium or potassium and thus should be nomenclated as the distinct ions) through compounds that are primarily covalent (such as when the metal is lead, tin, mercury or thallium for which the bonding is covalent)

†Older reports and textbooks always included the prefix n- (which was an abbreviation for "normal", when naming an unbranched chain. Nowadays this is usually omitted and the unprefixed name "Butyllithium" implies the "straight" [60] chain of four carbon atoms.

(1) The 1-dimensional relation is seen in tabulating, in sequence, the individual atoms, i.e., the structural formula.

This is sufficient only for that small subset in which all of the atoms are similarly connected in a linear or monocyclic path. As described above, this is as far as analogy with the dictionary can be carried. By the introduction of a second parameter, such as an arbitrary ordering rule for the respective languages, there is little to be gained as no words have letters in more than one alphabet. In chemistry, however, the value of the system can be greatly increased by introducing as a second parameter different modes of connection; viz., single, double and triple dashes to represent single, double and triple bonds, respectively.

At this point it should be noted that both conformers of Figure 1 have this "essentially 1-dimensional" (linear) character. However, when either:
(a) as in Figure 5, at least one of the ligands has a higher coordination; i.e., when the chain is branched, such as in the molecule with common name isobutane (IUPAC name 2-Methylpropane) or
(b) as in Figure 6, the chain of atoms forms a cycle (cyclopropane)
at least two dimensions are required for an accurate representation of the moiety. Nevertheless, various techniques may be introduced that give a convenient representation of such moieties in a linear formula. The desired linear representation of this planar model may sometimes be achieved by including some marks of punctuation, such as a pair of parentheses. For example, in Figure 5, either, or both, of the branches of a branched chain may be indicated by enclosing them inside parentheses. This may be written several ways, such as: $CH_3C(CH_3)HCH_3$; $CH_3CH(CH_3)CH_3$ or $CH_3C(CH_3,H)CH_3$. Similar ploys, using special symbols, have been created to represent selected connectivities that are commonly occurring. For example, a creative way to depict a monocyclic compound is via the Mars symbol, ♂ (a circle with an arrow, to indicate go back to the beginning) as the terminal character of the

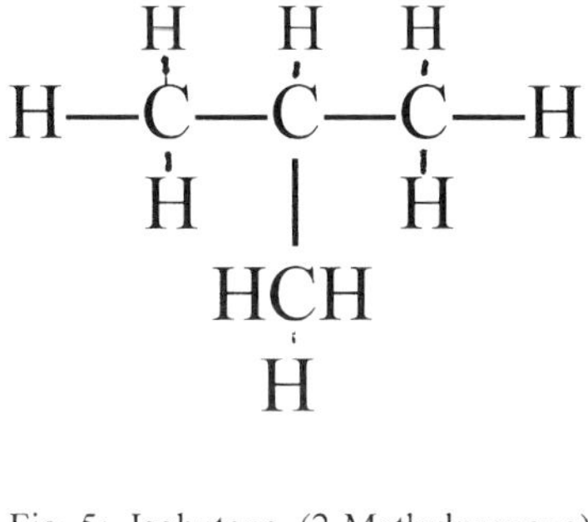

Fig. 5: Isobutane (2-Methylpropane)

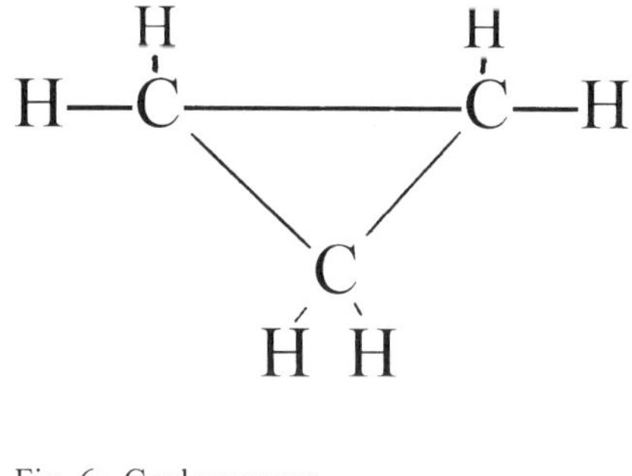

Fig. 6: Cyclopropane

name, etc. By this technique, one can represent cyclopropane (Figure 6) as: CH_2-CH_2-CH_2-♂. An entire system, called Wiswesser Line Notation, was formulated which uses only those characters that are part of the standard typewriter keyboard* [63]. Until the advent of the computer, this system, despite its complexity, had many proponents. In fact, one of the features of the Parent Compound Handbook is inclusion of the Wiswesser name for all compounds.

In contradistinction to the scenario depicted in Figures 5 and 6, there are many molecules (and their graphs) in which an inherent two-dimensionality can not be finessed, this will be described in the next section. Before examining them, however, it is desirable to return our focus to Figure 2. Here, one should note that had the chain of carbon atoms been one or more atoms longer in the cis conformation of n-Butane and had one of the ligands on a terminal carbon atom been a highly electronegative atom, such as fluorine, there would be Coulomb attraction between the fluorine atom and one of the hydrogen atoms at the other terminus. Because the attractive force between the fluorine and hydrogen atoms is weaker than a covalent single bond, it has been traditional to dismiss any interaction between these two atoms as immaterial. To the contrary, such a connectivity may be of great importance. This idea will be developed in Chapter 2, along with its effect on the nomenclature. Meanwhile, one should note that, although the GTDs between corresponding atoms is identical, there is a vast MD difference (about 800 nm in the trans conformer vs. 200 nm in the cis conformer), which produces a drastically different environment (Figure 7) in which these two conformers both exist and react. Consequently, it is disingenuous to treat them as "nearly similar".† This gross inadequacy, which is perpetuated in traditional nomenclatures, such as both IUPAC and nodal nomenclature, will be remedied in the proposed new system.

(2) Although above there was presented a simple representation technique that one could use for describing, and thus nomenclating, a monocyclic compound as though it were only one-dimensional, the presence of two or more rings in a compound makes such an evasion of the intrinsic planar geometry of

---

*Exotic symbols, such as the Mars sign, were not among those employed; rather one had only various combinations of numbers, letters that were not used as abbreviations of the elements, slashes, etc.

†That an oxymoron is created when one tries to combine "nearly" with "equal" (or any term involving equality, such as "similar", etc.) was demonstrated by S. Basak at the Skolnik Award Symposium at the American Chemical Society Meeting in Washington, D.C. on August 24, 1994. He defined such a relation as differing from its predecessor by a single letter, and then proceeded, using familiar words in the English language, to change black into white; namely: BLACK-SLACK-STACK-STARK-STORK-STORE-SHORE-SHARE-SHALE-WHALE-WHILE-WHITE.

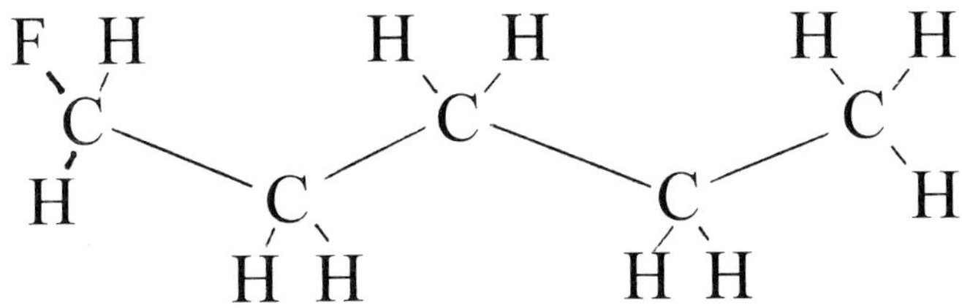

trans conformer

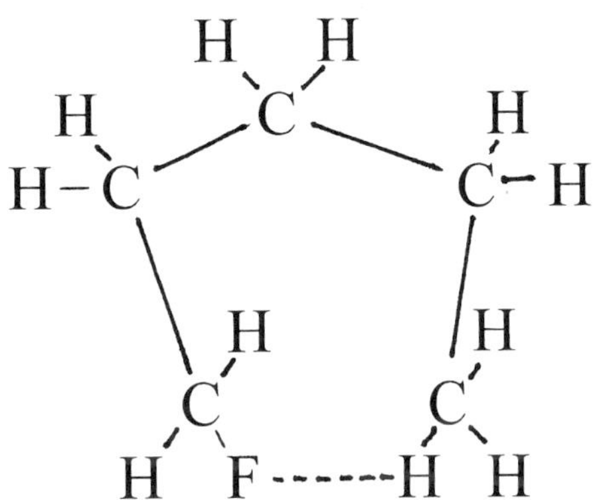

cis conformer

Fig. 7: The extreme conformers of 1-Fluoropentane

the moiety unacceptable. At this point it should be noted that in the traditional domain of organic chemistry, not only is the class of aromatic compounds primarily* two dimensional in it model representation, it is also two-dimensional in the physical world. This is in contradistinction to the sometimes allowable representation of aliphatic ring compounds using two dimensional projection and, more importantly, to the times that this projection introduces so much distortion that the scientific description is **WRONG**! For the former of these, in particular for most aromatic compounds, a planar representation is

---

*Some of the subtleties of precisely what the term "aromatic" denotes are presented in [64]. However, from a pragmatic, even though not *absolutely* accurate, perspective, for purposes of developing nomenclature, the connotation of this term in this treatise shall be the ability to assign a sequence of alternating single and double bonds that covers the moiety.

adequate. On the other hand, use of a planar description is sufficient for only a subset of the aliphatic multi-ring compounds. For others (what Goodson separated from Taylor's "reticular" class [65] and designated as "fisular" [66]) it is grossly inadequate. This latter set is described in the next section.

At this point the focus of developing nomenclature is returned to definition #14 given above. Note that although the term "locant" would not be created for another quarter century, the idea that underlies this term and its importance in nomenclature was evident despite that a term for this idea was not yet in the vocabulary. This may be seen in the "Proposed International Rules for Numbering Organic Ring Systems" [67]. In these rules, rings are subdivided into four categories: (A) Single rings; (D) Free Spiro Unions; (B) A selected, circumscribed set of two or more rings; and (C) none of the above. The inversion of order in this paragraph is intended to emphasize the ad hoc nature of Patterson's classification scheme; namely, only categories A and D (single rings and when "a single atom is the only common member of two rings", designated as "spiro"* compounds) are even relatively unambiguous; the various ways that rings may be joined (herein defined as "fused" or "bridged") are not. Category B compounds are those which have a certain heuristic of simplicity based on the common rings of his day. This subset eliminated: (a) rings considered as "strained" (rings of size 3 or 4), (b) rings whose intersection set contained more than a single edge which he called an "atomic bridge" (This was in contradistinction to his category B rings which were fused with only a single edge called a "valence bridge"), and (c) all bridges that crossed one another.† The eliminated combinations were considered as anomalies to be treated separately; i.e., his category C is "negatively-defined" ‡. In the taxonomy and nomenclature scheme being proposed to replace what is viewed as an antiquated system, there is no need for such an ad hoc (Category B) or an open-ended (Category C) taxonomy. Note that Categories B and C were most likely the model used by Taylor in devising his recticular and his two types of bridged subdivisions. Again, like Patterson, Taylor appears to have been

---

*And even this second category is not completely unambiguous. To try to accomplish his intended uniqueness, Patterson added the adjective "free" to emphasize that this is the only union between the rings.

†Unlike molecules that shall be described later in this study, one can assume from the example chosen that Patterson's idea of *bridges that crossed each other* was more a matter of assigning locant numbers and a canonical name to a wrong projection of a molecule than it was to there being an intrinsic non-planarity to the molecule. In other words, the mind set of planarity is deeply ingrained in IUPAC's rules.

‡"A term is positively defined if it belongs to a set that has some single characteristic or set of characteristics that can be used to test for inclusion in that set. Similarly a set is negatively-defined if it belongs to the complement of that set." The implications of such positive- or negative-definitions can be far-reaching, as described in [68].

preoccupied with planarity, as was the custom of that time.

Additional items of concern in Patterson's paper delineating IUPAC's system of organic chemical nomenclature include:

(a) the priority for ordering atoms; namely, they be "... as high a group in the periodic table ... and as low an atomic number in that group as possible." By this rule, not only would organometallic compounds be either permanently relegated to the category of inconsequential — after all of the nonmetal carbon combinations had been described or else forced to form their own new fiefdom — which they did.*

(b) a predilection for five and six member rings† that is so strong that it often overrides other, what nowadays is viewed as more important, considerations; namely, Dewar benzene is treated NOT as the fusion of two four member non-coplanar rings, but as an aberrant form of benzene and thus should be named exactly as though it were benzene. This is demonstrated in Note 9 and their description of what one might call "Dewar anthracene"‡ (Figure 8 is a copy of XII§ in [70]). This is followed by the assertion that what one has here is the traditional three fused hexagonal rings, rather than the four ring system of Figure 8. Such a supposition is deemed to be grossly erroneous and throughout this treatise all structural names and illustrations will be compatible. Their

---

*The existence of a large class of organometallic compounds was not anticipated at the time — only the Grignard compounds, discovered in 1900 (which may be viewed as the insertion of a magnesium atom between a carbon and a halogen atom in an alkyl halide), and a few lithium compounds, had found a place in the domain of organic chemistry. Although these compounds are an important tool for synthesis, they were initially regarded as an anomoly, rather than as a whole new subset of chemical compounds.

†This is predicated on the geometric property that the interior angle of a regular pentagon (108°) is very close to the tetrahedral angle (109° 28') while that of a regular hexagon (120°) is precisely that of the trigonal angle. Molecules having such interatomic angles are subjected to less internal strain and so have added stability. Consequently, these angles are chosen whenever possible. When smaller angles are required, such as in a cyclopropane or cyclobutane, the molecule is viewed as "strained" [69] and thus more likely to break some or all of the bonds holding it in that configuration.

‡Or more precisely b-Dewar anthracene, inasmuch as any or all of the "benzene" rings could have such a central bond. The logical extension to the linear fusion of six cyclobutane rings would by this criterion, according to IUPAC, be: a,b,c-tri-Dewar anthracene. On the other hand, in the nomenclature being developed, rather than as some artificially concocted three ring system, such a mathematically viable aggregation would be named as the six ring system that more accurately describes its structure. This is notwithstanding that such a structure would be **chemically** unstable.

§The ten hydrogen atoms and the two Robinson ring cycles were only implied in [67]; however, they are included in Figure 8. The two interior four member rings have no double bonds and for consistency should, by the convention prescribed (see section d below), have been drawn as squares, rather than either trapezoids or rectangles.

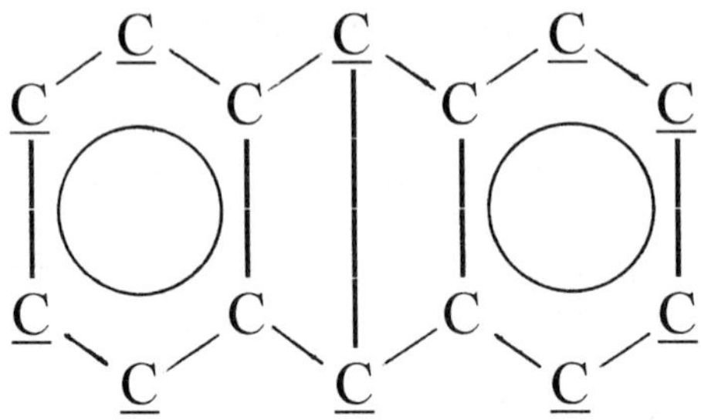

Fig. 8: A corrected version of figure XII in Patterson's 1925 article. An incorrect picture of anthracene

previous example, anthranil, in which a fused three and four member ring system is rearranged to form a five member conjugated ring is even more egregious. Whether such a rearrangement actually does or does not occur is irrelevant to the formation of nomenclature. A useful nomenclature must be able to name whatever structure has been presented for naming. There should further be the understanding that the capacity to name an actual moiety **does not exist**. Instead, what is being named is a mathematical model that **hopeful** approximates the moiety involved!

(c) the postulation of a prescribed orientation of each molecule so as to maximize the number of rings on a reference row, and to then center this reference line on the x-axis of a Cartesian coordinate system* in such a position that it maximizes the number of rings in the first quadrant. Nowadays, because of a general acceptance of a three dimension embedding space for all chemical moieties, any such positioning constraints are viewed as a liability, rather than a goal. Nevertheless, this aspect of the nomenclature has never been updated; consequently, according to the first of Patterson's rules:

> "Fixed orientations are an aid to memory and should not be neglected. Single rings should be oriented with Position 1 at the top and with numbers proceeding clockwise around the ring."

Such a rule, as well as prescribing how locant numbers are to be allocated, has a built-in bias favoring planarity. Such a pigeonholing of all considerations of the third dimension, which permeated chemical thinking at that time, continued unabated for nearly a century. This is despite the general acceptance of van't Hoff [70] and le Bel's [71] postulation of the tetrahedral carbon atom. Even Wells's short 1956 monograph "The Third Dimension in Chemistry" [72] only very superficially focused on chemistry, being more a treatise in geometry. It

*Even though neither the terminology nor any explicit reference to coordinate systems was ever made.

was not until the later part of the twentieth century that the "world-view" now referred to as "stereochemistry" came to the forefront.

(d) Not only was an orientation for multi-ring systems promulgated, additionally the shape to be used for illustrating each of the describing polygons was prescribed.

> "**Note 10**. Triangle ... should have one side vertical, other rings two sides vertical (this requires a deformation of the polygons with an odd number of sides....)"

Furthermore, **although it is not specifically spelled out,** for polygons with 2n+3 edges exactly one of the edges should be horizontal. This edge may be at either the top or the bottom of the figure. Moreover, for polygons with 4n edges, exactly two of the edges should be horizontal. All remaining edges beside those set vertically or horizontally should be as evenly distributed as possible. Observe that, in order to maintain its association with benzene, the center ring of Figure 8 does not subscribe to this implied convention. Additionally, this note acknowledges that, from IUPAC's perspective, selected deformations in the plane are desirable. Another item of note is that there is no description or extrapolation when there are multiple odd rings in the picture. The first three examples of Class B compounds, as well as many other on page 560 contradict the heuristic picture of "straightness". Here, fused pentagons that should have been depicted with one ring oriented "up" and the other "down" instead give preference to having the odd atom always drawn in the "up" position.

The logical development of the linear fusion of odd rings is described in [73]. That report presents both the IUPAC nomenclature for molecules formed by the fusion of tricyclo through heptacyclo pentane modules and a binary code that simplifies the description of the complete set of fused tricyclo through hexacyclo pentane aggregations. The latter examples in Table 3 of Chapter 2, shall focus on cyclopentane as the module of interest. In a similar manner, the extrapolation of linear chains of cyclopentane modules to form the counterpart of helicenes, henceforth referred to as "helicanes", is described in [73].

(3) Further expanding the representation of connectivity into three dimensions is traditionally accomplished by the use of various projective processes, supplemented by drawing techniques such as dashes and wedges, etc. Projection of the moiety onto a plane has been used irrespective whether the intrinsic topology of the moiety is one-dimensional, two-dimensional or three-dimensional. In IUPAC nomenclature the topological influence of the third dimension is greatly down-played, except in those cases in which there is "optical" isomerism wherein a complicated system of prefixes is employed to distinguish between such isomers. Discussion of the nomenclature of "optical isomerism" will appear in a follow-on treatise.

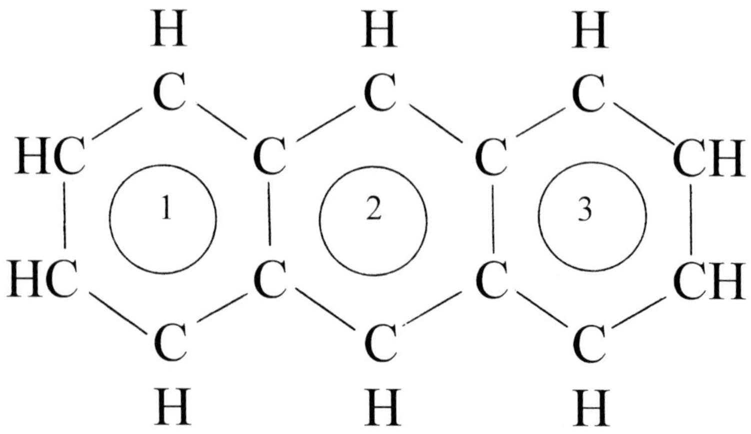

Fig. 9: Anthracene

What had not been taken into account (until Goodson's article [57], which was generally overlooked by even the mathematical chemistry community) in traditional chemical nomenclature was the fact that there is this inherent topological difference in two-dimensional vs. three-dimensional graphs [74]. When such graphs are used to model moieties, this will greatly impact what is desired in the nomenclature. For example, for those molecules that would be classified as reticular by Taylor (i.e., are "intrinsically planar"), an efficient method of description is in term of the smallest set of smallest rings SSSR [75-76]. In anthracene, for example, there are six rings {three six member carbon rings (1, 2 and 3 in Figure 9), two ten member rings (the fusion of 1-2 and of 2-3) and a circumscribing fourteen member ring}. Of these, only the set of three six member rings constitute the SSSR. The problem becomes much more interesting when the moiety is intrinsically three-dimensional, such as for cubane (Figure 10). For this molecule, there are 28 distinct cycles that could be formed using the various contiguous combinations of square faces; however, only six of these cycles (the six faces of the cube) are regarded as significant [77]*. Moreover, if one were to project this three-dimensional

*The remaining 22 cycles include: (a) 12 hexagonal rings formed from two abutting squares; (b) 4 heptagonal rings formed by three squares meeting at a vertex. Although there are eight vertices, the same boundary is traced out by pairs of trihedral angle triplets ; (c) 6 octagonal rings formed by three squares joined successively at opposite edges of a linear sequence of these squares. As in (b) the twelve edges of these rings are paired so that only six distinct octagons are formed. Although traditionally this is the total number of such rings [78], "spiro" (see Chapter 6) connected rings at each of the vertices, as well as multiply-spiro combinations are also possible.

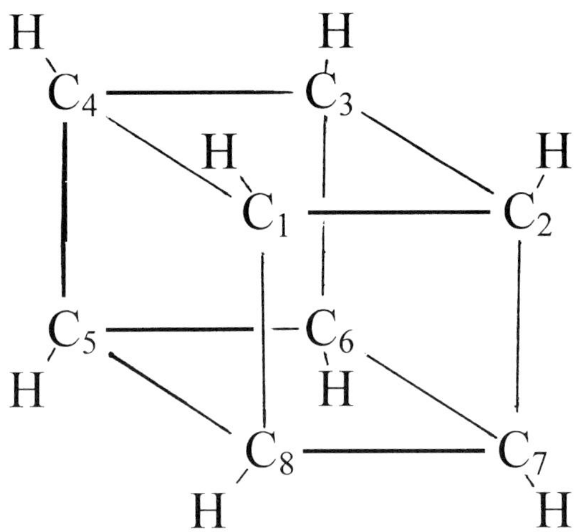

Fig. 10: Cubane

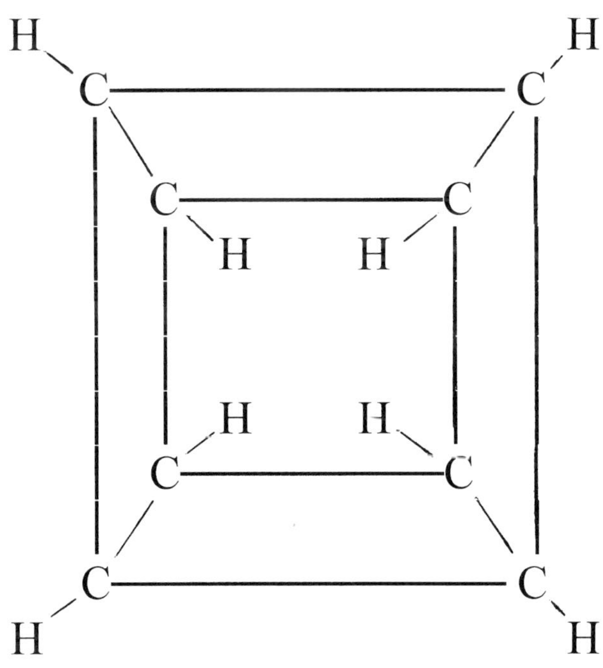

Fig. 11: Projection of cubane onto a plane

model onto a plane (Figure 11), one normally counts five closed regions — which is the number of SSSR that is traditionally used for purposes of nomenclating. On the other hand, even this number is large. The "minimum spanning set", the set which covers all of the edges of the molecule, is a set of only four squares.

Although cubane is three dimensional and its representation on a planar surface introduces some distortion, if one considers the outer perimeter as a ring to be counted, the correct number of "simple" faces has been formed. In other words, one of the inadequacies of the SSSR process is that there are different rules for molecules that are "co-planar" vs. those that are "three-dimensional".* In addition to the dimensionality of the model, one also must consider what one can called the dimensionality of the graph. All of the problems that are endemic to this phase of traditional nomenclature will be evaded in the nomenclature being proposed because the focus shall be strictly on the set of edges. The set of faces, in this system, has no significance. Consequently, the fact that Euler's Polyhedron Formula is applicable only to heuristically simple polytopes (of any dimension) is a problem that does NOT arise.

The topological question as to whether a particular graph could be inscribed on a planar surface or whether edges would have to cross was settled in 1930 when Kasimir Kuratowski proved that any graph containing the subgraphs K_5 or $K_{3,3}$ (see Figures 3 and 4 above) could not be drawn on a planar surface without the edges crossing [80]. Models of moieties that are represented by these two graphs are described in [81-82]. For such moieties, the graphical representation will require the use of the third dimension. A planar picture will have sufficiently great distortions that the basic geometry of the moiety will masked. Consequently, the task of correctly assigning the canonical name to such a moiety will be made much more difficult.

Along with the idea of greatly expanding the concept of connectivity into three dimensions, another major contribution of the proposed new nomenclature is to extend the means of connectivity (i.e., the bonds) to a larger, more flexible set. Although limiting attention to a small set of integer value bonds was, initially, consistent with a "simple" description in the newly developing field of quantum mechanics, creation of a more comprehensive nomenclature to apply to the emergent science (observations in the laboratory) was not deemed necessary. However, such success has been relatively short-lived. This inadequacy is the result of there being an ever-expanding set of moieties. Per www.cas.org/EO/casstats.pdf, there are, at latest count, over 42,000,000 moieties and sequences listed in the CAS Registry, and this number is

*Other problems, such as non-orientability, also contribute to making this a poor "fix" to a much deeper problem, especially for models whose geometry is not heuristically simple. See the discussion of "triangular prismane" vs. "triangular Moebiane" in [79].

increasing by over 2000 new additions daily. Moreover, these additions are often in ways that could not be anticipated at the time the original moieties were formulated and their names assigned. Furthermore, with each new entry there is the need to assign a canonical name to that moiety.

The advantages that will accrue by adopting this new systemic nomenclature include:

(1) The ability to more precisely correlate the various bonding types which have historically given rise to vastly different methods of assigning canonical names in the subdisciplines. This is done by standardizing on a larger set of bond descriptors than the traditional use of only integer (single, double, triple and occasionally, in the inorganic domain, quadruple) bonds. Additionally, the capacity to expand this bond set, whenever other, or more precise, bonding types are formulated.

(2) A complete dichotomy between bond order and other functionality.* This is in contrast to IUPAC's organic nomenclature, which uses morphemic suffixes to specify both degree of bond unsaturation (ane, -ene, and -yne), and also selected functional groups (-one, -al, etc.). In the proposed system, on the other hand, the individual chemical symbols for atoms alternate with bond descriptors (small integers or selected symbols). In particular, traditional single, double and triple bonds are represented by the integers (1, 2 or 3) as appropriate, rather than writing either neighboring atoms without any separation (assumed to be a single bond) or with a single, double or triple dash to denote the respective bond. Other functionality is NOT significant in the proposed nomenclature. Instead, all other consideration of functional groups is relegated to the atom and bond components which comprise them. In this way the nomenclature being developed is not dependent on any arbitrary priority rules† that are of historical, rather than of scientific or mathematical, origin. Also, much of the ambiguity that would, otherwise, arise from the use of selected words is eliminated. For instance, in most present systems determining what the phrase "longest chain"‡ denotes is influenced by the geometric character [83] of the modules making up that chain: IUPAC limits this definition to single

---

* Ability to maintain this dichotomy in "organic" (limited to sigma and pi bonding) chemistry seems to be unchallenged; however, when delta bonding is involved in "organo-metallic" chemistry, there arises a nebulousness as to what is the most useful designation of bond type. See Chapter 2.

† For example in the IUPAC system (Footnote #30), Rule A-3 and its sub-parts.

‡ It is interesting to note that although the term "chain" (or more precisely "0-chain" and "1-chain") has been given precise meaning in graph theory, only the heuristic meaning of this word is desired in developing the nomenclature of chemistry. The precision employed in formulating a formal mathematics may be stultifying in describing a science.

carbon atoms when nomenclating aliphatic compounds, but views benzene rings as the module of choice for many aromatic compounds [84]. Nodal nomenclature's concept of "the longest chain" includes not only heteroatoms (larger than hydrogen), but also shrinks selected sets of rings to a point and includes a "node" representing this ring as a member of the principal cycle. In the nomenclature being developed, the "longest chain" has as its **only** condition that the bond descriptors between successive atoms be greater than zero. Hydrogen is just as important as any other element.

(3) There is no need to use numeric prefixes in any language, (such as use of Latin and Greek in Taylor's seminal attempt [85] to add order to IUPAC organic nomenclature and its extension by Goodson [86]) to indicate the number of a given type of substituent group in a molecule. This information is given using only single, unambiguous integers ("2", "3", "4", etc.). In other words, rather than using "di" when naming two sets of a "simple" group, "bis" for two sets of a "complex" group, and "bi" for two sets of a ring assembly, where the line separating "simple" vs. "complex" is an unstated heuristic that may be interpreted differently by different persons, as well as similar prefixes of "tri" vs. "tris" vs. "ter" for three, "tetra" vs. tetrakis" vs. "quater" for four, etc., ONLY the simple integer set (which carries no additional connotations) is used.

(4) All ordering parameters have a chemical basis (usually atomic number). No alphabetizing of names is ever needed. Any use of any alphabet is language dependent, and, consequently, inherently capricious.

(5) There is no use of, no less dependence on, the admittedly ***empirical*** [87] concept of "oxidation number", as is the practice in IUPAC's inorganic nomenclature.

(6) By redirecting the focus from local to global, the concept of chelation is brought into consonance with that of graph theoretical cycles, and thus may be described in the same manner as the system that is used for organic molecules.

Turning now to some sample molecules and the names assigned in other nomenclature systems: From the discussion of item #2 above, it should be reiterated that the inclusion of hydrogen atoms often impacts the final geometry of a molecule. This is true for all purposes, including nomenclature. In the proposed systemic nomenclature, for example, the longest chain in methane is 3 atoms long; consequently, the coding (name) that describes this chain starts from one terminal atom, lists the bond order of the bond connecting it to the second atom of the chain, then the symbol of the second atom, etc. until the entire chain has been traversed; i.e.,

H1C1H (1)

Before describing how the remaining two hydrogen atoms are attached to this longest chain, observe that selected entire molecules (those that are strictly linear in the graph theoretical sense) can be nomenclated in an identical manner. For example, the carbon dioxide and water molecules are simply:

O2C2O (2)

and

H1O1H (3)

respectively*. In order to now incorporate the two remaining hydrogen atoms into the canonical name for methane, the algorithm is: at the end of the code of the principal chain place a colon followed by the locant number of the atom to which each secondary chain is to be attached. This locant number is written as a superscript inside a set of parentheses. Next the code of the ligand chain starting with the connecting bond, is included. For the methane molecule, this is:

H1C1H:$^{(3,3)}$(1H) (4)

Note that when there are two or more identical secondary codes, these may be combined in a single set of parentheses. Also, when multiple equal length longest chains exist, the priority is to choose the chain with the highest atomic number for the element or, if all equal, continue to the next entry in this chain, the bond order, when designating the "principal chain". This is in contradistinction to IUPAC, which lists the names of the substituents alphabetically [88], rather than in any mathematically or scientifically logical sequence. Moreover, the hydrogen atom is not included in their name! For example, IUPAC assigns chlorofluoroiodomethane as the name for that tri-substituted molecule. The proposed systemic name, on the other hand, selects the iodine as the lead atom in the principal chain and the second largest atom, chlorine, as the terminus. The canonical name of this molecule is thus†:

---

*The non-linearity of a chain of three atoms in space, such as the water molecule having a 105° angle vs. the geometrically linear carbon dioxide molecule (180°), is not a consideration; only that the chain is unbranched and thus can be represented by a linear graph.

†Although the similarity in font between the lower case letter **l** and the numeral **1** may be disconcerting to the human reader, it is of no importance to the computer. Nevertheless,

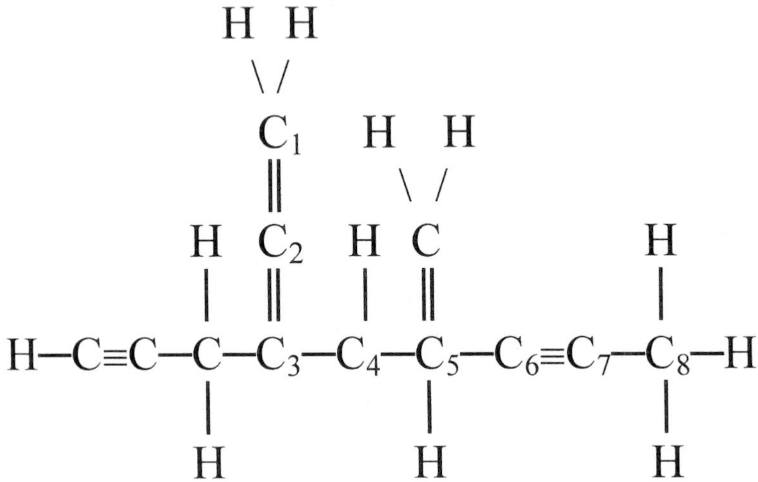

Fig. 12. A traditional example of an acyclic carbon compounds and its IUPAC name: 3(2-propynyl),5-methenyl-oct-1,2-diene,6-yne

$$\text{I1C1C}\ell\text{:}^{(3)}\text{(1F);}^{(3)}\text{(1H)} \qquad (5)$$

Observe that the systemic nomenclature being developed is a strictly analytic, vs. a synthetic, one. [89]

Next, Figure 12 is an example wherein the fiat of the IUPAC name is, at best, arcane. In IUPAC organic nomenclature, priority is given to chains which have the largest number of multiple (double and triple) bonds, even when there are longer chains with fewer multiple bonds. In this molecule, a shorter chain (8 carbon atoms long) is given precedence over a longer (9 carbon atom) chain. Additionally, using the IUPAC algorithm, when chains have an equal total number of multiple bonds, the one having more double bonds (thus fewer triple bonds) is given precedence. Note that the number of single bonds in either compound is immaterial. The effect of such a naming algorithm is that cumulenes, although rare in nature, have a disproportionately high priority when selecting the principal chain; viz., the extended (5 carbon long) cumulene CH_2=C=C=C=CH_2 is given priority over the longer (8 carbon long)

---

when the appropriate choice of fonts is available, when writing a name in the proposed system, avoid using the lower case letter l altogether. Instead, for those two letter symbols whose second letter is l (Aluminum, Chlorine and Thallium), use the manuscript capital letter and a script lower case l; e.g., Cℓ. This is illustrated in (5).

polyacetylene HC≡C—C≡C—C ≡C—C ≡CH. Furthermore, when assigning locant numbers to the atoms in a chain, per Rule A-3.3. [90]:

> "Numbers as low as possible are given to double and triple bonds even though this may at times give'-yne' a lower number than '-ene'."

This is seen in its examples of: 3-Penten-1-yne vs. 1-Penten-4-yne, where the combined numbers 1,3 are selected as the priority numbering in the first of these names; however, since the same numbering 1,4 would result in the second of these compounds, the "-ene" is given preference. Moreover, in writing the IUPAC canonical name, the suffix "-ene" always precedes the "-yne".

A more logical choice, incorporated into the proposed system, selects the longest contiguous, non-redundant path as the principal chain — without regard to **any** other parameters (such as maximizing the number of sites of bond unsaturation or of one degree of unsaturation over another).

Applying the proposed systemic nomenclature to the above cumulene and polyacetylene, their respective names are:

$$\mathrm{H1C2C2C2C2C1H{:}^{(3,11)}(1H)} \tag{6}$$

and

$$\mathrm{H1C3C1C3C1C3C1C3C1H} \tag{7}$$

Returning the focus to Figure 12, the canonical name will, unlike in IUPAC nomenclature, have along its principal chain nine carbon plus two hydrogen atoms (This was indicated in the figure along the horizontal line). Additionally, since the locant numbering could have started at either end, one notes the first difference in these two potential chains occurs at the second bond (locant #4).* Therefore, one chooses to have a triple bond at locant #4, rather than a single bond. This produces: H1C3C1C1C1C1C1C3C1C1H as the principal chain. The full name can now be formed by numbering each of the secondary chains starting from locant #2 as the connecting bond to the primary chain. Similarly, each tertiary chain will be located along a secondary chain enclosed in square brackets and have its own set of locant numbers; namely:

$$\mathrm{H1C3C1C1C1C1C1C3C1C1H{:}^{(7,7,11,11,19,19)}(1H);^{(9)}[2C2C1H{:}^{(5)}(1H)];\ ^{(13)}[2C1H{:}^{(3)}(1H)]} \tag{8}$$

---

*Remembering that every path can be traversed in two directions, excluding some trivial cases such as a single atom or a chain having a mirror plane through the center atom, there is more than one path that needs to be considered when selecting which of all possible paths to is to be selected as the "principal path".

Conversely, from the name, one can determine: (a) there are 11 atoms (9C and 2 H) in the principal chain; (b) there are secondary branches at locants 7, 9, 11, 13 and 19; (c) emanating from the secondary branches at locants 9 and 13 there are tertiary branches (of a single hydrogen atom) attached at positions 5 and 3 on the respective secondary chains. Moreover, had any of these atoms not been carbon or hydrogen, this would have been included by the appropriate chemical symbol without the need of either a set of substitutional affixes as in the extended Hantzsch-Widman system of IUPAC organic chemistry or the Greek letter affixes of IUPAC inorganic chemistry. See Chapter 2. Attention is further directed to the use of repeated superscripts, rather than the use of subscripts, to indicate that two atoms are attached to a common atom. The code $(1H)_2$ would have indicated a different (in this case incorrect) constitution; namely, that two hydrogen atoms were bonded to each other, as in C-H-H, rather than that each hydrogen was bonded to a common (carbon) atom.

Two important abbreviations that are introduced at this time are:

(1) Repeated groups can be condensed to a single repeat inside parentheses, with the number of repeats listed as a subscript. Caution should be exercised that subscripts, superscripts and in-line numbers are clearly distinguished from one another.

(2) Much of the tedium in (8) is introduced by naming **all** of the atoms in the molecule. Unlike other nomenclatures in which some atoms, namely the hydrogens, are inferred by default, no atoms are omitted in the proposed systemic nomenclature. Instead, noting the ubiquity of non-terminal CH and CH_2 groups in the various chains that comprise a complex "organic" molecule, a major simplification is introduced by underscoring chemical symbols to denote a grouping of that atom and the indicated number of non-terminal hydrogen atoms*; e.g., $\underline{C}$ and $\underline{\underline{C}}$ respectively denote non-terminal CH and CH_2 groups.

Next observe that in the primary chain there is no distinction between C, $\underline{C}$ and $\underline{\underline{C}}$; therefore, it is often convenient to mix abbreviated and unabbreviated symbols. Furthermore, in order to maintain consistency, terminal hydrogen atoms are not incorporated into any abbreviation. In particular, the carbon atom in a terminal methyl group does not have three underscores. Instead it is written as $1\underline{\underline{C}}1H$, etc. Similarly, use of $1\underline{O}$ for the hydroxyl group is precluded. The correct code contains the terminal hydrogen atom: 1O1H, etc.

Using these simplifications, (8) can be rewritten as:

$$H1C3(C1\underline{\underline{C}}1)_2C1C3C1\underline{\underline{C}}1H{:}^{9}[(2C)_2 1H{:}^{5}(1H)];\ ^{13}[2C1H{:}^{3}(1H)] \qquad (9)$$

---

* An identical simplification could be made in the domain of highly fluorinated [91-92] (or any other similarly selected parameter) molecules.

where parentheses around the superscripts is optional, but should be included if one thinks there is a chance of ambiguity. Next, comparing (9) to the IUPAC name (see legend of Figure 11), one observes a more straight-forward, but longer name; also a name that requires meticulous attention to detail but no memorized affixes. In other words, a name perfectly suited to a computer.

In a similar manner, (6) becomes either*:

$$H1(C2)_4C1H{:}\ ^{(3,11)}(1H) \qquad (10)$$

or

$$H1\underline{C}2(C2)_3\underline{C}1H \qquad (11)$$

As the next examples, attention is directed to the newer, mathematically more sophisticated, nodal nomenclature [93]. In this system all atoms (except hydrogen) are initially regarded as points (nodes) in a graph, which are all

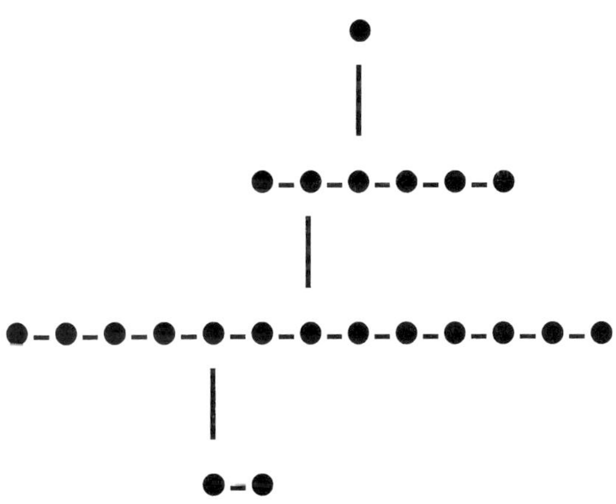

Fig. 13. A traditional example of a more highly branched acyclic carbon compound and its nodal name: $[13.5^7 2^5 1^{14} 1^{15}]$Docosanodane.

*Since simplifications are merely convenient ways of writing the full canonical name, any one is as good as any other. Thus one normally opts for the shortest way to write the name; (11) in this case.

Fig. 14: The graphic picture for many trianodanes in nodal nomenclature

treated equally. Although this is a major improvement over IUPAC nomenclature, it has some important shortcomings. One of these is that by labeling the nodes all equally, there is an increase in complexity of the nomenclature. Another layer of information needs to be addended in order to identify the atoms that comprise a given path. Furthermore, the names of the individual atoms are added **only after** all of the nodes of the graph have been named. Similarly, only after that, are the bond multiplicities indicated. For example, the same graph applies to both propane and to ethanol, as well as to dimethyl ether.[*] Some molecules which have this graph (Figure 14) along with their IUPAC and nodal canonical names are:

propane:	[3]-Trianodane [†]
propene:	[3]-Trianodene
propyne	[3]-Trianodyne
ethanol:	[3]-Oxatrianodane
diethyl ether:	[2]-Oxatrianodane
ethenol:	[3]-Oxatrianodene
ethynol:	[3]-Oxatrianodyne
carbon dioxide	[1,3]-Dioxatrianodiene
1,3-dichloromethane	[1,3]-Dichlorotrianodane

Observe that this system continues to assume that both the number and the location of hydrogen atoms can be inferred, rather than having to be named. This is a potential source of ambiguity that will be revisited shortly. Meanwhile, the name assigned by nodal nomenclature for a more highly branched alkane (Figure 13) is: $[13.5^{7}2^{5}1^{14}1^{15}]$ Docosanodane. Another major disadvantage of nodal nomenclature is that the assignment of locant numbers beyond the principal path is tenuous at best and ambiguous at worst. This is seen when closely examining the name assigned. Whereas the five carbon long chain at locant #7 and the two length chain at locant #5 are easily recognized, the designation of which atom is to be named locants 14 and 15 is less obvious.

---

[*]Whether the comparison of identical structural isomers as in Figure 1 or this comparison of identical nodal graphs is chemically more significant is a heuristic decision that each individual must make.

[†]Note the [3] is redundant for propane (having been included in the original report [18] but omitted by Gottlieb [53]). A similar comment applies to propene and propyne. In the succeeding examples it serves as a locant number. Also note the locant numbering selected gives highest priority to the lowest atomic number; i.e., oxygen is locant #3, rather than #1, in ethanol, ethenol and ethynol.

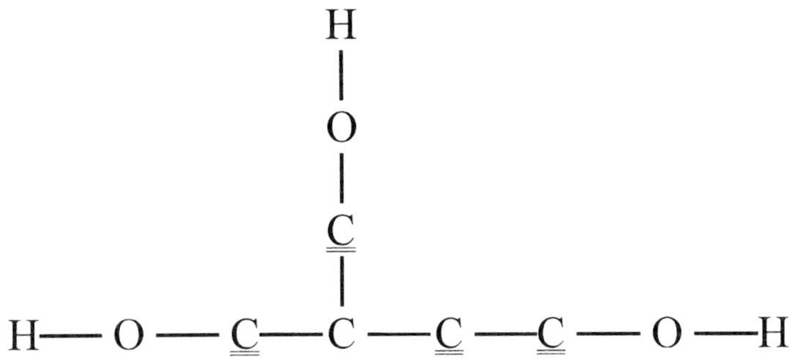

Fig. 15: A molecule having two equally acceptable nodal nomenclature names, rather than a single canonical name.

In this case there is only a single methyl group at each locations and so there is no ambiguity; however, there will be more problems with increasing branching, especially when rings are introduced. Meanwhile, upon comparing the above nodal name and the proposed systematic name,

$$H1(\underline{\underline{C}}1)_4\underline{\underline{C}}1\underline{\underline{C}}1\underline{\underline{C}}1(\underline{\underline{C}}1)_6H;^{11}[(1\underline{\underline{C}})_2 1H];^{15}[(1\underline{\underline{C}})_2(\underline{\underline{C}}1)_3H;^{(3,5)}(1\underline{\underline{C}}1H)] \quad (12)$$

one finds that the systemic name is again longer, but completely unambiguous.

Returning to another example in Gottlieb's follow-on paper, his Figure 9 is reproduced here as Figure 15. Note that two acceptable names are presented: 1,6,8-trioxa[6.2^3]octane and [4.1^2]pentane-1,4,5-triol. This allowable ambiguity follows the traditional approach of IUPAC in accepting alternate names, rather than maintaining a uniform set of priorities. In the proposed nomenclature, there is exactly one name that is **the** canonical name:

$$H1O1\underline{\underline{C}}1\underline{\underline{C}}1(\underline{\underline{C}}1)_2O1H:^{7}(1O1H) \quad (13)$$

A third historically important system of nomenclature created by Dyson in 1947, which was limited to organic compounds [94], is next examined. The name that would be generated in that system for the thirty-three carbon alkane illustrated in Figure 16 is given by Polton [95] as: Tetradecan,(3-heptan,(2-trian,monan-2)4)7,(trian,monan-2)5,(2-trian, monan-2)7. Rather than what seems to be a convoluted set of word and digit numbers, the longer but more direct systemic name would be:

$$H1(\underline{\underline{C}}1)_4\underline{\underline{C}}1\underline{\underline{C}}1C1(\underline{\underline{C}}1)_7H:^{11}[1\underline{\underline{C}}1\underline{\underline{C}}1\underline{\underline{C}}1H:^{5}(1\underline{\underline{C}}1H)];\ ^{15}\{1(C1)_2(\underline{\underline{C}}1)_3H:^{3}[1(\underline{\underline{C}}1)_2H];\ ^{5}[1(\underline{\underline{C}}1)_2H:^{(3,3)}(1\underline{\underline{C}}1H)]\};^{15}[1C1\underline{\underline{C}}1H]:^{(3,3)}(1\underline{\underline{C}}1H)] \quad (14)$$

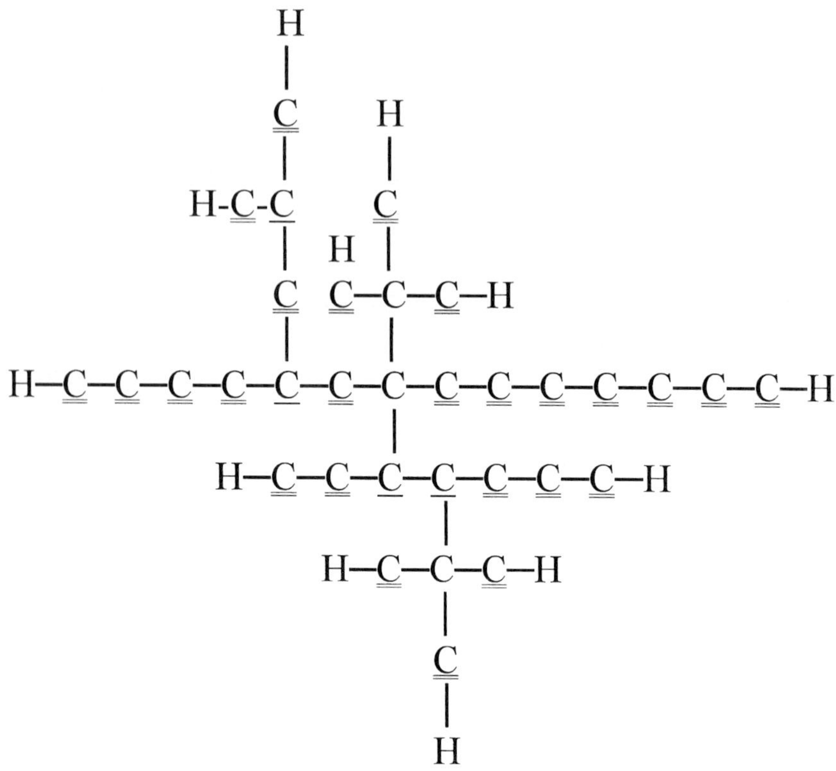

Fig. 16: A selected isomer of $C_{33}H_{68}$ that Polton used to demonstrate the Dyson system

A fourth system of mathematical interest, but of little historical or chemical interest, is the Matula system for naming rooted trees (acyclic alkanes) [10] and its extension to all graphic representation of moieties [11-15]. The output of this system is a single very large integer, which can be decoded into a unique acyclic graph. For example, the alkane depicted in Figure 16 has as its Matula name: 548,813,133,611.*

*This enormous number was obtained by starting from each of the leaves of the tree and working toward the center using the following algorithm. Label each leaf as 1 and on the edge joining that leaf to the next node affix the first prime number, namely 2. Label the node at GTD = 1 by the product of the numbers inscribed on the incoming edges. (If there were only one incoming edge, this node would be named 2. With 2 incoming edges the node is labeled 4 and with 3 incoming edges 8). The edge emanating from this node at GTD = 2 from the leaves is now labeled by either the second prime (=3) when only one edge was incoming, the fourth prime (=7) when two edges were incoming or the 8-th prime (=19). This process is continued until an agreed upon node (called the root) of the graph is reached.

Note that the examples selected in the two original nodal nomenclature reports ([19] and [20]) and in the Dyson system [49] reports are all from the much simpler class of alkanes. Moreover, both the nodal and the Dyson systems have an increase in complexity when naming molecules having multiple types of atoms, a disadvantage that the proposed nomenclature system does not have. To the contrary, because atomic symbols are included in the first "layer of information" about a chemical moiety, there is no need to create a second layer of information in which the names of the different atoms are listed. This inclusion of atom symbols immediately makes for a more user-friendly nomenclature, especially for rapidly scanning a name to see whether it is the moiety under consideration.

The extension of this coding system to monocycles is straightforward: Instead of an atom at the end of a path being labeled as locant #1, there is no end to a ring; consequently, the highest atomic number atom in a ring is designated as locant #1. Also, instead of there being only two paths to consider as the principal path, every atom of a carbocycle is a potential starting point. Similarly, when there is more than one heteroatom larger than carbon in a ring or more than one carbon atom in a carbon-boron ring, paths going in both directions from these atoms must be examined as candidates for the lead position in the principal ring. Furthermore, observe that the code for monocycles is readily distinguishable from that for paths; namely, the last item in the code of a ring is a bond, while it is an atom for a path.

Table 2.
Summary of morphemes used to nomenclate compounds

In-line numerals = bond (multiplicity) descriptors
Superscripts = locant numbering
- A pair of superscripts separated by a hyphen designates the initial and final locant numbers of a "bridging" chain.
- A pair of superscripts separated by a comma designates the existence of disjoint open chains at the indicated locant.
- Each new locant number beyond those in the principal chain are indicated by (=#). For example, see (9) Acenaphthylene, etc. in Table 1 of Chapter 3.

Subscripts = ligand (bond and atom) multiplicity.

For all but a very few small trees, the smallest number attained by this process will have the root at the center of the graph. The name assigned to that node by this protocol is a Matula number. Each number so achieved is unique to a particular rooted tree and can thus be called the name of that tree. Since each tree can have a root at any node, there are a maximum of n (= number of nodes) Matula numbers for a given alkane, and the minimum of these numbers is the canonical Matula name.

In the assignment scheme herein formulated, the name associated with any molecule is a purely machine-readable number/letter/punctuation string*. A summary of the morphemes used in assigning names to a compound is listed in Table 2.

Additional comments about the proposed systemic nomenclature include:

(1) Because of the method of selecting priorities, one seldom has to examine all potential candidates for the canonical name. Instead a cursory inspection of the principal rings (if there are any cycles in the compound) or chains (for acyclic compounds) is sufficient†. Only when several, "apparently equal" candidates for principal chain (or cycle) exist are secondary chains examined.

(2) In expanding the scope of molecules under consideration from being represented by paths to cycles, the convention adopted is that every cycle has a higher priority than any path; i.e., a ring of as few as three atoms takes precedence over a longer chain. This feature of the nomenclature is based on the perspective introduced by Harary [96], which is now the one generally accepted by the graph theory community. Namely, the smallest graph theoretical cycle has three nodes; i.e., a double bond is not a two-member ring. Similarly, neither is a triple bond a conjugated two member ring nor is an electron pair loop a one-member ring. This is notwithstanding that, under certain conditions, there is merit in adopting such an interpretation. For example, by the use of such a (pre-Harary) perspective, a nomenclature system that "pseudo-converted" polycyclic aromatic hydrocarbons into acyclic polyenynes was formulated [97-98]; additionally, bridges‡ of any length are listed before terminating chains.

---

* Because one of the motivations in creating the proposed nomenclature system was predicated on its interface with the computer, there is no need for a word-stem, such as the suffix "-nodane" created in nodal nomenclature. If one wishes to make the system more user-friendly, especially for the student user, such a word-stem could easily be included; however, a problem with such an inclusion is that it encourages the inclusion of other functionalities so that eventually several of the difficulties that the system has eliminated will be reintroduced.

† Remember that when making this cursory inspection, there is no distinction in the principle ring or path between underlined and not underlined atoms. Underlining indicates a difference in the secondary, but NOT the primary, ring/path. Consequently, for indexing, rather than nomenclating compounds, it is prudent to eschew the simplification and use the longer names in which all of the hydrogen atoms are individually spelled out.

‡ An important item of vocabulary relates to the term “bridge”. By the IUPAC definition in the domain of “organic chemistry”, a "bridge" is: “a valence bond or an atom or an unbranched chain of atoms connecting two different parts of a molecule” [99]. This, however, was shown to be an *ill-defined* word, rather than its intended status as a *primitive* word [100]. Meanwhile, in “inorganic chemistry”, the term “bridging group” was successfully incorporated as a primitive word with the denotation “a ligand attached to more

As a further word of introduction, attention is directed to organic compounds which, despite being of drastically different constitution, have their IUPAC names so closely connected that the only difference is the inclusion or exclusion of a blank space. E. W. Godley [104] in Chapter 1 of Thurlow's book, illustrates three molecules whose only difference in IUPAC name is the

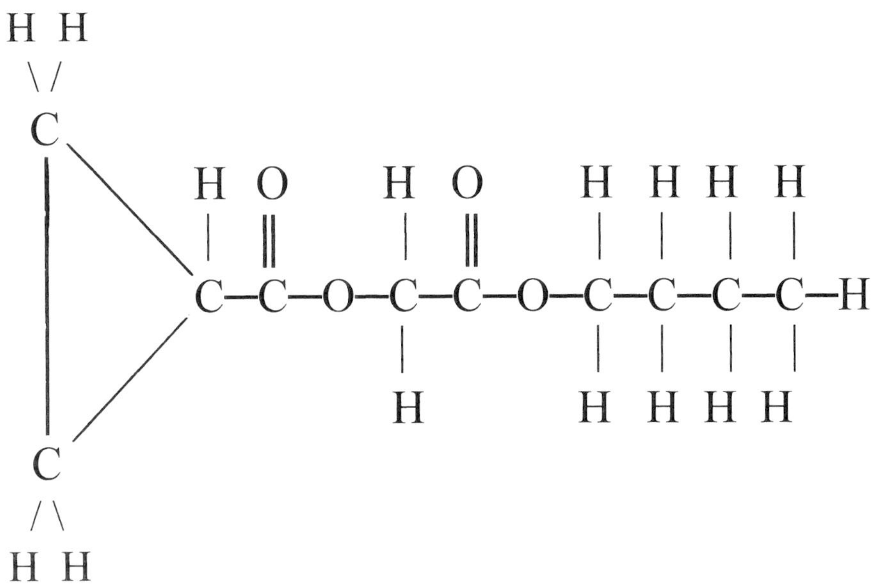

Fig. 17: IUPAC Name: Butyl cyclopropyl malonate

---

than one central atom." [101]. This distinction arose historically because of the equal importance assigned to each of the carbon atoms in a chain or ring in "organic chemistry" vs. the elevated importance for selected atoms (designated as "central") in "inorganic chemistry". In other words, the heuristics of relative importance determined the "clarity" of the definition. Consequently, in order to formulate a unified nomenclature, one must, at the expense of discarding historical precedent, choose the same perspective for both sub-domains. The milieu chosen is a graph theory based one. Harary [102] describes the term "bridge" as a single edge of a connected graph whose removal would disconnect that graph. This correlation to a chemical bond is the one that shall be selected for the purpose of nomenclating ring assemblies in Chapter 6. Meanwhile, the IUPAC organic chemistry definition is much broader in scope as it also includes a single atom, whose removal disconnects the graph; i.e., spiro compounds. For organizational purposes in formulating a canonical nomenclature, spiro compounds and ring assemblies have sufficient overlap that it is desirable to use a common approach for nomenclating them. Some additional limitations in graph theory, especially as relates to metric vs. topological properties, which become Goedelian [103], will be described at that time.

presence or absence of a space. These are presented here as Figures 17, 18 and 19. Note that this potential for ambiguity could not arise in the proposed nomenclature because of the postulation that any closed cycle is given priority over all open chains; i.e., that a three atom ring connected to a hundred atom chain will be the primary path and the ring is that part of the molecule which is to be named first; namely:

(a) butyl cyclopropyl malonate has as its systemic name:

$$\underline{C}1(\underline{\underline{C}}1)_2{:}^1[1C1O1\underline{\underline{C}}1C1O1(\underline{\underline{C}}1)_4H;^{(3,9)}2O] \qquad (15)$$

(b) butyl cyclopropylmalonate has as its systemic name:

$$\underline{C}1(\underline{\underline{C}}1)_2{:}^1\{[1\underline{C}1C1O1(\underline{\underline{C}}1)_4H{:}[^3(1C1O1H){:}^32O;^52O]\} \qquad (16)$$

and

(c) butylcyclopropyl malonate has as its systemic name:

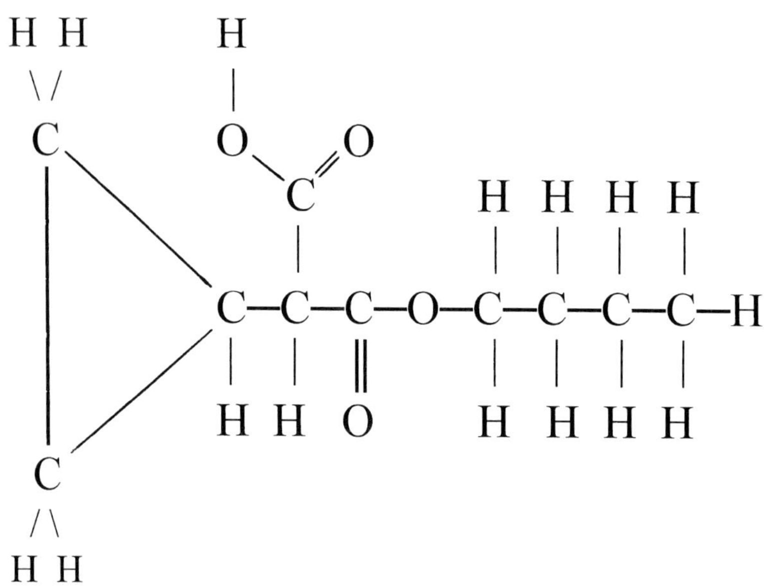

Fig. 18: IUPAC Name: Butyl cyclopropylmalonate

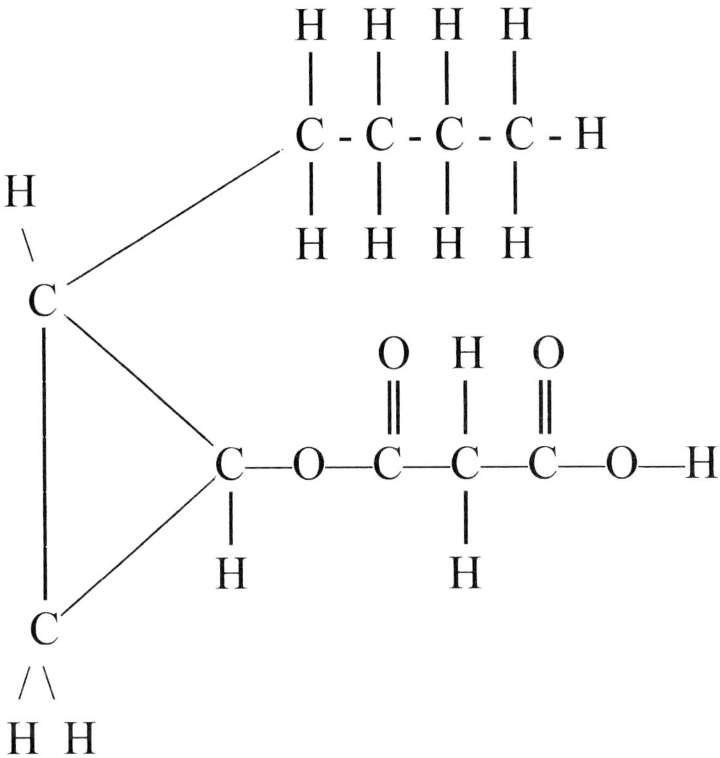

Fig. 19: IUPAC Name: Butylcyclopropyl malonate

$$(\underline{C}1)_2\underline{\underline{C}}1:^{1}\{[1O1C1\underline{\underline{C}}1C1O1H:^{(5.9)}(2O)];^{3}[1(\underline{\underline{C}}1)_4H)] \quad (18)$$

As the figures indicate, these molcecules are so sufficiently distinct from one another that upon comparing their structures there is no likelihood of confusion.

Next, one should note that the same protocol that had been developed in the proposed nomenclature for "organic" compounds is readily applied to coordination compounds in “inorganic” chemistry; namely, the monocyclic compound which I.U.P.A.C. calls: dichloro[N,N-dimethyl-2,2’-thiobis(ethylamine) -S,N’]platinum (II) (Figure 20), may be named, without resorting to different prefixes for the number two (di- and bis-) and without the prime symbol, as:

$$Pt1S1(\underline{\underline{C}}1)_2N1:^{(1,1)}1Cl;^{3}1(C1)_2N1\underline{\underline{C}}1H:^{7}(1\underline{\underline{C}}1H) \quad (19)$$

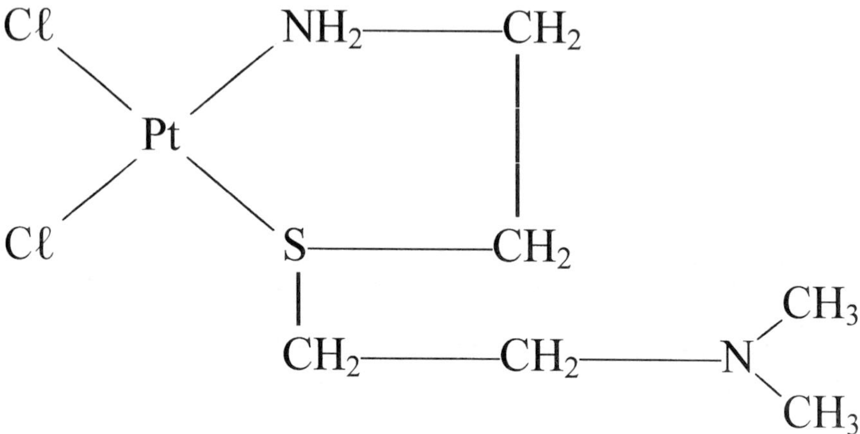

Fig. 20: IUPAC Name: Dichloro[N,N-dimethyl-2,2'-thiobis(ethylamine)-S,N']platinum(II)
Systemic Name: $\mathrm{Pt1S1(\underline{\underline{C}}1)_2\underline{\underline{N}}1{:}^{(1,1)}(1C\ell);\ {}^{3}[1(\underline{\underline{C}}1)_2N1\underline{\underline{C}}1H{:}^{7}(1\underline{\underline{C}}1H)]}$

Some further comments of significance are:

(1) In contrast to the proposed method of naming features of a graph, which is also the method used by the organic chemistry division of IUPAC [105] and in nodal nomenclature [106], the inorganic chemistry community has opted to name a set of seemingly disparate properties that are of special interest by affixing a Greek letter as an indication of this property.

Because of this difference in focus, this is a major area that needs to be finessed in formulating a common nomenclature. The first affix to be herein examined, λ, indicates a different coordination than the one anticipated by its "lowest absolute value place" in the periodic table. Namely, λ followed by a superscript indicates that the coordination is not 4 for an atom in column 14, not 3 for an atom in columns 13 or 15, and not 2 for an atom in column 16.* For example, λ^6-sulfane indicates that six hydrogen atoms are attached to a central sulfur. The proposed systemic name for this compound is:

$$\mathrm{H1S1H{:}^{(3,3,3,3)}(1H)} \qquad \text{or alternately} \qquad \mathrm{H1\underline{\underline{S}}1H{:}^{(3,3)}(1H)} \tag{20}$$

Note that although it would be consistent to do so, using more than two underlines to indicate more hydrogen atoms is avoided; not only because most computer programs do not have triple and higher underscore keys, but, more importantly, because such abbreviated forms would be prone to error in

---

* Covalent bonds for other coordinations are seldom, if ever, encountered.

reading by humans.*

Three other Greek affixes (η, μ and κ), all of which are used much more frequently in I.U.P.A.C. nomenclature shall be encounter in Chapter 2. Additionally, the inorganic chemistry community employs a set of shape affixes, which are abbreviations of various geometrical figures. For example, TBPY-5 denotes a Trigonal BiPYramid where the 5 redundantly advises that five atoms are at the vertices of this solid figure, etc. Naming of these latter affixes is easily done in the proposed system by focusing on the edge sets from which they are formed; e.g., the trigonal bipyramid has nine edges (out of the possible set of ten line segments that can be drawn between the five points. The two apical vertices are not connected.) Such a polyhedron containing five like atoms "X" joined by single bonds (Figure 21) would be named:

$$(X1)_5{:}^{(1\text{-}3,1\text{-}4,2\text{-}4,2\text{-}5)}(1) \qquad (21)$$

Many of these polyhedral shapes will be important when nomenclating boranes. (See Chapter 5).

(2) The traditional idea of the number of "bonds" versus the fact that a covalent bond is normally the union of two electrons forms the basis for describing the connection between atoms. An alternative representation, which does have more than just a modicum of merit, would be to present (9) above as:

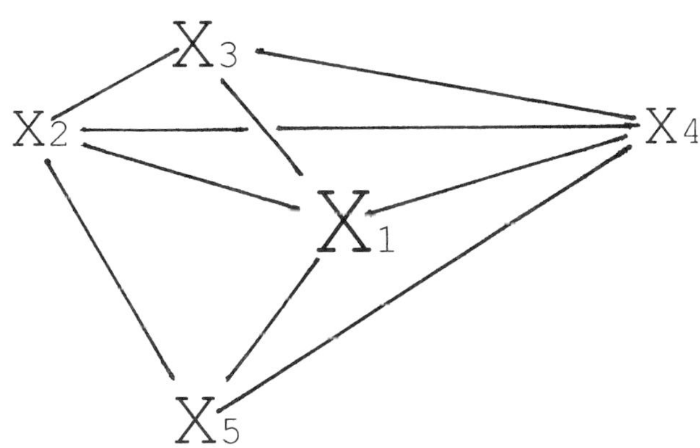

Fig. 21: IUPAC Name: TBPY-5
Systemic Name: $(X1)_5{:}(1)^{(1\text{-}3,1\text{-}4,2\text{-}4,2\text{-}5)}$

---

* In Chapter 6, a simplification that could be applied to (18) using spherical nomenclature is introduced; however, discussion of this alternate nomenclature shall be delayed until that time.

$$\mathrm{H2C6(C2\underline{\underline{C}}2)_2C2C6C2\underline{\underline{C}}2H{:}^{9}((4C)_2 2H){:}^{5}(2H));^{13}(4C2H{:}^{3}(2H))} \qquad (22)$$

(22) counts the number of electrons shared by atoms, rather than on the number of bonds between them. The virtue of such a scheme is that **much** of the nomenclature that shall be developed in the next chapter, relating to "alpha" and "beta" bonds, could be replaced by integers — especially in the domain of organic chemistry, as well as in boron compounds in inorganic chemistry. In these areas the bond orders of alpha bonds frequently are one-half while beta bonds are one and a half. **However**, because this is not always the case, the disadvantages outweigh the advantages. Those instances in which a false picture would be created, such as when hydrogen atoms bridge more than two atoms in the inorganic domain and when ring formation is best described using aleph bonds (See Chapter 2), make adopting such a scheme undesirable.

REFERENCES

[1] D.H. Rouvray, Endeavor, 1 (1997) 23.
[2] A. Lavosier, Mem.Acad.Roy.Soc.,Paris 1782, p.202.
[3] J.H. Hassenfratz and P.A. Adet, Methode de Nomenclature Chiumique, A. Lavosier, (Ed.), Cuchet, Paris, 1787, p.253.
[4] R.S. Cahn and O.C. Dermer, Introduction to Chemical Nomenclature, 5. Ed., Butterworth, London, 1979, p.2.
[5] Ibid.
[6] S.B. Elk, THEOCHEM, 358 (1995) 119.
[7] S.B. Elk, J.Chem.Inf.Comput.Sci., 37 (1997) 835.
[8] Ibid #4.
[9] Ibid #7.
[10] D.W. Matula, S.I.A.M. Rev., 10 (1968), 273.
[11] S.B. Elk, Graph Theory Notes of N.Y., XVIII (1989) 40.
[12] S.B. Elk, J.Math.Chem., 4 (1990) 55.
[13] I. Gutman, A. Ivic and S.B. Elk, J.Serb.Chem.Soc., 58 (1993) 193.
[14] S.B. Elk & I. Gutman, J.Chem.Inf.Comput.Sci., 34 (1994) 331.
[15] S.B. Elk, J.Chem.Inf.Comput.Sci., 35 (1995) 233.
[16] S.B. Elk, J.Chem.Inf.Comput.Sci., 34 (1994) 942.
[17] S.B. Elk, J.Chem.Inf.Comput.Sci., 34 (1994) 637.
[18] International Union of Pure and Applied Chemistry, Nomenclature of Organic Chemistry: Section A, Pergamon Press: Oxford, U.K., 1979.
[19] N. Lozac'h, A.L. Goodson and W.H. Powell, Angew.Chem.Int.Ed.Engl., 18 (1979) 887.
[20] N. Lozac'h and A.L. Goodson, Angew.Chem.Int.Ed.Engl., 23 (1984) 33.
[21] S. H. Bertz, Chem.Appl.Topology & Graph Theory, 28 (1983) 206.
[22] F. Harary, Graph Theory, Addison-Wesley, Reading, Ma., 1969, p.37.
[23] K. Goedel, On Formally Undecidable Propositions, Basic Books, New York, 1962.

[24] R.C. Read and R.S. Milner, Research Report CORR-78-42, Dept. Combinatorics and Optimization, Univ. Waterloo, 1978.
[25] A.L. Goodson, J.Chem.Inf.Comput.Sci., 20 (1980) 167.
[26] S.B. Elk, J.Chem.Inf.Comput.Sci., 38 (1998) 54.
[27] C.U. Kim et al, J.Am.Chem.Soc. 119 (1997) 681.
[28] Oxford English Dictionary, Clarendon Press, Oxford, VII (1993) 204.
[29] S.B. Elk, THEOCHEM, 313 (1994) 199.
[30] S.B. Elk, J.Chem.Inf.Comput.Sci., 34 (1994) 325.
[31] Ibid #6.
[32] S.B. Elk, J.Chem.Inf.Comput.Sci., 36 (1996) 385.
[33] Ibid #7.
[34] S.B. Elk, MATCH, 36 (1997) 157.
[35] S.B. Elk, THEOCHEM, 489 (1999) 177.
[36] S.B. Elk, THEOCHEM, 589-90 (2002) 27.
[37] The American Heritage Dictionary of the English Language, 3-rd Ed. Houghton-Mifflin Co., Boston, Mass, 1992, p.849.
[38] Brian Pfaffenberger, The Webster New World Dictionary of Computer Terms, 8-th Ed., IDG Books Worldwide, Inc., Foster City, Calif, 2000, p.256.
[39] Webster's New Collegiate Dictionary; G&C Merriam Co.; Springfield, Mass, 1981, p.734.
[40] Ibid, p. 28.
[41] Ibid, p. 608.
[42] Ibid, p. 234.
[43] Ibid, p. 235.
[44] S.B. Elk, MATCH 31 (1994) 89.
[45] Ibid #22, p.13.
[46] Ibid, p. 64.
[47] Ibid, p. 65.
[48] S.B. Elk, THEOCHEM 431 (1998) 237.
[49] G.M. Dyson, A New Notation and Enumeration System for Organic Compounds; Longmans, Green and Co., London, 2-nd Ed., 1949.
[50] W. Prenowitz and M. Jordan, Basic Concepts of Geometry, Blaisdell Publ. Co., Waltham, Mass. 1965, p.4.
[51] A.M. Patterson, L.T. Capell and D.F. Walker, The Ring Index, (1-st Ed.), Am.Chem. Soc., Washington, D.C., 1940; 2-nd Ed. 1960; Supplements 1963, 1964, 1965.
[52] J.E. Blake et al, J.Chem.Inf.Comput.Sci., 20 (1980) 162
[53] O.R. Gottlieb and M.A.C. Kaplan J.Chem.Inf.Comput.Sci., 26 (1986) 1.
[54] S.B. Elk, J.Chem.Inf.Comput.Sci., 37 (1997) 696.
[55] Ibid #22, p. 17
[56] Ibid #18.
[57] A.L. Goodson, J.Chem.Inf.Comput.Sci., 20 (1980) 172.
[58] Ibid #4, p. 45.
[59] T.W.G. Solomons, Organic Chemistry, 5-th Ed., John Wiley, New York, 1992, p.458.
[60] S.B. Elk, MATCH, 23 (1988) 19.
[61] www.iupac.org/publications/pac/71_08_pdf/7108salzer_1557.pdf
[62] S.B. Elk, J.Chem.Inf.Comput.Sci., 25 (1985) 17.
[63] W.J. Wiswesser, A Line Formula Chemical Notation, Thomas Y. Crowell Co, New York, 1954

[64] P.vR. Schleyer and H. Jiao, Chem.Int. 18 (1996) 205.
[65] F.L. Taylor, Ind.Eng.Chem., 40 (1948) 734.
[66] Ibid #57.
[67] A.M. Patterson, J.Am.Chem.Soc., 47 (1925) 543.
[68] S.B. Elk, THEOCHEM, 431 (1998)237; footnote on 239..
[69] A. Greenberg and J.F. Liebman, Strained Organic Molecules, Academic Press, New York, 1978.
[70] J.H. Van't Hoff, Arcxh. Neerland. Sci, 9 (1874) 445.
[71] J.A. LeBel, Bull.Soc.Chim.France 22 (1874) 337.
[72] A.F. Wells, The Third Dimension in Chemistry, Clarendon Press, Oxford, U.K. 1956.
[73] Ibid #66.
[74] S.B. Elk, THEOCHEM, 201 (1989) 75.
[75] E.J. Corey and G.A. Peterson, J.Am.Chem.Soc., 94 (1972) 460.
[76] A. Zamora, J.Chem.Inf.Comput.Sci., 16 (1976) 40.
[77] S.B. Elk, J.Chem.Inf.Comput.Sci., 25 (1985) 11.
[78] Ibid #72.
[79] S.B. Elk, J.Chem.Inf.Comput.Sci., 24 (1984) 203.
[80] K. Kuatowski, Fund.Math., 15 (1930) 271
[81] S.B. Elk, J.Chem.Inf.Comput.Sci., 30 (1990) 69
[82] S.B. Elk, NYAS GTD Notes, XIV (1987) 44
[83] Ibid #54.
[84] Ibid #7.
[85] Ibid #65.
[86] Ibid #57.
[87] International Union of Pure and Applied Chemistry, Nomenclature of Inorganic Chemistry, 2-nd Ed., Definitive Rules, 1970, Butterworths, London, p.5.
[88] Ibid #4, p. 63.
[89] S.B. Elk, J.Chem.Inf.Comput.Sci., 37 (1997) 162.
[90] Ibid #18, p. 14.
[91] J.A. Young, J.Chem.Doc., 7 (1967) 82.
[92] J.A. Young, J.Chem.Doc., 14 (1974) 98.
[93] Ibid #19,20 and 53.
[94] Ibid #47.
[95] D.J. Polton, Chemical Nomenclatures and the Computer, Research Studies Press, Taunton, Somerset, England, 1993, p.19.
[96] Ibid # 22, p.13.
[97] S.B. Elk, J.Chem.Inf.Comput.Sci., 26 (1986) 126.
[98] S.B. Elk, Graph Theory Notes of N.Y., XVI (1988) 29.
[99] Ibid #18, p.32, footnote.
[100] S.B. Elk, J.Chem.Inf.Comput.Sci., 27(1987) 70.
[101] Ibid # 4, p.17.
[102] Ibid # 22, p.26.
[103] Ibid # 23.
[104] E.W. Godly, Chemical Nomenclature, K.J. Thurlow (Ed.); Kluwer Academic Publishers, Dordrecht, The Netherlands, 1998, p.9.
[105] Ibid #18.
[106] Ibid #19.

Chapter 2

Non-integer bonds

CHAPTER ABSTRACT:

In addition to the traditional integer (single, double, triple, and, in inorganic chemistry, quadruple) bonds, some additional bonds having selected standardized bond orders are introduced. Three of these play a major role in establishing a unified system of nomenclature:

(1) "alpha" (α) bond: a bond intermediate between no bond and a traditional single bond

(2) "beta" (β) bond: a bond intermediate between a traditional single and a double bond

(3) "aleph" (ℵ) bond: a bond with bond order near to 1, but with certain properties more closely related to a beta bond than to the traditional single bond.

Two other standardized bonds, one an intermediate bond characterized by non-integer bond orders higher than 2 and the other with bond order near to 2, **but** with certain properties distinct from the traditional double bond, are also postulated. These, however, are of lesser importance. Instances where α bonds are used include nomenclating compounds having 3-center-2-electron bonds, such as the bond between a boron and a hydrogen atom in diborane, and also in systems of atoms in which there is a strong presence of hydrogen bonding. β bonds are of major importance in aromatic compounds, especially in situations wherein one wishes to designate "resonance" versus "fixed" bonds. Both α and β bonds are used for canonically naming most of the oxy (inorganic and organic) acids. ℵ bonds are used primarily in nomenclating organo-metallic compounds in which the bond between a metal atom and a carbon is substantially longer than traditional single bonds, but in which a degree of conjugation (which is usually indicative of a shortening of bonds) is implied. Because of the introduction of these standardized intermediate bonds, many fundamental chemical properties are more clearly delineated. Thirty examples of problems with ambiguity or inconsistency in traditional inorganic and organic chemistry nomenclature that would either never have arisen or else would be solved by use of the proposed system are presented, along with a detailed discussion of the synergy with chemical structure that has been introduced by the proposed nomenclature.

Two important properties of the interconnection between pairs of atoms which chemists call a "chemical bond"* are the "bond length" and the "bond strength". Bond length is defined as the distance from the center of one atom to the center of a neighboring atom. Bond strength refers to the mean disassociation enthalpy between two atoms. Because both the bond length and bond strength are approximately the same between identical atoms in different compounds (e.g., the length and strength of an oxygen to hydrogen "bond" in water is nearly equal to a similar "bond" in methanol, as well as most other compounds in which these two atoms are covalently bonded), a means of quantifying the description of bonds exists.

Meanwhile, however, a third property of great importance in describing chemical moieties, bond angle [both between three atoms (coplanar) and between four or more atoms (in 3-dimensional space)], is not **as** amenable to standardization. Nevertheless, because some degree of standardization is better than none, a partial utilization of the system based on G. N. Lewis's description of bonding [2] that defines "bond order" as: "one-half the number of electrons in bonding orbitals minus one-half the number of electrons in anti-bonding orbitals" has been adopted.

At this point a major mathematical shortcoming in Lewis's definition is noted: namely, his measurement of bond order uses a discrete variable, rather than a continuous one. To the contrary, the empiricism of the entire concept of bond order should be noted; namely, what has been designated as a single bond is NOT precisely bond order = 1.000... Not only is such accuracy unattainable, there is not exact equality of bond order for **all** permutations of atoms that have been designated as having a single bond between them. Similarly, neither are all double bonds precisely bond order = 2.000... , etc. In actuality, bond order is a gross oversimplification that depends on which two (or more) atoms are being bonded together, as well as the environment in which this bonding is taking place. This is especially

---

* Pauling [1] advises "there is a chemical bond between two atoms or groups of atoms in case that the forces acting between them are such as to lead to the formation of an aggregate with sufficient stability to make it convenient for the chemist to consider it as an independent molecular species".

true with a system of alternating single and triple bonds, as is illustrated by"In many molecules containing systems of alternating single and multiple bonds (*conjugated systems*) there is very considerable shortening of formal single bonds. In molecules such as

$HC{\equiv}C\text{-}C{\equiv}H$ and $H_3C\text{-}C{\equiv}C\text{-}C{\equiv}C\text{-}CH_3$

not only is the central bond (1.38Å) nearly as short as a double bond but the terminal $C\text{-}CH_3$ bonds are also shortened (to about 1.46Å)."

In this example, as well as in many other cases, the line that separates the individual bond orders in adjacent bonds is not clearly drawn. A partial solution occurred in 1925 with the introduction of the Robinson ring as a better descriptor for benzene [4]. Although this representation corrected the false picture of benzene having localized single and double bonds and thus being nomenclated as 1,3,5-cyclohexatriene, no other significant use of non-integer bonds or the effect of "bonding" between non-adjacent atoms (e.g., the Dewar benzenes) has been accepted. Moreover, even this one notable success has had no influence on chemical nomenclature. Today, one still encounters many similar types of inconsistency when nomenclating selected molecules. This will be described in detail in a later part of this chapter.

One of the most significant differences between the system of nomenclature being developed in this treatise and the internationally agreed upon IUPAC system, as well as other proposed systems that have appeared (and then disappeared through neglect), is an introduction and a selective use of bonds having non-integer bond orders. For purposes of formulating a coherent nomenclature, it is important to differentiate between **adjacent integer** and **non-adjacent integer** bond orders: Sequential polyacetylenes (molecules having an alternating sequence of triple and single bonds) are distinct from and thus should be distinguishable from the corresponding length cumulenes (molecules having successive double bonds). For both such systems continued use of these traditional bonds is desired.

In contrast to **non-adjacent integer** bond orders, alternation between **adjacent integer** bond orders presents an opportunity to better describe the observed chemistry. This is especially true for systems in which the alternation in bond order is between single and double bonds. Because such an alternation occurs so frequently in molecules, the proposed nomenclature system shall regard selected cases of this alternation as a standardized bond with a bond order intermediate between the two integer values, while retaining the traditional single and double bond in others. In other words, rather than Well's extended usage of the term "conjugated", the nomenclature being proposed accepts the traditional connotation associated with that term; namely that it be limited to alternation between single and double bonds. Additionally, the sequence of single bond with no bond is of special interest. For these two scenarios, standardized bond symbols, such as letters of the Greek alphabet, are introduced to describe those bonds having these particular "bond orders".

Quantification of the concept referred to as "bond order" is more easily undertaken in "organic chemistry", where it is limited to sigma and pi bonds. However, in order to be able to finesse the differences that have historically (and some might say even logically) evolved in their respective fiefdoms and to establish a common nomenclature across **all** chemistry, this idea must be extended to all types of bonds. In particular, it must be applied to pi-delta and delta-delta bonds in the traditional domain of "organo-metallic" chemistry. In making such an extension, some ambiguity will be unavoidable. Consequently, although acknowledging that the foundation on which the entire concept of bond order is built is an oversimplification, the immediate goal is to use this concept, where appropriate, to maximize order while minimizing contradiction and/or overlap.

Because of the incongruity in the desired extension of "bond order" (using a discrete variable as the measure for a concept that requires a continuous variable), any nomenclature will have to lump together structures having "similar" bond orders. Here the heuristics of "similar" is assumed — with all of the consistency problems that ***could*** arise by such an assumption[*]. In the proposed system,

bond orders greater than 0 and less than 1 are designated by the symbol α,

bond orders greater than 1 and less than 2 by the symbol β,

and non-integer bond orders greater than two by γ [†].

---

[*]See footnote on page 20.

[†] For pragmatic purposes, because the concept of bond order is an empirical one, all values "near" to an integer may be regarded as though it were that integer; consequently, α will connote a real number with bond order somewhere between, say, 0.2 and 0.8 and β between 1.2 and 1.8. This wide range of deviation from integer value bonds was selected in order that single bonds continue to be used in formulating the nomenclature of peroxides, as well as having an equal interpretation in both alcohols and phenols. (Note that in hydrogen peroxide, as well as in alkyl peroxides, the bond between the oxygen atoms is much more easily broken than the bond between an oxygen atom and its hydrogen or carbon neighbor — both of which are still represented using single bonds. Similarly, there is a preferential breaking of the O-C bond in alcohols vs. the O-H bond in phenols).

Note the compatibility of this choice with the Robinson ring indicating a bond and a half between each pair of carbon atoms in benzene.

Continuing in this vein, it might seem desirable to have selected different symbols to designate fractional bond orders between 2 and 3, and between 3 and 4. However, the effects of non-integer bond orders greater than two are, at present, only evident in the classical domain of organo-metallic compounds. For this domain there is no presently-known advantage to the nomenclature in making such a distinction. To the contrary, the differences in the chemistry between the classical carbon-carbon triple bond in ethyne vs. the molybdenum-molybdenum triple bond in $[Mo_2O_6PO_4H]^{2-}$ are greater than the difference between the bonds in that molybdenum compound and in other molybdenum and ruthenium compounds such as $[Mo_2SO_{10}]^{3-}$and $[Ru_2C_2O_8Cl_2H_3]^-$, which have Lewis bond orders of $2^1/_2$ and $3^1/_2$ respectively [5]. Consequently, in the proposed system, such bonds are lumped together and designated by the symbol γ. Whether it is better to also view the "triple" bond between the two molybdenum atoms in $[Mo_2O_6PO_4H]^{2-}$ as a γ bond is left as ambiguous — comparable to the use of ℵ bonds later in this section. Without further study, judgment is reserved as to which bond order (γ or 3) is a better descriptor for this compound. In either case, no uncertainty is introduced into the nomenclature regardless which of these choices is made.

At this point it should be noted that although the pi-delta bond between a carbon and a metal atom is both weaker and longer than a traditional sigma or pi bond between two carbon atoms, it often seems to provide a continuity of "aromatic character" throughout an entire aggregate of atoms which forms a graph theoretical block [6]. In other words, it is desirable to postulate a new type of standardized bond. This bond has a bond order greater than an α bond and *usually* less than or equal to a traditional single bond. On the other hand, it can be viewed as extending aromaticity and thus has properties associated with the previously defined β bond -- which was defined above as having a bond order greater than 1. This seeming contradiction can be finessed by choosing a symbol from a second alphabet (Hebrew) ℵ ("aleph"). In other words, the ℵ bond is not ordered using the same sequential relationship that one uses for the traditional integer and the postulated Greek bonds. The prescribed ordering relationships are thus: $\alpha < \aleph < \beta$ and $\alpha < 1 < \beta$. Also, **usually but not guaranteed**, $\aleph \leq 1$.

Now, just as the introduction of a "modified single bond" (the ℵ bond) is of value in describing the chemistry of a large set of compounds (mostly in the organo-metallic domain), the recent discovery of a trisilaallene (-Si=Si=Si-) with a bond angle of 136°, instead of the 180° normally associated with sp-hybridization, might be better served by designating these "modified double bonds" by a specialized symbol, say **ב,** the second letter of the Hebrew alphabet ("bet"). The details associated with the nomenclature of this compound will be developed in Chapter 6. So far, this is the only compound in which the bet symbol seems advantageous; however, it should be noted that this is a very recent discovery, which, most likely, will lead to many other examples.

Returning focus back to the above discussion as to whether a traditional triple bond or a gamma bond better describes the chemistry of $[Mo_2O_6PO_4H]^{2-}$, one might also opt for the postulation of a ג (gimel, the third letter of the Hebrew alphabet) bond; thereby being compatible with א and ב bonds correlating to traditional single and double bonds.

Before tabulating specific examples wherein there is advantage in the proposed nomenclature, two features of this system should be reiterated:

(1) There is an alternation of atoms and bonds throughout the entire canonical name being formulated

(2) Coding (for both chains and rings) begins with an atomic symbol.

Consequently, a principal chain *of a given length* will always end with another atomic symbol, while a principal ring *(which is always a specified length)* will end with a bond descriptor. Thus, there is no need for the prefixes "catena" and "cyclo" in IUPAC inorganic nomenclature [7], as well as no use of the prefix "cyclo" in IUPAC organic nomenclature [8].

Many of the benefits that the choice of standardized intermediate bonds introduces comprise the remainder of this chapter. Additional advantages shall be introduced in succeeding chapters along with the development of alternate schemes of nomenclating. Meanwhile, it should be noted that continued usage of traditional single bonds in those places where the proposed systemic nomenclature would use aleph bonds, while inferior from the perspective of chemical precision, presents no major difficulty in communicating the desired connectivity in such molecules. This is true even though, especially in the organic case, it often masks the delocalization of bonds; i.e., the aromaticity.

A tabulation of examples wherein the proposed nomenclature either corrects or bypasses problem areas endemic to other nomenclature system, especially IUPAC, follows:

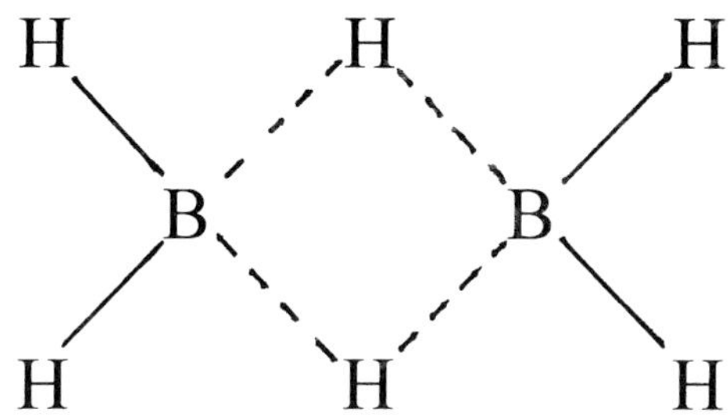

Fig. 1: Diborane

(1) The connectivity of atoms in the central part of the diborane molecule is by means of 3-center-2-electron bonds, which have a bond order substantially less than a single bond. This is represented in both Figure 1 and (1) by using α bonds in the proposed, systemic nomenclature. On the other hand, the four remaining hydrogen atoms are singly bonded to boron atoms in the traditional manner; consequently, the canonical name for this molecule (which has been further shortened by using the standardized underscore abbreviation for the boron hydrogen module) is:

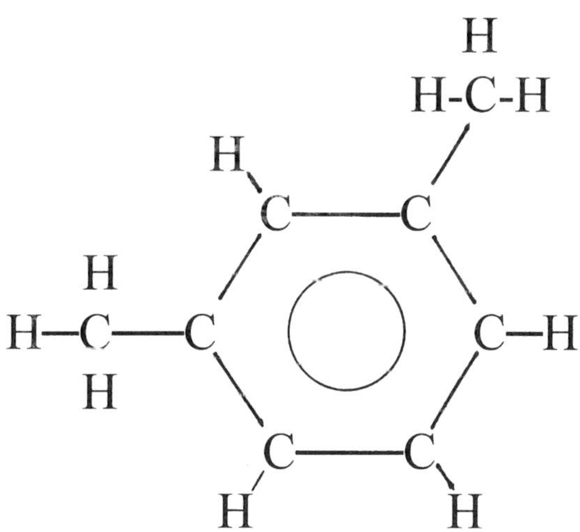

Fig. 2: 1,3-Dimethylbenzene

$(\underline{\underline{B}}\alpha H\alpha)_2$ (1)

(2) In a similar manner, using the proposed, systemic β bond, 1,3-Dimethylbenzene, (Figure 2) would be nomenclated as:

$(C\beta\underline{C}\beta)_2(\underline{C}\beta)_2{:}^{(1,5)}(1\underline{\underline{C}}1H)$ (2)

Observe that in (2) there was no need to assign locant numbers beyond those on the principal chain. (Unfortunately, this may not be possible in all cases, as will be noted below.)

(3) Unlike Well's use of the term "conjugated" [9], what this word is to connote in a unified nomenclature is in accordance with emphasizing the difference in functional groups between isolated, conjugated, and cumulenic double bonds; namely, use integer bond orders (1, 2, 3) for isolated bonds and bond order = 2 for all cumulenic bonds, HOWEVER **always** designate by β conjugated bonds, *even those not in a ring*. For example, name the four carbon chain having adjacent double bonds, which is a member of the functional group called "cumulenes"*, and which IUPAC names as 1,2-butadiene:

$H1\underline{C}2C2\underline{C}1\underline{\underline{C}}1H$ (3)

vs. the conjugated 1,3-butadiene:

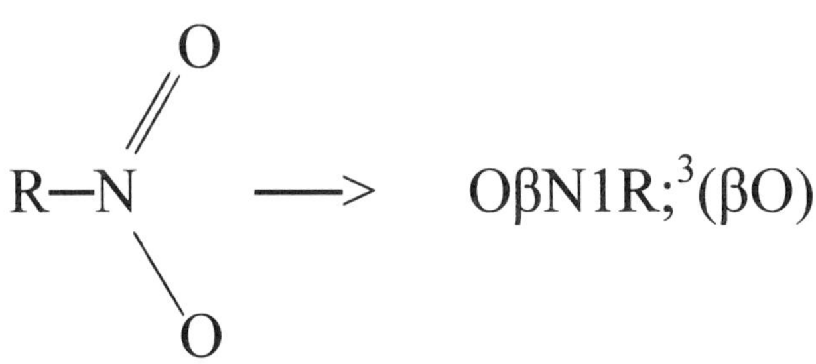

Fig. 3: Traditional and systemic representation of a nitroalkane

*The sequence C2C2C is the "signature" for cumulenes, just as C1O1C was the signature for ethers and C1O1H was the signature for either alcohols or phenols (See page 13 in Chapter 1).

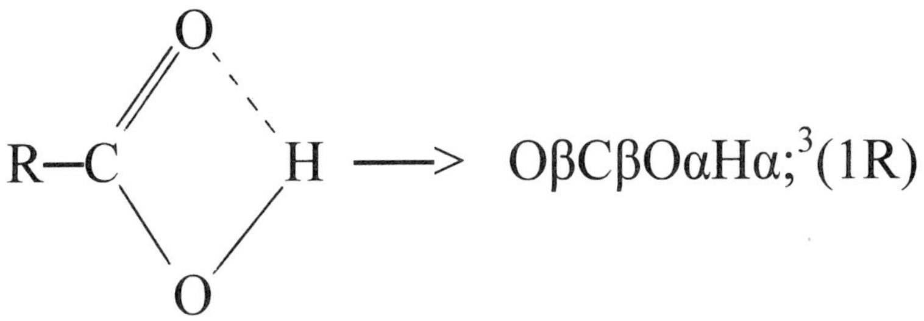

Fig.4: Traditional and systemic representation of a carboxylic acid monomer

$$\mathrm{H1(\underline{C}\beta)_3\underline{C}1H} \qquad (4)$$

(4) For **all** groupings of atoms, the nomenclature should accurately describes the observed chemistry. For example, consider the names that should be assigned to a nitroalkane (Figure 3) and to an alkanecarboxylic acid (Figure 4). An initial, superficial impulse might be to assign the names:

$$\mathrm{R1N(\beta O)_2} \qquad (5)$$

and

$$\mathrm{R1C(\beta O\alpha)_2H} \qquad (6)$$

However, there is a problem with each of these formulations: (5) advises that the second oxygen atom is bonded to the first oxygen atom, rather than forming the desired bond to the nitrogen atom. Such a code is indicative of a "peroxide" functional group.* Instead of this incorrect constitution, the desired systemic name† of this molecule is:

$$\mathrm{O\beta N1R{:}^{3}(\beta O).} \qquad (7)$$

---

*Although such a code might suggest a peroxide, this is most improbable since the bond between oxygen atoms in a peroxide are weaker, rather than stronger, than the standard single bond. Consequently, a beta bond, which has bond order greater than 1, is contraindicated.

†The generic symbol R traditionally denotes only carbon and hydrogen atoms; therefore, the oxygen atom has priority in determining which end of the chain is the beginning.

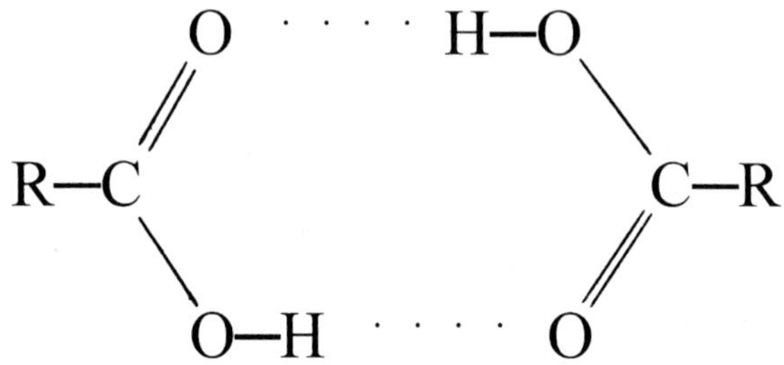

Fig. 5: Traditional picture of a carboxylic acid dimer

Although (6) does not refer to a different constitution, as did (5), it is incompatible with the coding techniques that have been used *so far*.* The math model developed up to this point does not contain structures with two bonds together – without an intervening atom; consequently, until this part of the nomenclature is developed at a later time, alternation of atoms and bonds is required. Furthermore, such a scheme is not needed. A simple cyclic repetition conveys the desired bonding pattern of the monomer; namely:

$$O\beta C\beta O\alpha H\alpha{:}^{(3)}(1R). \qquad (8)$$

At this point, it should be noted that Figure 4 portrays a carboxylic acid monomer. This is notwithstanding that in most instances there is hydrogen bonding between two carboxylic acid groups, thereby creating an eight member hydrogen bonded "ring". A logical extrapolation from "monomer"† to "dimer" suggests that the dimer be nomenclated as:

$$(O\beta C\beta O1H\alpha)_2{:}^{(3,11)}(1R) \qquad (9)$$

This, however, is not what is found in the laboratory. Instead, the measured bond lengths in formic acid, HCOOH, are: C=O = 120.2 nm and C-O = 134.3 nm [10]. These bond lengths are consistent

*In Chapter 3, such a coding shall be introduced for "cylindrical" molecules.
†The terms "monomer" and "dimer", as well the logical extension to "polymer", have the denotation of a cohesive group of atoms acting once, twice or many times as a congruent unit (see [1] of this chapter). This idea is carried over to the term "isomer", see definition #9 on page 9 (Chapter 1).

with the usual values for such double vs. single bonds respectively. In other words, in the monomer one set of bond lengths are measured; however, when two monomers join to form the dimer, there is a reversion back to a state wherein the single vs. double bonds are distinct (Figure 5). In this state, each of the carbon atoms of the carboxylic acid is double bonded to one oxygen atom and single bonded to the other while each of the hydrogen atoms forms a traditional intra-molecular covalent single bond with one neighboring oxygen atom and an inter-molecular hydrogen bond with the other. This is reflected in the nomenclature as:

$$(O2C1O1H\alpha)_2{:}^{(3,11)}(1R) \quad (10)$$

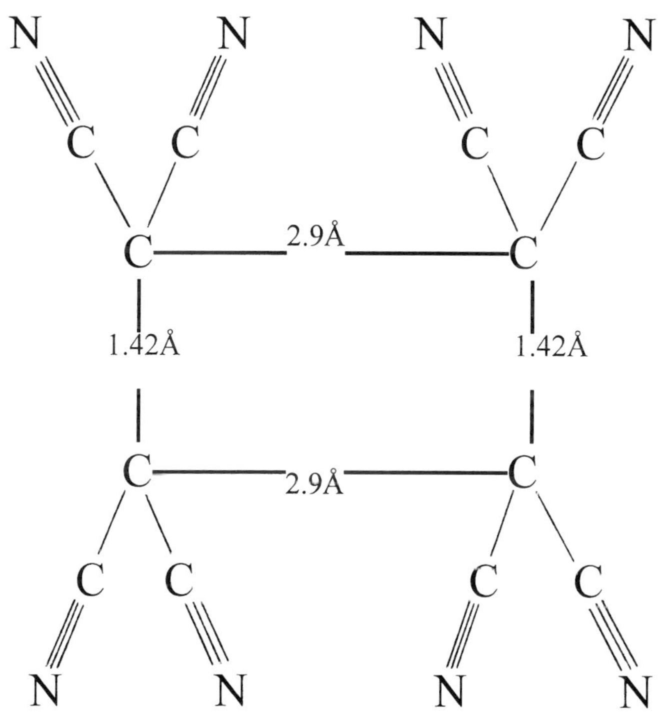

Fig. 6: Dimer formed from two tetracyanoethene modules

Furthermore, as (10) illustrates, multiplicity of the singly-connected R groups is indicated in the superscripts ***only.*** A subscript after the R would imply a repeat of the same R group, such as in an ethenyl vs. a methenyl group, etc.

(5) A logical extrapolation of the nomenclature used with the molecules described in Item #4 is next applied to the "extremely long" *single* bond (290 nm vs. the traditional 154 nm) [11] that is formed when two molecules of tetracyanoethene interact to form a four member carbon ring (Figure 6). Contrary to the explanation given, one can view this bond **NOT as "extremely long"**, but precisely consistent with the explanation of four carbon atoms sharing two electrons. Namely, let the 142 nm bond be designated by a beta bond and the bond with length 290 nm by an alpha bond.

$$(C\beta C\alpha)_2{:}^{(1,1,3,3,5,5,7,7)}(1C3N) \tag{11}$$

In other words, the nomenclature describes the relevant physics and one need not be concerned with the nature of the orbitals (π^*). On the other hand, had this been merely a transition state that is followed by a rearrangement to form a skew rhombus with traditional single bonds, this would also be reflected in the nomenclature; namely the final product would then be nomenclated as:

$$(C1)_4{:}^{(1,1,3,3,5,5,7,7)}(1C3N) \tag{12}$$

(6) The use of alpha and beta bonds simplifies describing several familiar "inorganic" anions. Figure 7 illustrates the traditional Lewis structure of the sulfate ion, with an explanatory note that a more realistically picture (which will involve metric, as well as graph theoretical, distances [12]) is a tetrahedral resonance hybrid [13]. In other words, there are four identical oxygen atoms, each bonded by 1 and 1/2 bonds to the central sulfur atom, rather than two single and two double bonds. The name for this aggregation:

$$[S(\beta O\alpha)_4]^{2-} \tag{13}$$

anticipates the introduction of spherical nomenclature in Chapter 6. It has been included at this time because of the importance of these ions.

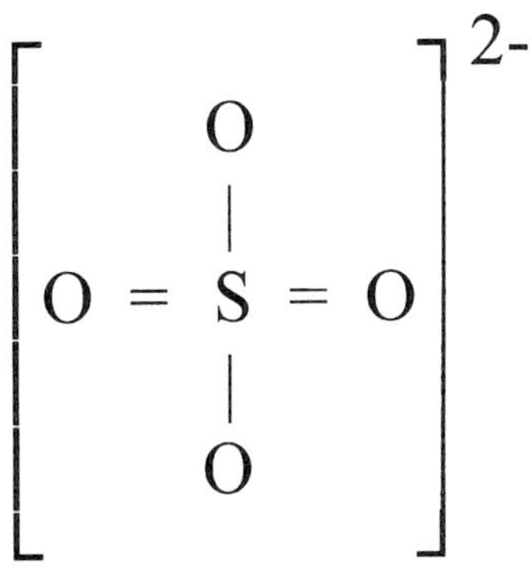

Fig. 7: Lewis structure of a sulfate ion

(7) Similar to the above description for the sulfate ion, in the thiosulfate ion, the distinction between the constitutionally different sulfur atoms is emphasized by the formula:

$$[S(\beta S\alpha),(\beta O\alpha)_3)]^{2-} \tag{14}$$

Additionally, unlike as suggested in the Lewis structure, this proposed systemic nomenclature correctly describes that neither the ligand sulfur nor any of the three oxygen ligands should be viewed as having priority to being "doubly bonded" to the central sulfur atom.

(8) A "nearly equivalent" picture to that of the sulfate ion describes the phosphate ions. For the orthophosphate ion, the phosphorus atom forms a completed outer shell by using its five valence electrons to form four double bonds with the four oxygen atoms, leaving a tetrahedral symmetric ion with a charge of minus 3, while the metaphosphate ion has a coplanar trigonal geometry using these same five phosphorus valence electrons double bonded with three oxygen atoms and a net charge of minus 1. Using the criteria described above, this is compatible with the respective anticipated range of bond orders for α and β bonds i.e., for ortho $\alpha = 3/4$, $\beta = 5/4$, and for meta $\alpha = 1/3$, $\beta = 5/3$. Consequently, the name for the ortho-phosphorate ion would be:

$$[P((\beta O\alpha)_4]^{3-} \tag{15}$$

while the meta-phosphate ion's name is:

$[P(\beta O\alpha)_3]^-$; (16)

The compatibility of these bond orders with the value for β in (4) is noted.*

(9) Incorporation of alpha bonds directly into the nomenclature allows various isomorphisms to be more accurately described. For example [14], hydrogen fluoride in the solid (and probably in the liquid) state consists of a long (unspecified length) chain of hydrogen bonded H-F molecules in a zig-zag ($\angle$HFH = 120.1°) chain. This would be nomenclated as:

$(F1H\alpha)_n$ (17)

where the subscript **n** indicates this ambiguity of length. On the other hand, the gaseous phase consists of a mixture of single HF molecules (canonical name = F1H) and puckered ring hexamers (with $\angle$HFH = 104°) which are nomenclated as:

$(F1H\alpha)_6$ (18)

(10) Because of the incorporation of beta bonds directly into the nomenclature, for conjugated monocyclic compounds having molecular formula $C_{2n}H_{2n}$ there is no need to add a declaration of odd vs. even in order to distinguish between compounds which are aromatic vs. anti-aromatic; namely, when **n** is odd the compound is aromatic vs. when **n** is even it is anti-aromatic. This is spelled out directly in the nomenclature: The bonds between carbon atoms of a monocyclic aromatic compound have bond order of 1.5; thus the compound is nomenclated by the formula

$(\underline{C}\beta)_{4n+2}$ (19)

Anti-aromatic compounds, on the other hand, have traditional single and double bonds and have as their canonical names:

---

*The three bonds which are traditionally represented as double, single, double bonds between the four carbon atoms each had approximate bond order of 5/3.

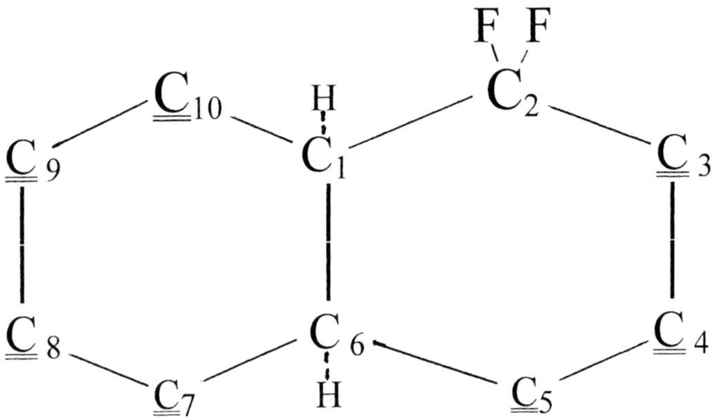

cis

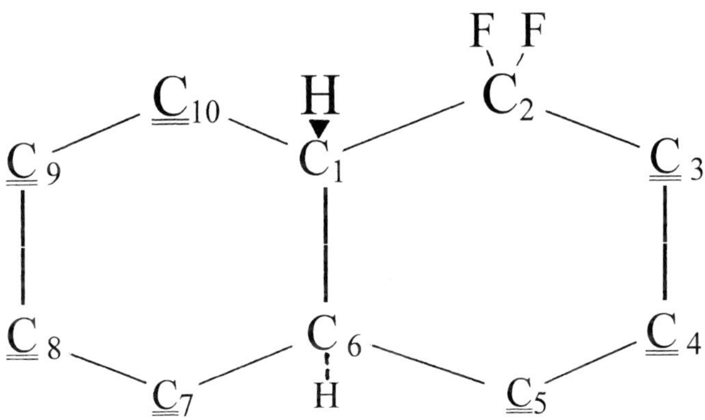

trans

Fig. 8: Traditional representation of geometrical Isomers of 2,2-Difluoro[4.4.0]decane with pertinent hydrogen atoms expressed

$$(\underline{C}2\underline{C}1)_{2n.} \qquad (20)$$

(11) A fundamental difference in the chemistry of certain selected geometrical isomers, which, is every bit as significant as the difference between constitutional isomers, may also be spelled out in the nomenclature. Consider 2,2-diflouro-cis-bicyclo[4.4.0]decane

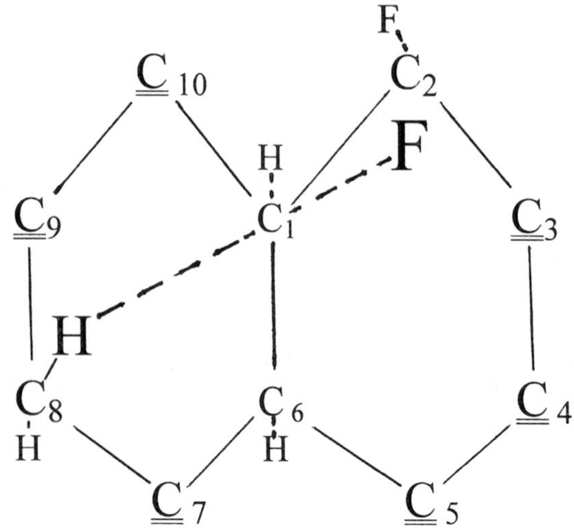

cis

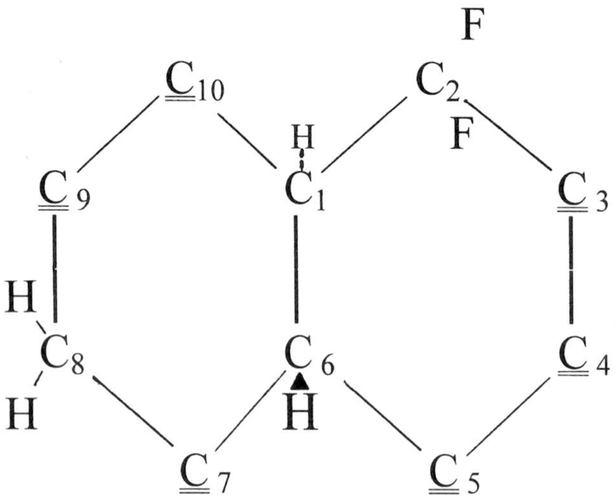

trans

Fig. 9: A geometrically realistic representation of geometrical isomers of 2,2-Difluoro[4.4.0]decane with pertinent hydrogen atoms expressed

(common name 2,2-diflouro-decalin) vs. its trans isomer. Figure 8 is the traditional (non-pertinent hydrogen-suppressed) picture of the type that is usually included in textbooks [15-17]. In this particular figure, for purposes of nomenclating these two geometric isomers, the locant numbers which IUPAC assigned have been used, rather than those locant numbers which are appropriate for the proposed system. Special attention should be paid to the fact that **there is not the slightest indication in such a picture** that one of the hydrogen atoms at IUPAC locant # 8 is close enough to a flourine atom at locant # 2 to form a hydrogen bond (which, in the perspective applicable to the proposed nomenclature scheme, is envisioned as a bridge) in the cis isomer*. In the trans isomer, on the other hand, the metric distance between the corresponding atoms is several times longer -- so long, in fact, that no interaction of significance is recognized. This difference is noted in the nomenclature by including an α bond (dotted line in top half of Figure 9) between those two atoms in the cis isomer. This compound is nomenclated as:

$$\underline{C}1C1(\underline{\underline{C}}1)_3\underline{C}1(\underline{\underline{C}}1)_4{:}\,^{(3\text{-}15)}(1F\alpha H1);\,^{(1\text{-}11)}1;\,^{(3,3)}(1F) \qquad (21)$$

In a follow on report, the difference between the class of molecules that Taylor designated as "reticular"[18] vs. that class which Goodson recognized as having essentially a three dimensional character, which he dubbed "fisular" [19] shall be related to whether an α bond between two atoms that traditionally were non-adjacent in the graph theoretical sense can be drawn. The topological implications of this reticular-fisular dichotomy are far-reaching and profoundly affect the perception of these molecules and schemes for nomenclating them [20]. However, before progressing further, it is important to note that the prefixes *cis* vs. *trans* are metric vs. graph theoretic descriptors [21], and metric properties have not, as yet, been introduced into the nomenclature. Nevertheless, from a strictly nomenclature perspective, observe that such a bond is not desired in the trans isomer and thus no negation of such a bond is required. On the other hand, inclusion of a symbol of negation for such a bond would **not** be pedantic. In other words, although

---

* The difluoro compound was selected for this example so as not to have to introduce the metric property of syn vs. anti at this time. In actuality, the monofluoro compound would suffice to illustrate this effect.

$$\underline{C}1C1(\underline{\underline{C}}1)_3\underline{C}1(\underline{\underline{C}}1)_4{:}^{(1\text{-}11)}1;^{(3,3)}(1F) \qquad (22)$$

would suffice,

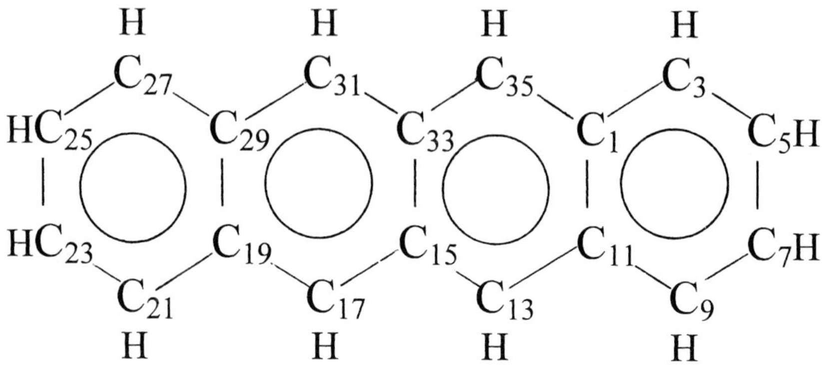

$$(C\beta)_{18}{:}^{(1\text{-}11,15\text{-}33,19\text{-}29)}\beta;^{(3,5,7,9,13,17,21,23,25,27,31,35)}(1H)$$

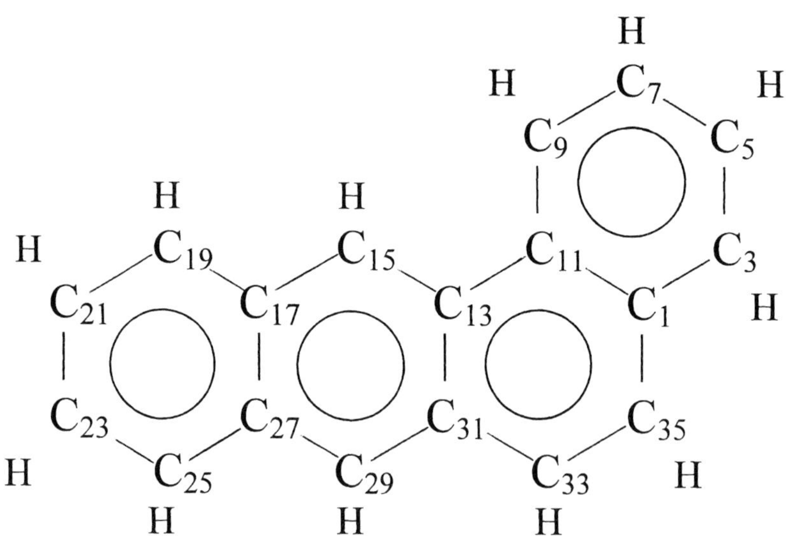

$$(C\beta)_{18}{:}^{(1\text{-}11,13\text{-}31,17\text{-}27)}\beta;^{(3,5,7,9,15,19,21,23,25,29,33,35)}(1H)$$

Fig. 10: Nomenclating tetrabenzenes (parts 1 and 2 of 7)

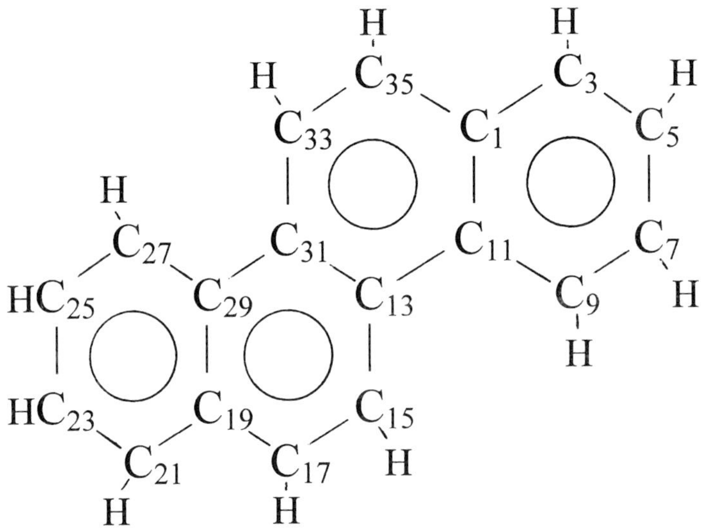

$$(C\beta)_{18}:^{(1\text{-}11,13\text{-}31,19\text{-}29)}\beta;^{(3,5,7,9,15,17,21,23,25,27,33,35)}(1H)$$

Fig. 10: Nomenclating tetrabenzenes (part 3 of 7)

$$\underline{C}1C1(\underline{\underline{C}}1)_3\underline{\underline{C}}1(\underline{\underline{C}}1)_4:^{(1\text{-}11)}1;^{(3\text{-}15)}(1F0H1);^{(3,3)}(1F) \qquad (23)$$

is preferred in order to emphasize that, ***in this particular case***, there is a major difference, which even shows up in the connectivity matrix, between the trans vs. the cis isomers.

(12) The nomenclature applicable to cata-condensed (no atom is common to three or more rings) benzenoid (all carbon rings of size 6) molecules comprised of n rings, that are not also corona-condensed (forms a ring of rings) is one of simplicity and organization. Every member of this set has the same principal part in their name:

$$(C\beta)_{4n+2} \qquad (24)$$

All that is needed to complete the canonical name is to locate the remaining bonds, as well as the location of the hydrogen atoms, by the above described system. For example, Figure 10 illustrates the names of the seven tetrabenzenes (5 cata-condensed, 1 peri-condensed and 1

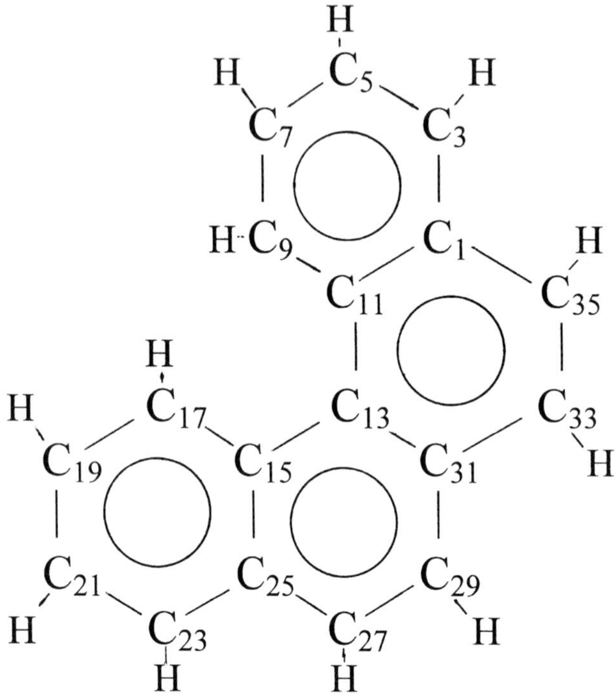

$$(C\beta)_{18}{:}^{(1\text{-}11,13\text{-}31,15\text{-}25)}\beta{;}^{(3,5,7,9,17,19,21,23,27,29,33,35)}(1H)$$

Fig. 10: Nomenclating tetrabenzenes (part 4 of 7)

non-viable) and invites comparison with names generated in the past using an augmenting set of prefixes [22], as well as some esoteric number theory based codes [23-25]. Note that, although such may be easily determined from the structure, excluding the principal (graph theoretical) cycle, there is no other reference to "ring" in the canonical name. This idea is in concert with being able to fuse multi-dentate "inorganic" structures with those of rings and ring assemblies in "organic" molecules.

(13) In the previous item, one notes the similarity in systemic name for the first five of the tetrabenzenes — all of which are stereoisomers. In a similar manner, attention is now directed to a common grouping of structural isomers — those having <u>C</u> as a single congruent module.

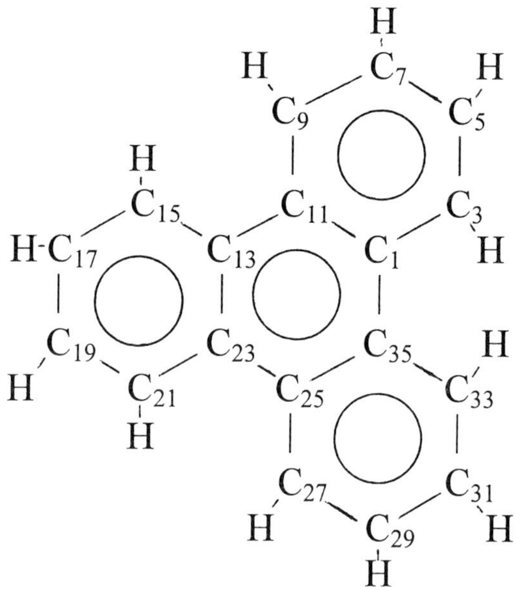

$(C\beta)_{18}{:}^{(1\text{-}11,13\text{-}23,25\text{-}35)}\beta;^{(3,5,7,9,15,17,19,21,27,29,31,33)}(1H)$

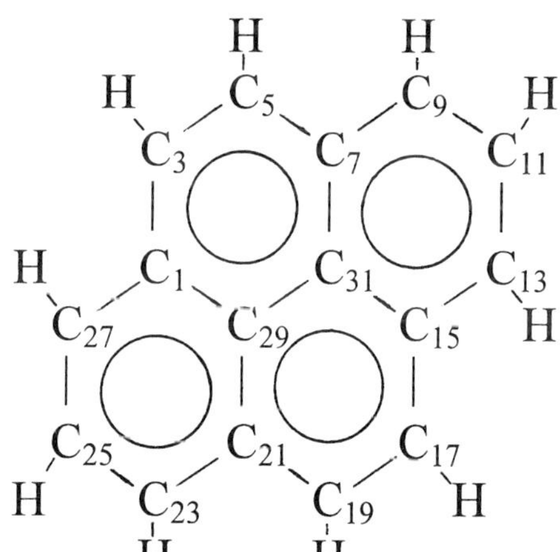

$(C\beta)_{14}{:}^{(1\text{-}7)}[\beta C^{(=29)}\beta C^{(=31)}\beta];^{(7\text{-}31,21\text{-}29)}(\beta);\ ^{(3,5,7,11,13,17,19,23,25,27)}(1H)$

Fig. 10: Nomenclating tetrabenzenes (parts 5 and 6 of 7)

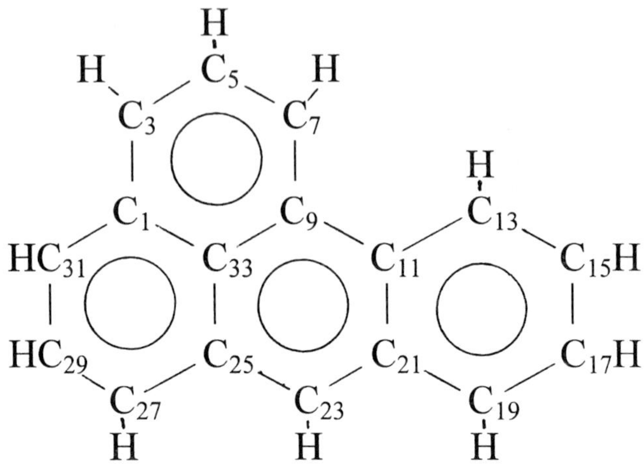

$(C\beta)_{16}{:}^{(1\text{-}9)}(1C^{(=33).}1);^{(25\text{-}33)}(1);\ ^{(3,5,7,13,15,17,19,23,27,29,31)}(1H)$

Fig. 10: Nomenclating tetrabenzenes (part 7 of 7)

One special subset of this group, contained in an advertisement by the software company JEOL Limited of Tokyo, Japan entitled "What is C_6H_6? Benzene?" appeared in several chemistry journals about 15 years ago. This ad was analyzed and then made into the subject of a report showing the limitations of any such computer program [26]. The first six of the 217 structures listed in this ad, which were repeated at the bottom of the ad and which have $\underline{C}_6$ as the principal part of the formula, are included at this point for nomenclating. The IUPAC name (including a diagram (Figure 11) with locant numbers, where needed) as well as the systemic names for these six compounds are:

(a) IUPAC name: Benzene [27]
Systemic name: $(\underline{C}\beta)_6$

(b) IUPAC name: 1-Bicyclo-[3.1.0.0^{3,5}.0^{4,6}]-hexene (part b1)
Systemic name: $\underline{C}2(\underline{C}1)_4{:}^{(5\text{-}7)}(1\underline{C}^{(=11)}1);^{(1\text{-}7,9\text{-}11)}(1)$ (part b2)

(c) IUPAC name: 2,5-Bicyclo-[2.2.0]hexadiene (part c1)
Systemic name: $[\underline{C}2(\underline{C}1)_2]_2{:}^{(5\text{-}11)}(1)$ (part c2)

(d) IUPAC name: Bicyclo-[2.2.0.0^{2,6}.0^{3,5}]hexane (part d1)
Systemic name: $(\underline{C}1)_{6:}{}^{(1\text{-}7,3\text{-}11,5\text{-}9)}(1)$ (part d2)

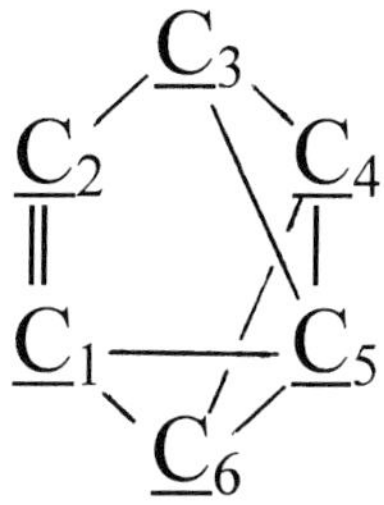

b1. IUPAC name

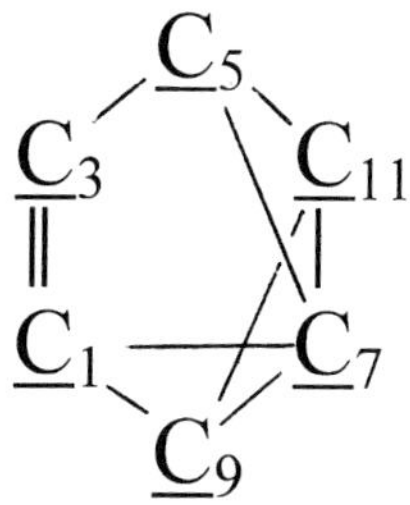

b2. Systemic name

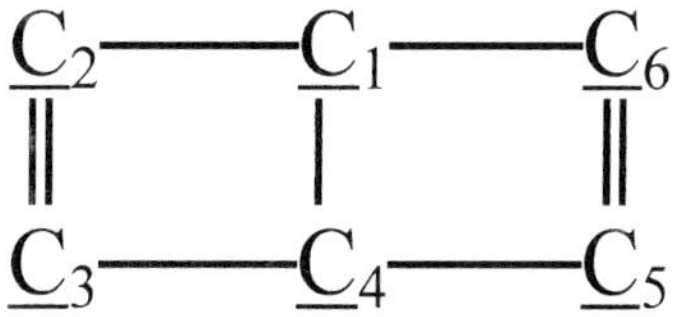

c1. IUPAC name

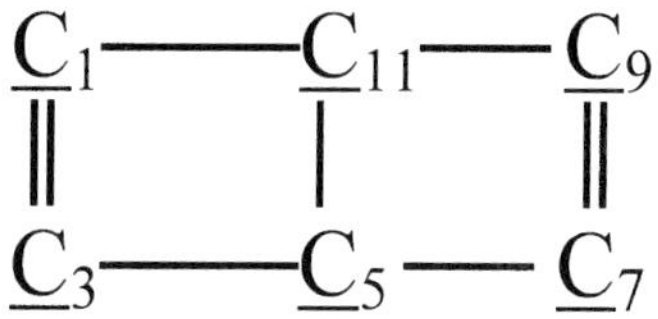

c2. Systemic name

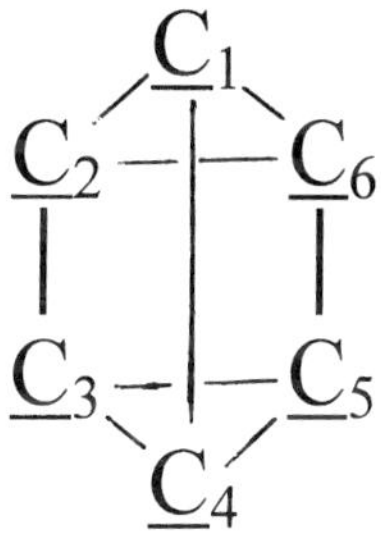

d1. IUPAC name

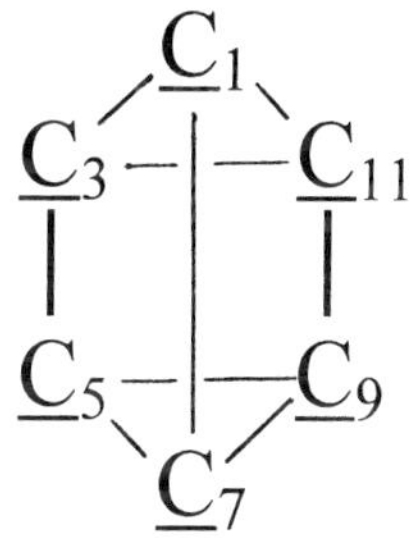

d2. Systemic name

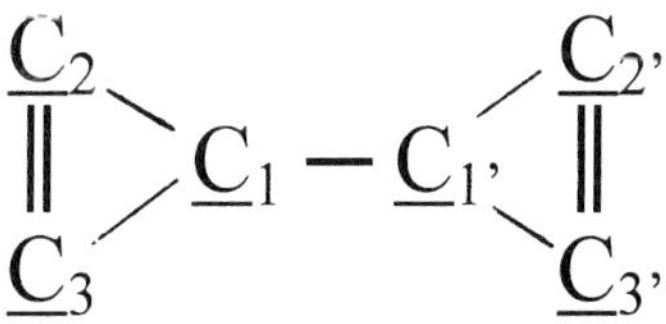

e1. IUPAC name

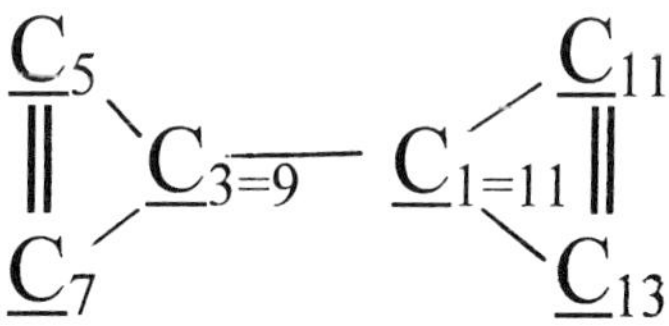

e2. Systemic name

Fig. 11: Locant numbering for selected structural isomers of $\underline{C}_6$ – part 1

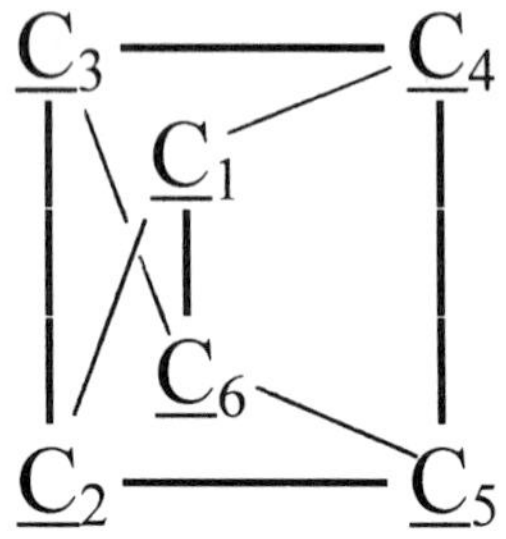

f1. IUPAC name

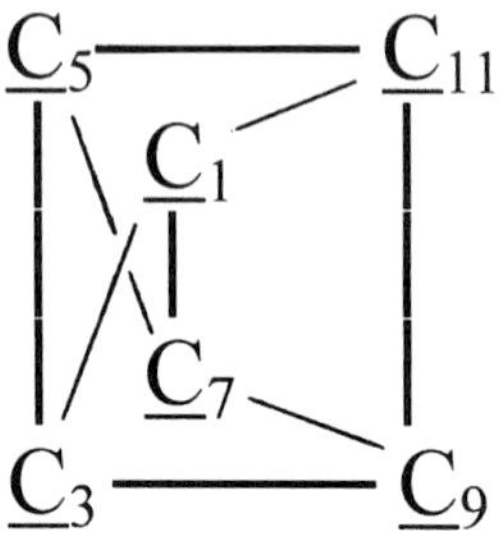

f2. Systemic name

Fig. 11: Locant numbering for selected structural isomers of $\underline{C}_6$ – part 2

(e) IUPAC name: 1,1'-Bicycloprop-2,2'-diene (part e1)
Systemic name*: $(\underline{C}1)_2\underline{C}2\underline{C}1C^{(9=3)}1^{(10=2)}C^{(11=1)}1\underline{C}2\underline{C}1$(part e2)

(f) IUPAC name: Bicyclo-$[2.2.0.0^{2,5}.0^{3,6}]$hexane (part f1)
Systemic name: $(\underline{C}1)_6{:}^{(1\text{-}7,3\text{-}9,5\text{-}11)}(1)$ (part f2)

(14) In some compounds specific bonds are either single or double in all viable resonance structures [28]. This important property is explicitly stated in the nomenclature system being developed by using the bond orders **1** or **2** respectively for such bonds, and β for those bonds which are single in one primary† resonance structure and double in another. This is in contradistinction to IUPAC's rules, which lump together conjugated systems and those in which the conjugation is broken. Instead, a table of 35 "reference" compounds (Rule A21.2) forms the basis that one is supposed to use for naming all fused polycyclic hydrocarbons [29]. Because most of the compounds toward the end of this table are only very slightly different from smaller molecules included earlier, only the first 26 of these compounds (all ring systems up through five rings) have been included in Table 1. Along with the illustration of each and its IUPAC reference name, this table includes:

(a) the proposed systemic locant number
(b) the systemic name in both full and abbreviated form,
(c) an arene name that had been proposed earlier [30].

Some additional smaller fused ring systems are next included as Table

---

*Details concerning how this name was generated will be supplied in Chapter 6.

†A primary resonance structure includes only the two main structures of benzene, not the Dewar benzene forms, etc.

Table 1
Systematic names for the IUPAC's reference compounds per Rule A-21.2

#	IUPAC Name/Illustration w/ systemic locants	Full Systemic Name Abbreviated SystemicName	Arene Name
1.	Pentalene (C_8H_6) 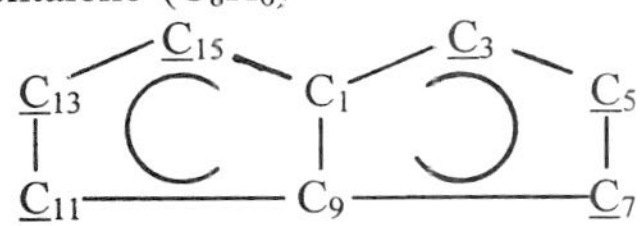	$(C\beta)_8$:$^{(1-9)}$(1);$^{(3,5,7,11,13,15)}$(1H) $[C\beta(\underline{C}\beta)_3]_2$:$^{(1-9)}$(1)	5,5-Diarene
2.	Indene (1H isomer) (C_9H_8) 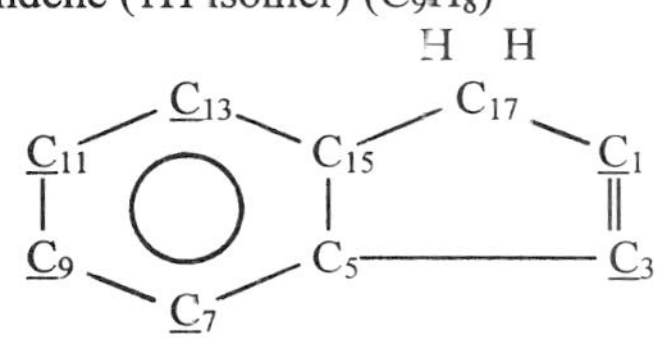	$C2C1(C\beta)_5(C1)_2$:$^{(5-15)}$(β); $^{(1,3,7,9,11,13,17,17)}$(1H) $\underline{C}2\underline{C}1C\beta(\underline{C}\beta)_4C1\underline{\underline{C}}1$:$^{(5-15)}$(β)	6,5-Diarene
3.	Naphthalene ($C_{10}H_8$) 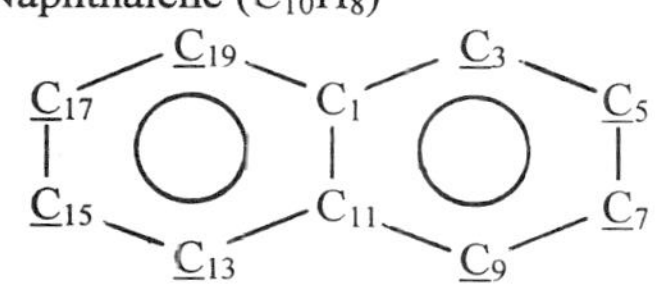	$(C\beta)_{10}$:$^{(1-11)}$(β);$^{(3,5,7,9,13,15,17,19)}$(1H) $[C\beta(\underline{C}\beta)_4]_2$:$^{(1-11)}$(β)	6,6-Diarene
4.	Azulene ($C_{10}H_8$) 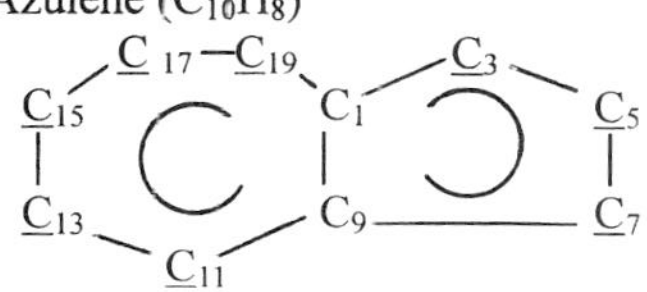	$(C\beta)_{10}$:$^{(1-9)}$(1);$^{(3,5,7,11,13,15,17,19)}$(1H) $C\beta(\underline{C}\beta)_3C\beta(\underline{C}\beta)_5$:$^{(1-9)}$(1)	7,5-Diarene

5. Heptalene ($C_{12}H_{10}$)

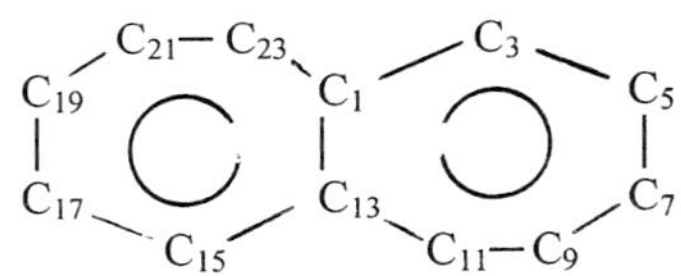

$(C\beta)_{12}:^{(1-13)}(1);^{(3,5,7,9,11,15,17,19,21,23)}(1H)$

$[C\beta(\underline{C}\beta)_5]_2:^{(1-13)}(1)$

7,7-Diarene

6. Biphenylene ($C_{12}H_8$)

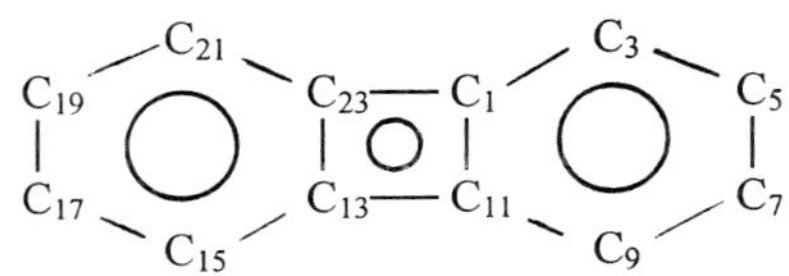

$(C\beta)_{12}:^{(1-11,13-23)}(\beta);\ ^{(3,5,7,9,15,17,19,21)}(1H)$

$[C\beta(\underline{C}\beta)_4]_2:^{(1-11,13-23)}(\beta)$

6,4,6a-Triarene

7. as-Indacene ($C_{12}H_8$)

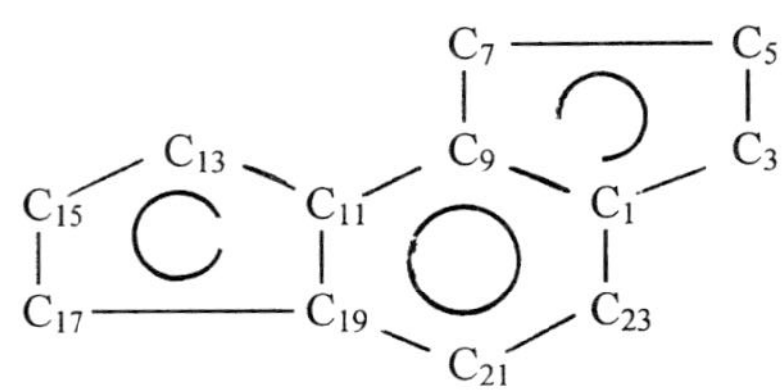

$(C\beta)_{12}:^{(1-9,11-19)}(1);\ ^{(3,5,7,13,15,17,21,23)}(1H)$

$[C\beta(\underline{C}\beta)_3C\beta]_2(\underline{C}\beta)_2:\ ^{(1-9,11-19)}(1)$

5,6,5b-Triarene

8. s-Indacene ($C_{12}H_8$)

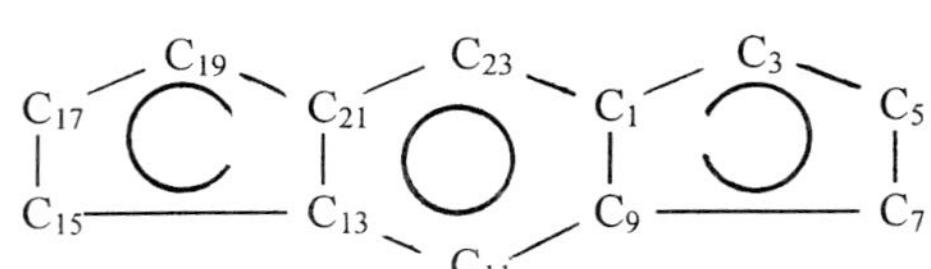

$(C\beta)_{12}:^{(1-9,11-19)}(1);\ ^{(3,5,7,13,15,17,21,23)}(1H)$

$[C\beta(\underline{C}\beta)_3C\beta\underline{C}\beta]_2:\ ^{(1-9,13-21)}(1)$

5,6,5a-Triarene

9. Acenaphthalene ($C_{12}H_8$)

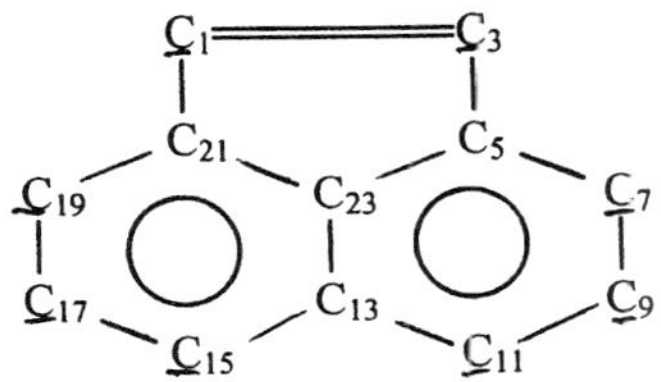

$C2C1(C\beta)_8C1:^{(5-13)}(\beta C^{(=23)}\beta);^{(13-23)}(\beta);^{(1,3,7,9,11,15,17,19)}(1H)$

$\underline{C}2\underline{C}1C\beta(\underline{C}\beta)_3C\beta(\underline{C}\beta)_3C1:^{(5-13)}(\beta C^{(=23)}\beta);^{(13-23)}(\beta)$

6,6,5c-Triarene

10. Fluorene (9H isomer) ($C_{13}H_{10}$)

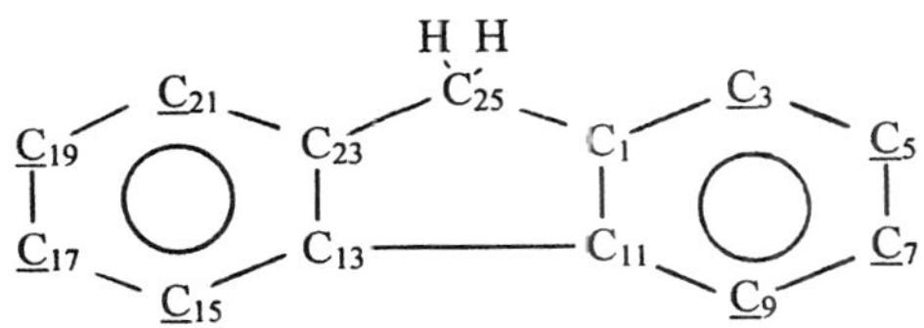

$(C\beta)_5C1]_2C1:^{(3,5,7,9,15,17,19,21,25,25)}(1H)$

$[C\beta(\underline{C}\beta)_4C1]_2\underline{\underline{C}}1:^{(1-11,13-23)}(\beta)$

6,5,6a-Triarene

11. Phenalene (1H isomer) ($C_{13}H_{10}$)

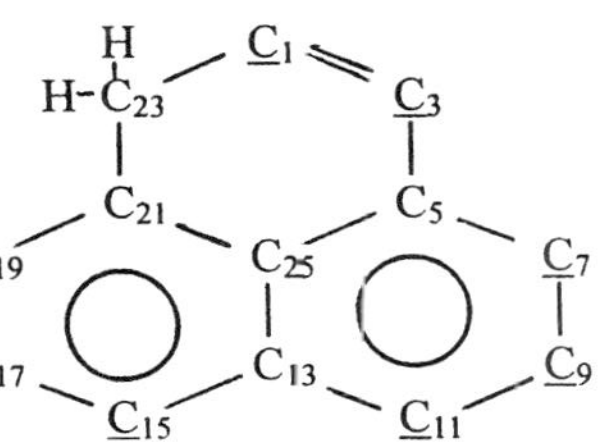

$C2C1(C\beta)_8(C1)_{2:}^{(5-13)}(\beta C^{(=25)}\beta);^{(21-25)}(\beta):^{(1,3,7,9,11,15,17,19,23,23)}(1H1)$

$\underline{C}2\underline{C}1[C\beta(\underline{C}\beta)_3]_2C1\underline{\underline{C}}1:^{(5-13)}(\beta C^{(=25)}\beta;^{(21-25)}\beta$

6,6,6c-Triarene

12. Phenanthrene ($C_{14}H_{10}$)

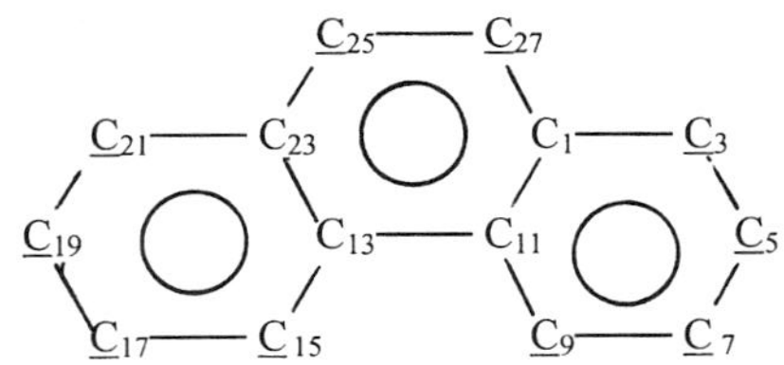

$(C\beta)_{14}:^{(1\text{-}11,13\text{-}23)}(\beta);^{(3,5,7,9,15,17,19,21,25,27)}(1H)$

$[C\beta(\underline{C}\beta)_4C\beta]_2(\underline{C}\beta)_2:^{(1\text{-}11.13\text{-}23)}(\beta)$

6,6,6b-Triarene

13. Anthracene ($C_{14}H_{10}$)

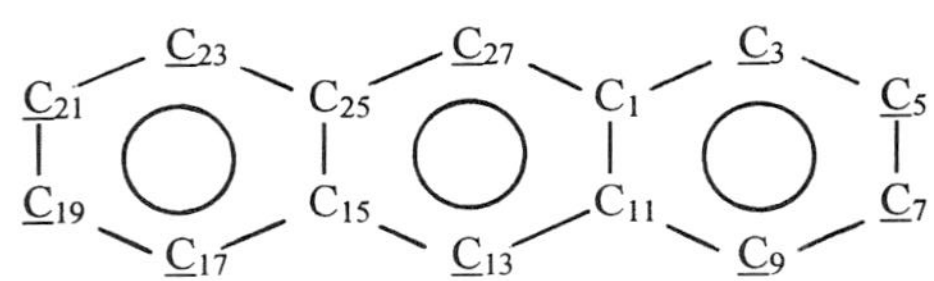

$(C\beta)_{14}:^{(1\text{-}11,15\text{-}25)}(\beta);^{(3,5,7,9,13,17,19,21,23,27)}(1H)$

$[C\beta(\underline{C}\beta)_4C\beta\underline{C}\beta]_2:^{(1\text{-}11,15\text{-}25)}(\beta)$

6,6,6a-Triarene

14. Fluoranthene ($C_{16}H_{10}$)

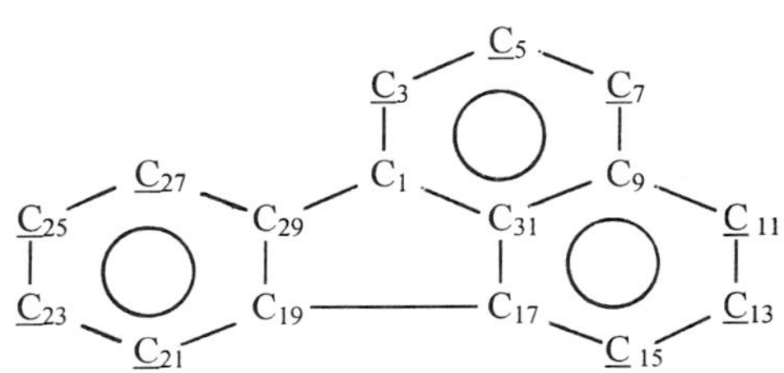

$(C\beta)_8C1(C\beta)_5C1:(\beta C^{(=31)}\beta;^{(17\text{-}31,19\text{-}29)}(\beta);^{(3,5,7,11,13,15,21,23,25,27)}(1H)$

6,5,6a,6c-Tetraarene

$[C\beta(\underline{C}\beta)_3]_2C1C\beta(\underline{C}\beta)_4C1:^{(1\text{-}9)}(\beta C^{(=31)}\beta);^{(17\text{-}31,19\text{-}29)}(\beta)$

15. Acephenanthrylene ($C_{16}H_{10}$)

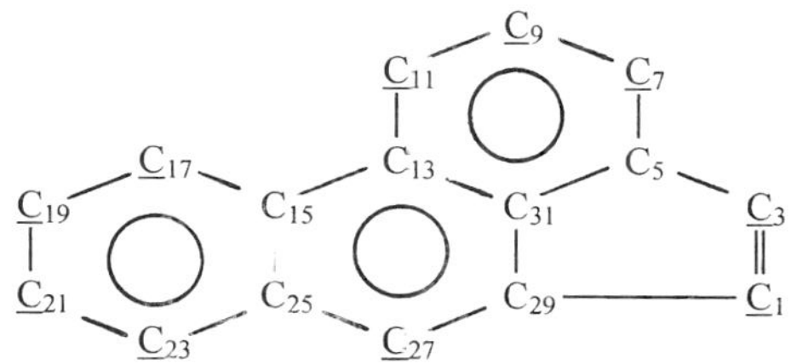

$C2C1(C\beta)_{12}C1{:}^{(5\text{-}13)}(\beta C^{(=31)}\beta;\ ^{(15\text{-}25,29\text{-}31)}(\beta);^{(1,3,7,9,11,17,19,21,23,27)}(1H)$

6,6,5a,6c-Tetraarene

$\underline{C}2\underline{C}1C\beta\underline{C}\beta C\beta(\underline{C}\beta)_4(C\beta)_2(\underline{C}\beta)_3C1{:}^{(5\text{-}13)}(\beta C^{(=31)}\beta);^{(15\text{-}25,29\text{-}31)}(\beta)$

16. Aceanthrylene ($C_{15}H_{10}$)

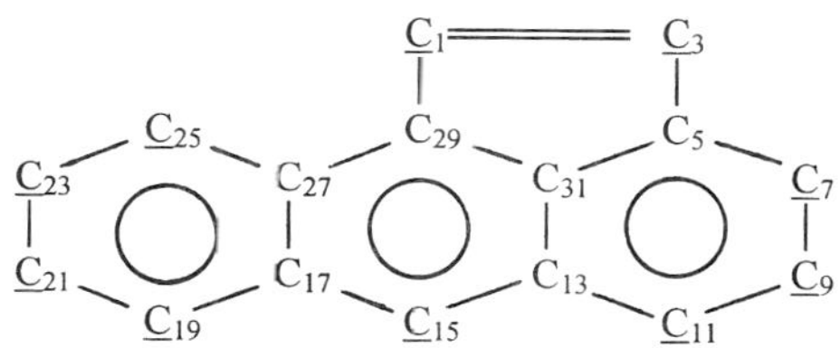

$C2C1(C\beta)_{12}C1{:}^{(5\text{-}13)}(\beta C^{(=31)}\beta;\ ^{(17\text{-}27,29\text{-}31)}(\beta);^{(1,3,7,9,11,15,19,21,23,25)}(1H)$

6,6,6a,5c-Tetraarene

$\underline{C}2\underline{C}1C\beta(\underline{C}\beta)_3C\beta\underline{C}\beta C\beta(\underline{C}\beta)_4C\beta C1{:}^{(5\text{-}13)}(\beta C^{(=31)}\beta);^{(17\text{-}27,29\text{-}31)}(\beta)$

17. Triphenylene ($C_{18}H_{12}$)

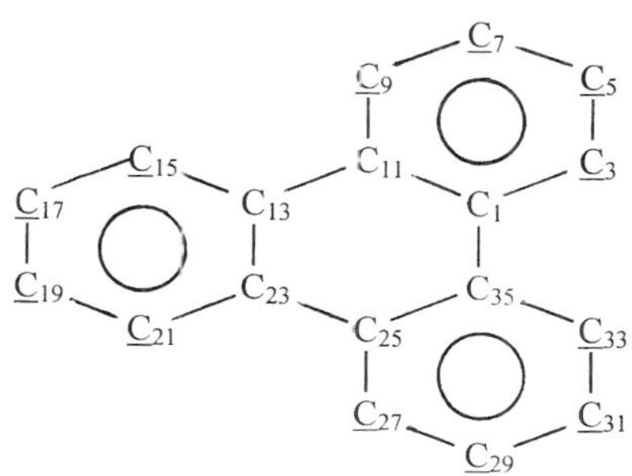

$[(C\beta)_5C1]_3{:}^{(1\text{-}11,13\text{-}23,25\text{-}35)}(\beta);\ ^{(3,5,7,9,15,17,19,21,27,29,31,33)}(1H)$

6,6,6b,6j-Tetraarene

$[C\beta(\underline{C}\beta)_4C1]_3{:}\ ^{(1\text{-}11,13\text{-}23,25\text{-}35)}(\beta)$

18. Pyrene ($C_{16}H_{10}$)*

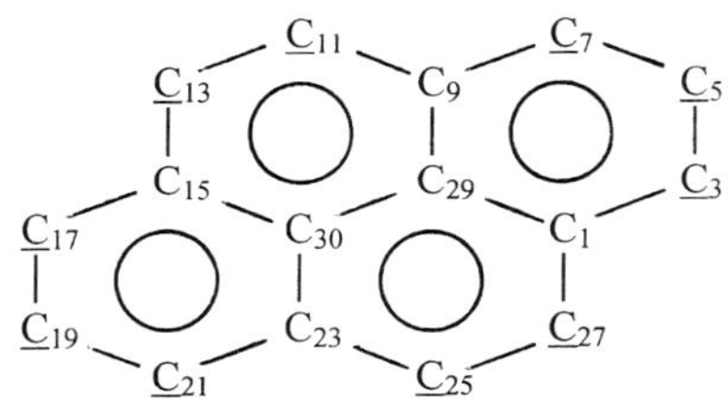

$(C\beta)_{14}:^{(1\text{-}9)}(\beta C^{(=29)}\beta);^{(15\text{-}23)}(\beta C^{(=30)}\beta);^{(29\text{-}30)}(\beta);^{(3,5,7,11,13,17,19,21,25,17)}(1H)$

6,6,6b,6e-Tetraarene

$[C\beta(\underline{C}\beta)_3C\beta(\underline{C}\beta)_2]_2:^{(1\text{-}9)}(BC^{=29}\beta);^{(15\text{-}23)}(\beta C^{=30}\beta);^{(29\text{-}30)}(\beta)$

19. Chrysene ($C_{18}H_{12}$)

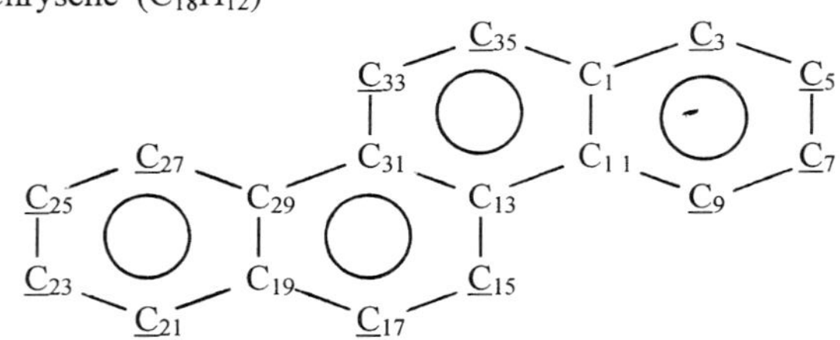

$(C\beta)_{18}:^{(1\text{-}11,13\text{-}31,19\text{-}29)}(\beta);^{(3,5,7,9,15,17,21,23,25,27,33,35)}(1H)$

6,6,6b,6b-Tetraarene

$[C\beta(\underline{C}\beta)_4C\beta(\underline{C}\beta)_2C\beta]_2:^{(1\text{-}11,13\text{-}31,19\text{-}29)}(\beta)$

20. Naphthacene ($C_{18}H_{12}$)

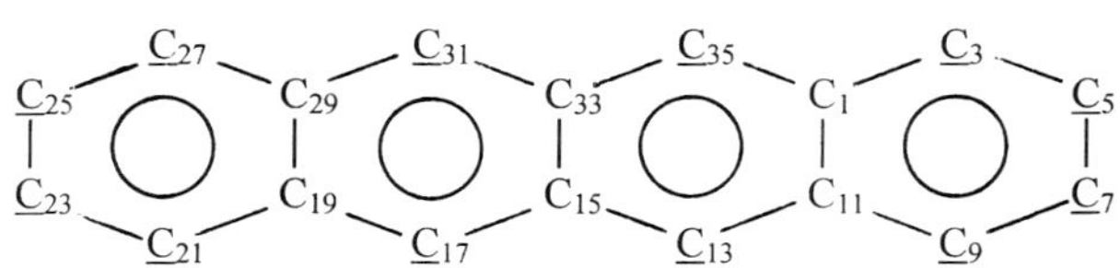

$(C\beta)_{18}:^{(1\text{-}11,15\text{-}33,19\text{-}29)}(\beta);^{(3,5,7,9,13,17,21,23,25,27,31,35)}(1H)$

6,6,6a,6a-Tetraarene

$[C\beta(\underline{C}\beta)_4(C\beta\underline{C}\beta)_2]_2:^{(1\text{-}11,15\text{-}33,19\text{-}29)}(\beta)$

---

*Although this is the "standard" locant numbering and name that the proposed system might assign to this molecule, there is another, *in some, but not all, ways better,* name that needs consideration. This subject shall be revisited in Chapter 3, when a comparison with another name will be investigated.

21. Pleiadene ($C_{18}H_{12}$)

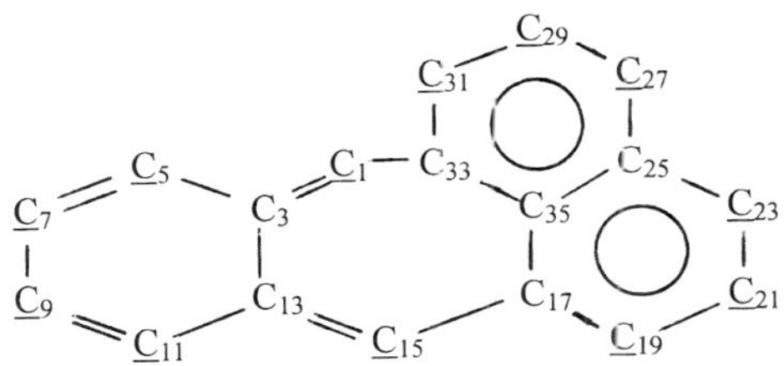

$(C2C1)_4[C\beta(\underline{C}\beta)_3]_2C1{:}^{(17\text{-}25)}(\beta C^{(=35)}\beta);^{(33\text{-}35)}(\beta);^{(3\text{-}13)}(1);^{(1,5,7,9,11,15,19,21,23,27,29,31)}(1H)$

6,7,6a,6c-Tetraarene

$\underline{C}2C1(\underline{C}2\underline{C}1)_2C2\underline{C}1[C\beta(\underline{C}\beta)_3]_2C1{:}^{(17\text{-}25)}(\beta C^{(=35)}\beta);^{(33\text{-}35)}(\beta);^{(3\text{-}13)}(1)$

22. Picene ($C_{22}H_{14}$)

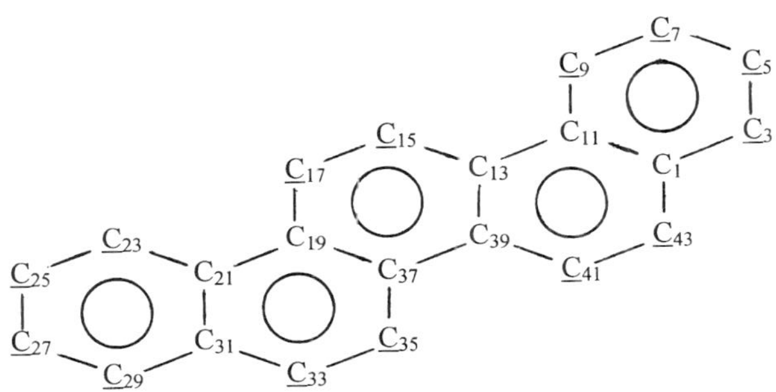

$(C\beta)_{22}{:}^{(1\text{-}11,13\text{-}39,19\text{-}37,21\text{-}31)}\beta;^{(3,5,7,9,15,17,23,25,27,29,33,35,41,43)}(1H)$

6,6,6b,6b,6c-Pentaarene

$C\beta(\underline{C}\beta)_4(C\beta)_2(\underline{C}\beta)_2(C\beta)_2(\underline{C}\beta)_4C\beta(\underline{C}\beta)_2(C\beta)_2(\underline{C}\beta)_2{:}^{(1\text{-}11,13\text{-}39,19\text{-}37,21\text{-}31)}\beta$

23. Perylene ($C_{20}H_{12}$)*

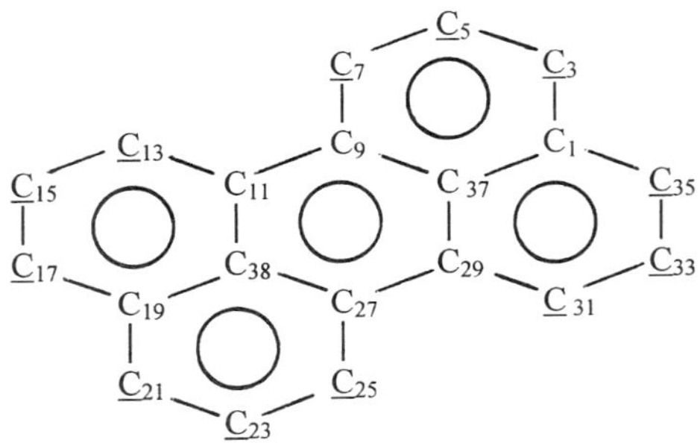

$(C\beta)_{18};^{(1\text{-}9)}(\beta C^{(=37)}\beta);^{(11\text{-}19)}(\beta C^{(=38)}\beta);^{(27\text{-}38,29\text{-}37)}(\beta);^{(3,5,7,13,15,17,21,23,25,31,33,35)}(1H)$

6,6,6a,6c.6i-Pentaarene

$C\beta(\underline{C}\beta)_3(C\beta)_2(\underline{C}\beta)_3C\beta(\underline{C}\beta)_3(C\beta)_2(\underline{C}\beta)_3{:}\ ^{(1\text{-}9)}(\beta C^{(=37)}\beta);^{(11\text{-}19)}(\beta C^{(=38)}\beta);^{(27\text{-}38,29\text{-}37)}(\beta)$

24. Pentaphene ($C_{22}H_{14}$)

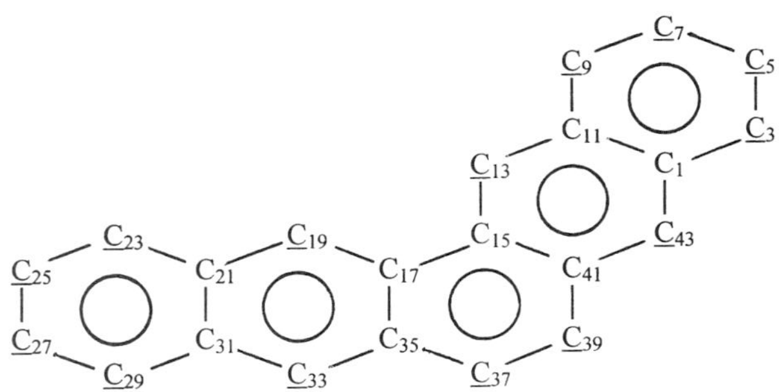

$(C\beta)_{22}{:}^{(1\text{-}11,15\text{-}41,17\text{-}35,21\text{-}31)}\beta;^{(3,5,7,9,13,19,23,25,27,29,33,37,39,43)}(1H)$

6,6,6a,6b,6c-Pentaarene

$C\beta(\underline{C}\beta)_4C\beta\underline{C}\beta(C\beta)_2\underline{C}\beta C\beta(\underline{C}\beta)_4C\beta\underline{C}\beta C\beta\underline{C}\beta\underline{C}\beta C\beta\underline{C}\beta{:}\ ^{(1\text{-}11,15\text{-}41,17\text{-}35,21\text{-}31)}\beta$

*This molecule will be examined in more detail in Chapter 3. The particular nomenclature given here is for the largest eulerian cycle

25. Pentacene ($C_{22}H_{14}$)

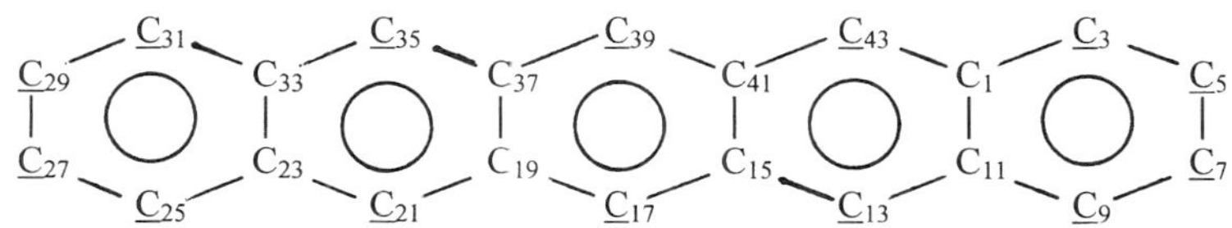

$(C\beta)_{22}:^{(1\text{-}11,15\text{-}41,19\text{-}37,23\text{-}33)}(\underline{\beta})$; $^{(3,5,7,9,13,17,21,25,27,29,31,35,39,43)}(1H)$

6,6,6a,6a-Tetraarene

$[C\beta(\underline{C}\beta)_4(C\beta\underline{C}\beta)_3]_2:^{(1\text{-}11,15\text{-}41,19\text{-}37,23\text{-}33)}(\beta)$

26. Tetraphenylene ($C_{24}H_{16}$)

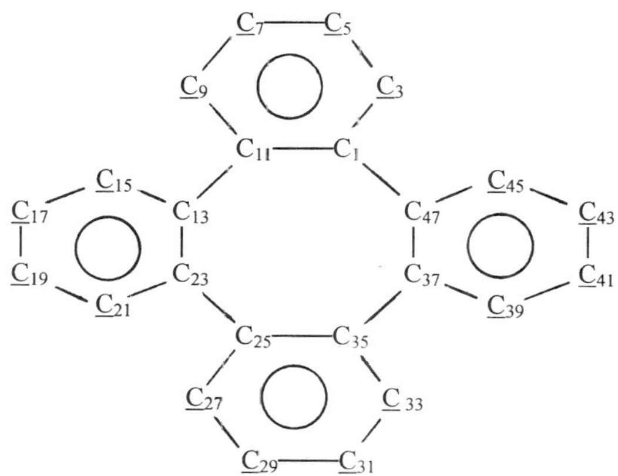

$[(C\beta)_5C1]_4:^{(1\text{-}11,13\text{-}23,25\text{-}35,37\text{-}47)}(\beta)$; $^{(3,5,7,9,15,17,19,21,27,29,31,33,39,41,43,45)}(1H)$

6,8,6a,6d,6ℓ-pentaarene

$[C\beta(\underline{C}\beta)_4C1]_4:^{(1\text{-}11,13\text{-}23,25\text{-}35,37\text{-}47)}(\beta)$

2. Unlike the lack of logic in the IUPAC system that results when additional "fused polycyclic hydrocarbons" are named by addending rings to one of these "privileged" 35 reference compounds at specified locations, it is highly desirable that the nomenclature for all ring systems uses the same criteria. This is another virtue of proposed system. Meanwhile a close scrutiny of the molecules depicted in Tables 1 and 2 reveal:

(a) In the first entry of Table 1, pentalene has a "fixed" single bond; namely, the bond that is common to both rings. The other eight bonds will be single in one resonance form and double in another. The specification of one or more fixed bonds will occur for most polycyclic hydrocarbons in which there is a ring having an odd number of edges, as well as in some selected molecules having an even number of edges. In other words, a necessary, but NOT a sufficient, condition for a compound to be classified as an arene (an aromatic hydrocarbon) is that it have all of its bonds variable. This will occur only when there is be an even number of bonds. Figure 12 shows the two "geometrically-simple"* viable resonance structures for pentalene. Other resonance structures comparable to the Dewar benzenes in which the "fixed" single bond is, in fact, a double bond, can be constructed , however, these are beyond the purview of establishing a system of nomenclature. In a similar manner, for molecules with unbroken conjugation (i.e., did not need to have an extra hydrogen atom), all terminal odd rings will be connected to the remainder of the molecule by a fixed single bond. This means azulene (#4), heptalene (#5), both indacenes (#6 and 7), etc.

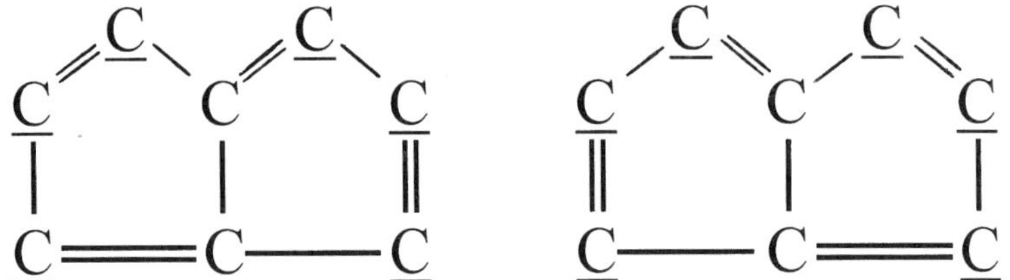

Fig. 12: The geometrically simple resonance forms of pentalene

*The heuristics of what qualifies as "geometrically simple" would require a detailed study of topology. Some of the *more familiar* properties subsumed by this term include: "simply-connected"; "simply-closed"; "orientable"; "iso-dimensional", etc.

Table 2
IUPAC and systematic names for other selected four-ring polycyclic hydrocarbons

1. Benz[a]anthracene ($C_{18}H_{12}$)

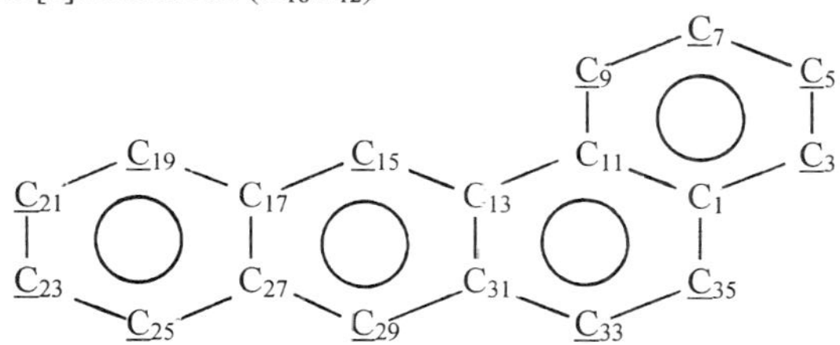

$(C\beta)_{18}:^{(1\text{-}11,13\text{-}31,17\text{-}27)}(\beta);^{(3,5,7,9,15,19,21,23,25,29,33,35)}(1H)$

6,6,6a,6b-Tetraarene

$C\beta(\underline{C}\beta)_4(C\beta)_2\underline{C}\beta C\beta(\underline{C}\beta)_4(C\beta\underline{C}\beta)_2\underline{C}\beta:^{(1\text{-}11,13\text{-}31,17\text{-}27)}(\beta)$

Note that IUPAC's locant designation for anthracene (Rule A-21.5) starts with locant **a** being at the upper left corner of the anthracene kernel. One then progress clockwise to allocate the remaining locant letters: b, c, d, etc. It is for this reason that this molecule was designated as benz[a], rather than the more logical extension of the molecule linearly, which, by IUPAC fiat, is designated as locant b. Moreover, in the process two locant letters are allocated at every carbon atom that is common to two benzene rings.

2. Benz[d]anthracene ($C_{17}H_{12}$)
Note that such a molecule, although mathematically possible, is not included in IUPAC nomenclature inasmuch as it can not be described with a completely conjugated system. Instead, as in phenalene (#11 in Table 2) an extra hydrogen atom is required and one will nomenclate this molecule starting from a carbon atom, followed by a double bond.

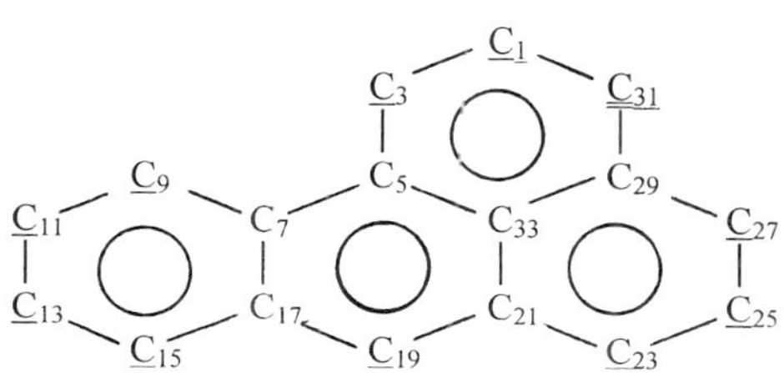

$C2C1(C\beta)_{12}(C1)_2:^{(5\text{-}21)}(\beta C^{=33}\beta);^{(7\text{-}17.29\text{-}33)}(\beta);$
$^{(1,3,9,11,13,15,19,23,25,27,31,31)}(1H)$

6,6,6a,6c-Tetraarene

$\underline{C}2\underline{C}1(C\beta)_2(\underline{C}\beta)_4C\beta\underline{C}\beta C\beta(\underline{C}\beta)_3C\beta\underline{C}1:^{(5\text{-}21)}(\beta C^{=33}\beta);^{(7\text{-}17.29\text{-}33)}(\beta)$

3. Benzo[c]phenanthrene ($C_{18}H_{12}$)

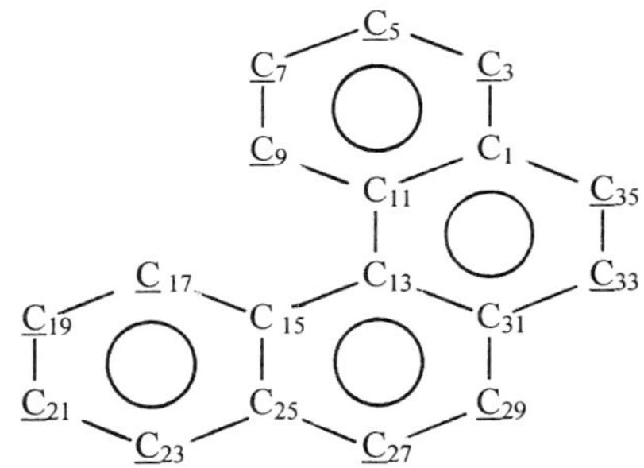

$(C\beta)_{18}:^{(1\text{-}11,13\text{-}31,15\text{-}25)}(\beta):^{(3,5,7,9,17,19,21,23,27,29,33,35)}(1H)$

6,6,6b,6d-Tetraarene

$C\beta(\underline{C}\beta)_4(C\beta)_3(\underline{C}\beta)_4[C\beta(\underline{C}\beta)_2]_2:^{(1\text{-}11,13\text{-}31,15\text{-}25)}(\beta)$

4. Cyclopent[fg]acenaphthylene ($C_{14}H_8$)

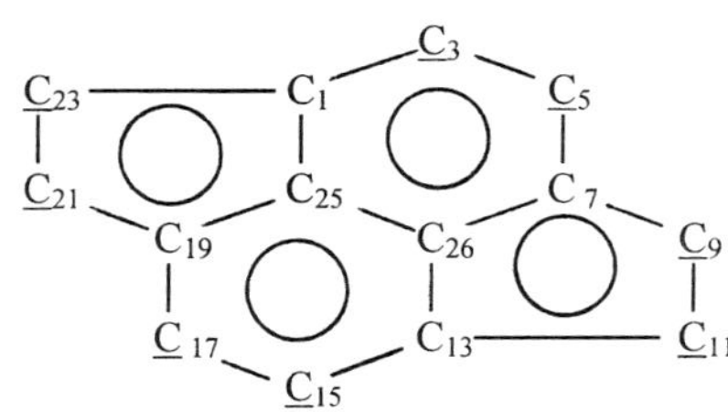

$(C\beta)_{12}:^{(1\text{-}19)}(\beta C^{(=25)}\beta);\ ^{(7\text{-}13)}(\beta C^{(=26)}\beta):^{(25\text{-}26)}(\beta),$ $^{(3,5,9,11,15,17,21,23)}(1H)$

5,6,5b,6d-Tetraarene

$[C\beta(\underline{C}\beta)_2]_4:\ ^{(1\text{-}19)}(\beta C^{(=25)}\beta);\ ^{(7\text{-}13)}(\beta C^{(=26)}\beta):^{(25\text{-}26)}(\beta)$

5. Pentalen[2,3-a]pentalene $C_{14}H_8$

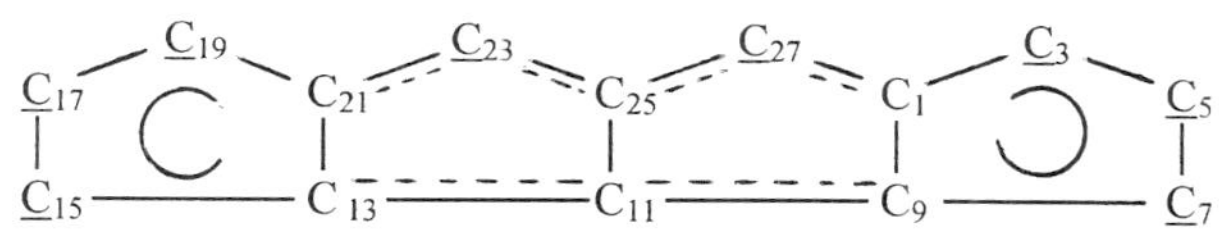

$(C\beta)_{14}{:}^{(1\text{-}9,11\text{-}25,13\text{-}21)}(1);\ ^{(3,5,7,15,17,19,23,27)}(1H)$

5,5,5a,5a-Tetraarene

$C\beta(\underline{C}\beta)_3(C\beta)_3(\underline{C}\beta)_3(C\beta\underline{C}\beta)_{2:}\,^{(1\text{-}9,11\text{-}25,13\text{-}21)}(1)$

Note that, despite the Patterson drawing convention [31] which resulted in the above picture, this molecule is not a straight line compound; instead these four pentagons form an arc with the "base" of each succeeding pentagon 36° higher than its predecessor.

6. Pentalen[2,3-b]pentalene $C_{14}H_8$

A geometrically accurate picture of a "straight" chain [32] of pentane modules would have the pentagons oriented in an alternating up and down pattern.

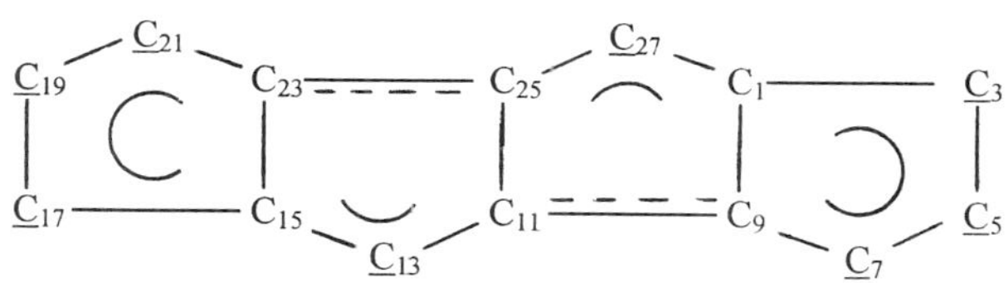

$(C\beta)_{14}{:}^{(1\text{-}9,11\text{-}25,15\text{-}23)}(1);\ ^{(3,5,7,13,17,19,21,27)}(1H)$

5,5,5a,5b-Tetraarene

$[C\beta(\underline{C}\beta)_3(C\beta)_2(\underline{C}\beta)]_{2:}\,^{(1\text{-}9,11\text{-}25,15\text{-}23)}(1)$

7. Pentalen[fgh]pentalene $C_{12}H_6$

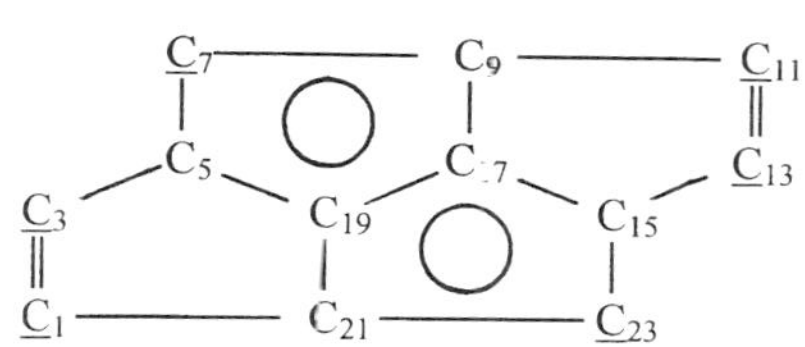

$C2C1(C\beta)_2C1C2C1(C\beta C1)_2;\ ^{(1,3,7,11,13,23)}(1H)$

5,5,5b,5c-Tetraarene

$\underline{C}2\underline{C}1C\beta\underline{C}\beta C1\underline{C}2\underline{C}1(C\beta C1)_{2:}\,^{(15\text{-}21)}(\beta\underline{C}^{=23}\beta);^{(5\text{-}19,9\text{-}17)}(\beta)$

8. Azuleno[2,3a-cd]azulene ($C_{16}H_{10}$)

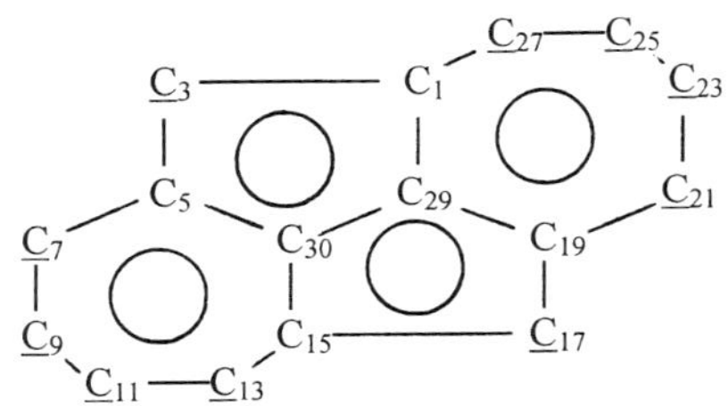

$(C\beta)_{14}:^{(1\text{-}5)}(\beta C^{(=29)}\beta C^{(=30)}\beta);\ ^{(3,7,9,11,13,17,,21,23,25,17)}(1H)$

7,5,7b,5f-Tetraarene

$[C\beta(\underline{C}\beta)_3C\beta(\underline{C}\beta)_2]_{2:}{}^{(1\text{-}9)}(\beta C^{=29}\beta);^{(15\text{-}23)}(\beta C^{=30}\beta);^{(29\text{-}30)}(\beta)$

9. Azuleno[3,4-ef]azulene ($C_{16}H_{10}$)

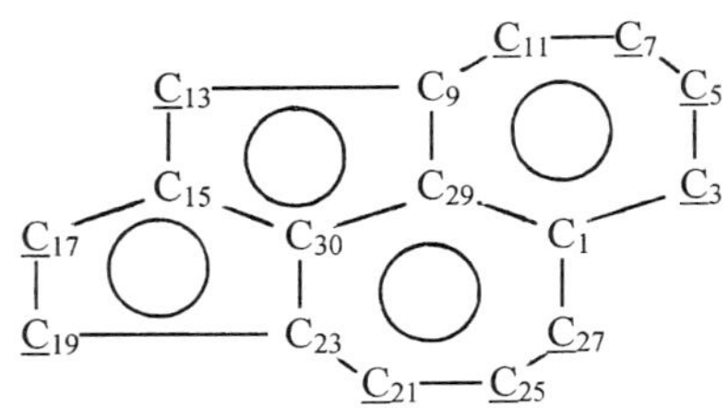

$(C\beta)_{14}:^{(1\text{-}5)}(\beta C^{(=29)}\beta C^{(=30)}\beta);\ ^{(3,7,9,13,15,17,21,23,25,27)}(1H)$

7,7,5c,5e-Tetraarene

$[C\beta\underline{C}\beta C\beta(\underline{C}\beta)_2C\beta(\underline{C}\beta)_3C\beta(\underline{C}\beta)_{4:}{}^{(1\text{-}5)}(\beta C^{=29}\beta C^{=30}\beta)\ ;^{(11\text{-}30,19\text{-}29)}(\beta)$

10. Azuleno[3,5-cde]azulene ($C_{16}H_{10}$)

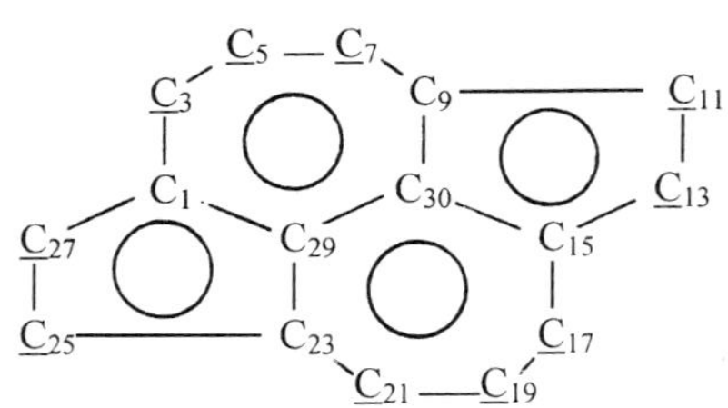

$(C\beta)_{14}:^{(1\text{-}9)}(\beta C^{(=29)}\beta C^{(=30)}\beta);^{(15\text{-}30,23\text{-}29)}(\beta);\ ^{(3,5,7,11,13,17,19,21,25,27)}(1H)$

5,7,5b,7d-Tetraarene

$[C\beta(\underline{C}\beta)_3C\beta(\underline{C}\beta)_2]_{2:}{}^{(1\text{-}9)}(BC^{=29}\beta C^{=30}\beta);^{(15\text{-}30,23\text{-}29)}(\beta)$

(b) Although there **may** have been **some** semblance of logic in IUPAC's reference table when using addended benzene modules, there is less, *if any,* advantage using other, especially odd, size rings. In particular, note that the fixed single bond in pentalene introduces inconsistency into the nomenclature that is not evident when using a saturated compound as the parent and then augmenting this parent with four double bonds; namely: Bicyclo[3.3.0]octa[1,3,5,7]tetraene. However, instead of having to resort to either a new memorized name as in Rule A-21.2 or to the logically more consistent, but still undesirable, categorization of this molecule as derived from an unsaturated bicyclic parent compound, the proposed nomenclature introduces simplicity by using beta bonds. See example 1 in Table 1. Such inconsistency in the IUPAC system will be greatly increased as more cyclopentadienyl modules are added to a cyclopentadienyl core. See molecules 5 though 7 in Table 2. Note the tedium for IUPAC nomenclating vs. the simplicity for the proposed system

(c) In three of IUPAC's reference compounds, (#2 indene, #10 fluorene, and #11 phenalene). a spanning set of conjugated single and double bonds is not possible. To classify these compounds, which have interrupted aromatic character, the best that can be done is to select one specific location for that "extra" hydrogen atom. This subject will be amplified in Chapter 9, when examining various molecular rearrangements.

(d) From the perspective of assigning canonical names to these compounds using the above established guidelines, the locant numbering for any hydrocarbon compound having a fixed double bond will start with a carbon atom adjacent to one of the double bonds and progress so as the next double bond will appear in the name as soon as possible; for example, indene (#2), acenaphthalene (#9), phenalene (#11), acephenanthrylene (#15), and aceanthrylene (#16). The locant numbering of all other combinations will have locant #1 assigned so that it starts the longest contiguous sequence of beta bonds as well as being a member of two rings on the boundary.

(e) The IUPAC rules for establishing the canonical orientation of acenaphthylene (#9) are, at best, confusing. Noting that only the number of rings and not ring sizes are considered in determining the "horizontal" line, it appears that the pentagonal ring could be either as shown in Table 1 or with the molecule rotated through either $\pm 120^{o}$. Such a rotation would influence the choice of locant numbers for IUPAC. It is only through the imposition of a convoluted set of additional rules that IUPAC's decision has been made. Such ambiguity does not arise in the proposed system which does not consider orientation when assigning locant numbers.

(f) For fluorene (#10), all three bond that are exclusively in the five member ring are constrained to be single bonds, thus the locant numbering will start from one of the carbon atoms common to the five and one of the six member rings and progress through that hexagon first; thereby leaving the carbon atom that is bonded to two hydrogen atoms last.

(g) The orientation and consequently the locant numbering selected for phenanthrene (#12) violates the rules established by IUPAC [31]. By note 10, especially figure XVI, there should have been five vertical sides in the depiction of phenanthrene; instead there are six horizontal sides and ZERO vertical sides.

(h) The IUPAC locant numbering for all of the reference compounds is sequential **only** for atoms that are members of a single ring. When rings are fused together, per Rule 22.2 "atoms common to two or more rings are designated by adding roman letters "a", "b", "c", etc. to the number of the position immediately preceding. Interior atoms follow the highest number, taking a clockwise sequence whenever there is a choice." In the proposed system, on the other hand, the locant numbers are sequential integers on the principal path and higher integers on bridging paths. There are no other symbols, such as "a", "b", etc., which IUPAC has inconsistently interspersed with these integers whenever either two or three rings have an atom in common.

(i) In pleiadene (#21), of the 21 bonds only the right most naphthalene segment supports conjugation. As in (a) above, the heptagonal ring forces there to be fixed single and double bonds at the remaining edges; namely there are four fixed double bonds, six fixed single bonds and eleven beta bonds; consequently, the systemic name begins with a sequence of four pairs of double bond – single bond.

(j) As illustrated in Table 1, with two to four rings, the abbreviated (underscored carbon atom) name is shorter and thus advantageous; however, if one continues to addend rings, building up to five or more rings, the abbreviated name often becomes longer and more tedious than the canonical name that would result had one listed all of the hydrogen atoms.

(15) An attribute of the arcanum of the IUPAC system is the importance

placed on later, rather than earlier, entries in the various tables, such as in Rule 21.3: "The base component should contain as many rings as possible (provided it has a trivial name) and should occur as far as possible from the beginning of the list of Rule A-21.1." [32] One particular ring system of interest is the fused seven ring system shown in Figure 13. This molecule* has tetrabenz[a,c,h,j]anthracene as its IUPAC name. Note that the choice of anthracene (#13 on IUPAC's list) as the "base component" to which the four benzene rings are to be addended is a curious one. The preference for anthracene over phenanthrene (#12) can be supported by this rule, even though use of phenanthrene would require addending only two modules (a second phenanthrene and a benzene); however, triphenylene (#17) would be a more logical choice — both on the basis of containing four, rather than just three rings and also being later on the list. This is but another

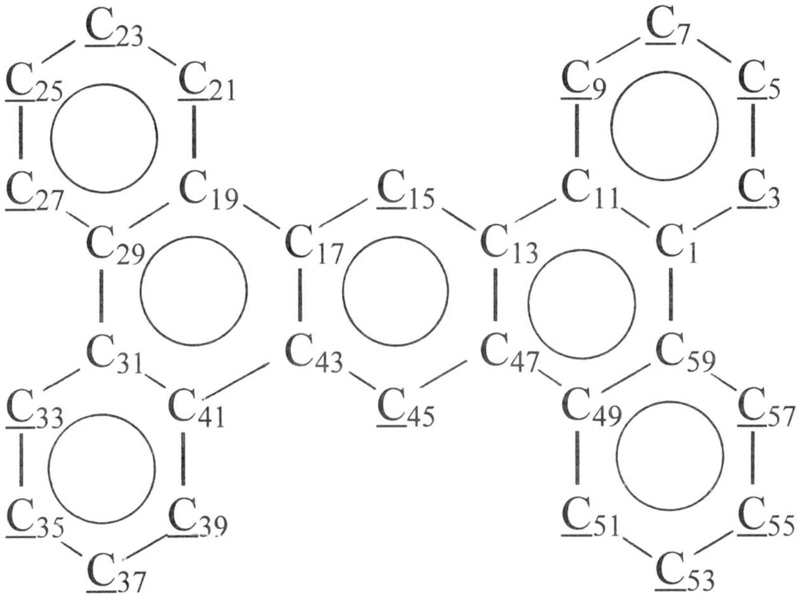

Fig. 13: Bond assignment and locant numbering for a selected seven ring aromatic compound

---

*This co-planar compound, when addended with two additional benzene rings. thereby forcing the molecule out of the plane, shall be the subject of further examination in Chapter 3

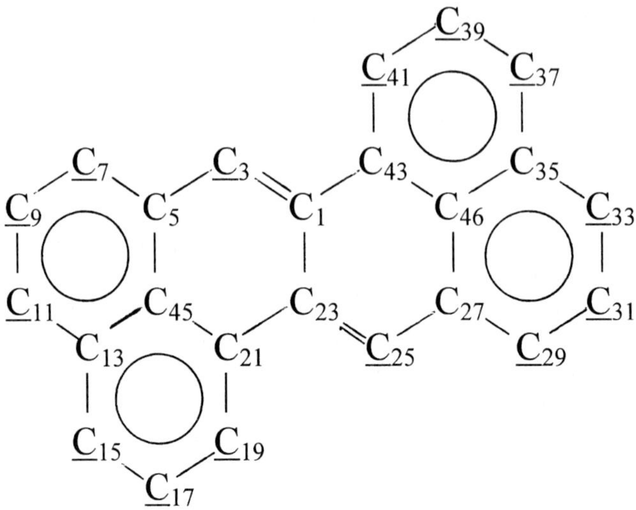

Fig. 14: Bond assignment and locant numbering for zethrene

example of where the IUPAC system obscures the finding of better nomenclature choices.

(16) In the six ring benzenoid (common name zethrene) shown in Figure 14, of the 29 bonds that comprise this molecule, exactly 5 single and 2 double bonds are fixed, while the remaining 22 bond positions have a viable resonance structure. In other words a chemically accurate name should contain five1's, two 2's and twenty-two β's as the bond descriptors. To create the desired name, select as locant #1 a perimeter carbon atom adjacent to a double bond and then note the name would be optimized by going through the second double bond as soon as possible. All paths, traversing either clockwise or counterclockwise around the maximum length perimeter, position the two double bonds at locants #2 and 24 and leave two nodes (to which the name "triple points" had been assigned in [33]) and six edges uncovered. The longest bridges that can be formed to cover these triple points are βCβ bridges from locants #5 to 13 and from #27 to 35. But this still creates the need for two tertiary bridges to cover the last two edges and locant numbers have not yet been assigned to these "triple points". This is done by labeling the node with a superscripted equal sign and the next sequential locant number.

Observe that the locant number convention (odd = atom vs. even = bond) is no longer maintained **beyond the principal cycle**. The canonical name for the molecule portrayed in Figure 14 is:

$$(C2\underline{C}1[C\beta\,(\underline{C}\beta)_3)]_2C1)_2{:}^{(5\text{-}13)}\,(\beta C^{(=45)}\beta)\,^{(27\text{-}35)}(\beta C^{(=46)}\beta);^{(21\text{-}45),(43\text{-}46)}(\beta);\,^{(1\text{-}23)}(1) \quad (25)$$

(17) So far in the treatment of polycyclic compounds, attention has been focused on the simpler class of hydrocarbons. As the scope of molecules being examined is increase by including heteroatoms as members of 'traditional "organic" chemistry' rings, several of the problems endemic to IUPAC nomenclature grow worse. Unlike for benzene, wherein the use of the Robinson ring [34] is standard, even highly meticulous textbooks in both inorganic [35] and organic [36] chemistry revert to the use of single and double bonds when illustrating hetero compounds. One special case of concern is the diazabenzenes. Although this presents no problems for 1,3- and 1,4-diazabenzene, it is disconcerting for pyridazine (1,2-diazabenzene) wherein the false impression of a localized single bond between the nitrogen atoms is not as readily recognized. The nomenclature system described herein encounters no such problem; instead, this compound unambiguously is nomenclated as:

$$(N\beta)_2(\underline{C}\beta)_4 \quad (26)$$

This is but one of many examples of rampant inconsistency in the IUPAC system, to say nothing of the tedium of having to memorize not only uncoordinated names but also inconsistent locant number assignments in these names. This particular compound is #18 in a tabulated list of 47 reference parent compounds containing heteroatoms. In place of the six pages of specialized names tabulated in Rule B-5 (page 54 of [8]), all of these will be handled in the proposed system by a simple substitution of atom type, without requiring the decreeing of any special locant numbering that varies from one atom to another in similar compounds. Table 3 contains: the IUPAC accepted common name, the corresponding IUPAC long name and the proposed systemic name. Similarly, the 14 trivial and semi-trivial names in Rule B-12 (page 60 of [8]) comprise Table 4.

Table 3
IUPAC Rule B-2.11: parent heteroatom compounds

	IUPAC common name	IUPAC alternate name	Systemic name
1.	Theophene	Thia-2,4-cyclopentadiene	$S\beta(\underline{C}\beta)_4$
2.	Benz[b]thiophene	1-Thiaindene	$S\beta(\underline{C}\beta)_2C\beta(\underline{C}\beta)_4C\beta{:}^{(7\text{-}17)}\beta$
3.	Naptho[2.3-b]thiophene	1-Thiabenz[5.6-e]indene	$S\beta(\underline{C}\beta)_2C\beta\underline{C}\beta C\beta(\underline{C}\beta)_4C\beta\underline{C}\beta C\beta{:}^{(7\text{-}25,11\text{-}21)}\beta$
4.	Thianthrene*	9,10-Dithiaanthracene	$[S\beta C\beta(\underline{C}\beta)_4C\beta]_2{:}^{(3\text{-}13,17\text{-}27)}(\beta)$
5.	Furan	Oxa-2,4-cyclopentadiene	$O\beta(\underline{C}\beta)_4$
6.	Pyran†	Oxa-3,5-cyclohexadiene	$O\beta(\underline{C}\beta)_5{:}^{(3)}(1H)$
7.	Isobenzofuran	2-Oxaindene	$O\beta\underline{C}\beta C\beta(\underline{C}\beta)_4C\beta\underline{C}\beta{:}^{(5\text{-}15)}\beta$
8.	Chromene‡	Oxa-2-hydronaphthalene	$O\beta C\beta(\underline{C}\beta)_4C\beta(\underline{C}\beta)_3{:}^{(3\text{-}13)}\beta\,;^{(19)}(1H)$
9.	Xanthene§	10-Oxa-9-hydroanthracene	$O\beta C\beta(\underline{C}\beta)_4C1\underline{C}1C\beta(\underline{C}\beta)_4\,C\beta{:}^{(3\text{-}13,7\text{-}17)}\beta$
10.	Phenoxathun	9-Thia-10-Oxaanthracene	$S\beta C\beta(\underline{C}\beta)_4C\beta O\beta C\beta(\underline{C}\beta)_4C\beta{:}^{(3\text{-}13,7\text{-}17)}\beta$
11.	2H-Pyrrole**	1-Aza-1,3-pentadiene	$N\beta(\underline{C}\beta)_4{:}^{(3)}(1H)$
12.	Pyrrole	1-Aza-2,4-pentadiene	$\underline{N}\beta(\underline{C}\beta)_4$
13.	Imidazole	1,3-Diaza-2,4-pentadiene	$\underline{N}\beta\underline{C}\beta N\beta(\underline{C}\beta)_2$
14.	Pyrazole	1,2-Diaza-2,4-pentadiene	$\underline{N}\beta N\beta(\underline{C}\beta)_3$

*The locant numbering for thianthrene is clockwise sequential starting from the upper rightmost atom. This gives the two sulfur atoms locant numbers 5 and 10 respectively, which is in contradistinction to the numbering in the parent compound, anthracene, which assigns locant numbers 9 and 10 to these locations.

†"The parent neutral ring system is that of pyran-2-H or pyran-4-H. In this nomenclature the term "pyran" refers to the hypothetical neutral aromatic ring system; the real molecules must have an extra hydrogen ... in the name [33]. The systemic name included here is for pyran-2-H. For pyran-4-H, the systemic name is $O\beta(\underline{C}\beta)_5{:}^{(7)}(1H)$

‡As in pyran above, a chromene-4-H, as well as the 2-H variety herein named is also existent, whose systemic name would be: $O\beta C\beta(\underline{C}\beta)_4C\beta(\underline{C}\beta)_3{:}^{(3\text{-}13)}\beta\,;^{(15)}(1H)$.

§As in thianthrene above, the locant numbering of anthracene prevails for this molecule also.

**Note that because of the use of beta bonds, rather than fixed single and double bonds, there is not inconsistency between the molecule depicted and the common vs. the alternate IUPAC names. The double bonds are NOT localized, as the IUPAC picture implies. To the contrary, 5H-Pyrrole is, because of symmetry just another (non-canonical) name for 2H--Pyrrole. Additionally, one should note that a 3H-Pyrrole can also be formed. This would have as its systemic name: $N\beta(\underline{C}\beta)_4{:}^{(5)}(1H)$.

15.	Pyridine	Azabenzene	$\underline{N}\beta(\underline{C}\beta)_5$
16.	Pyrazine	1,4-Diazabenzene	$[\underline{N}\beta(\underline{C}\beta)_2]_2$
17.	Pyrimidine	1,3-Diazabenzene	$[\underline{N}\beta(\underline{C}\beta)]_2(\underline{C}\beta]_2$
18.	Pyridazine	1,2-Diazabenzene	$(N\beta)_2(\underline{C}\beta)_4$
19.	Indolizine*	Dehydro-3a-azaindene	$N\beta(\underline{C}\beta)_3C\beta(\underline{C}\beta)_4{:}^{(1-9)}(\beta)$
20.	Isoindole	Dehydro-2-azaindene	$N\beta\underline{C}\beta C\beta(\underline{C}\beta)_4C\beta\underline{C}\beta{:}^{(5-15)}(\beta)$
21.	3H-Indole	3H-1-dehydoazaindene	$N\beta C\beta(\underline{C}\beta)_4C\beta\underline{C}\beta C\beta{:}^{(3-13)}(\beta)$
22.	Indole	Azaindene	$N\beta C\beta(\underline{C}\beta)_4C\beta(\underline{C}\beta)_2{:}^{(3-13)}(\beta)$
23.	1H-Indazole	1,2-Diazaindene	$(N\beta\underline{C}\beta N\beta C\beta)_2\underline{C}\beta{:}^{(7-15)}(\beta)$
24.	Purine†	1,3,4,6-Tetraazaindene	$(N\beta\underline{C}\beta N\beta C\beta)_2\underline{C}\beta{:}^{(7-15)}(\beta)$
25.	4H-Quinolizine	5-Aza-4-hyrdonaphthalene	$N\beta(\underline{C}\beta)_4C\beta(\underline{C}\beta)_3\underline{C}\beta{:}^{(1-11)}\beta$
26.	Isoquinoline	2-Azanaphthalene	$N\beta\underline{C}\beta C\beta(\underline{C}\beta)_4C\beta(\underline{C}\beta)_2{:}^{(5-15)}(\beta)$
27.	Quinoline	1-Azanaphthalene	$N\beta C\beta(\underline{C}\beta)_4C\beta(\underline{C}\beta)_3{:}^{(3-13)}(\beta)$
28.	Phthalazine	2,3-Diazanaphthalene	$(N\beta)_2\underline{C}\beta C\beta(\underline{C}\beta)_4C\beta\underline{C}\beta{:}^{(7-17)}(\beta)$
29.	Naphthyridine	1,8-Diazanaphthalene	$\underline{N}\beta\underline{C}\beta N\beta(\underline{C}\beta)_3C\beta(\underline{C}\beta)_3{:}^{(3-13)}(\beta)$
30.	Quinoxaline	1,4-Diazanaphthalene	$\underline{N}\beta(\underline{C}\beta)_2N\beta C\beta(\underline{C}\beta)_4C\beta{:}^{(9-19)}(\beta)$
31.	Quinazoline	1,3-Diazanaphthalene	$N\beta\underline{C}\beta N\beta C\beta(\underline{C}\beta)_4C\beta\underline{C}\beta{:}^{(7-17)}(\beta)$
32.	Cinnoline	1,2-Diazanaphthalene	$(N\beta)_2C\beta(\underline{C}\beta)_4C\beta(\underline{C}\beta)_2{:}^{(5-15)}(\beta)$
33.	Pteridine	1,3,5,8-Tetranaphthalene	$N\beta C\beta N\beta\underline{C}\beta N\beta\underline{C}\beta C\beta N\beta(\underline{C}\beta)_2{:}^{(3-13)}(\beta)$
34.	4aH-Carbazole	9-Aza-4a-hydrofluorene	$N\beta C\beta(\underline{C}\beta)_4(C\beta)_2(\underline{C}\beta)_4\underline{C}\beta{:}^{(1-11,13-23)}\beta{;}^{(11)}(1H)$
35.	Carbazole	9-Azafluorene	$N\beta[C\beta(\underline{C}\beta)_4C\beta]_2{:}^{(3-13,15-25)}\beta$
36.	β-Carboline	2,9-Diazafluorene	$N\beta C\beta\underline{C}\beta N\beta(\underline{C}\beta)_2(C\beta)_2(\underline{C}\beta)_4C\beta{:}^{(3-13,15-25)}\beta$
37.	Phenanthridine*	9-Azaphenanthrene	$N\beta[C\beta(\underline{C}\beta)_4C\beta]_2\underline{C}\beta{:}^{(3-13,15-25)}\beta$

*Note that in the IUPAC system this parent compound allocates a locant number to the nitrogen atom at the position of ring fusion, whereas in the corresponding hydrocarbon, indene, this location is assumed not to have either a heteroatom or a ligand. Also the hydrocarbon has an extra hydrogen atom , thus the need for the "dehydro-" All of this is obviated in the proposed system.

†There are exactly two sequences of NβCβ repeated four times that have to be examined when selecting the principal cycle. All of the other sequences have a carbon atom ahead of one or more of the nitrogen atoms and may thus be ignored. The decision as to which of these to choose is made based on the locant numbering on the bond between the two rings; namely, by traversing the hexagon first the bond between the rings is numbered 7-15, while traversing the pentagon first given an inferior numbering for this bond: 7-17.

38.	Acridine[†]	9-Azaanthracene	$N\beta C\beta(\underline{C}\beta)_4C\beta\underline{C}\beta C\beta(\underline{C}\beta)_4C\beta:^{(3-13,17-27)}\beta$
39.	Perimidine[‡]	1,3-Diazaphenalene	$\underline{N}\beta\underline{C}\beta N\beta[C\beta(\underline{C}\beta)]_2C\beta:^{(7-15)}(\beta C^{(=25)}\beta;^{(23-25)}\beta$
40.	Phenanthroline	1,5-Diazaphenanthrene[§]	$N\beta(C\beta)_2(\underline{C}\beta)_3N\beta C\beta(\underline{C}\beta)_2C\beta(\underline{C}\beta)_3:^{(3-21,5-15)}\beta$
41.	Phenazine[**]	9,10-Diazaanthracene	$[N\beta C\beta(\underline{C}\beta)_4C\beta]_2:^{(3-13,17-27)}\beta$
42.	Phenarsazine[††]	9-Arsa-10-azaanthracene	$As\beta C\beta(\underline{C}\beta)_4C\beta N\beta C\beta(\underline{C}\beta)_4C\beta:^{(3-13,17-27)}\beta$
43.	Isothiazole[‡‡]	1-Thia-2-aza-cyclopentadiene	$S\beta N\beta(\underline{C}\beta)_3$
44.	Phenothiazine	9-Thia-10-azaanthracene	$S\beta C\beta(\underline{C}\beta)_4C\beta N\beta C\beta(\underline{C}\beta)_4C\beta:^{(3-13,17-27)}\beta$

---

[*] Note the orientation and the locant numbering of this compound follows the standard IUPAC system, rather than the special orientation and numbering that had been used for phenanthrene; viz., the nitrogen atom in this reference compound is designated as being locant #5, rather than locant #9 — which is the locant number it would be assigned as a derivative of phenanthrene

[†] Despite that there is complete equivalence between locations 9 and 10, IUPAC has illustrated this compound with the nitrogen atom at location #10. This is in contradistinction to the standard choice of lowest locant numbers. There is also the frequently used comment: "Denotes exceptions to systemic numbering".

[‡] As in xanthene above because of the symmetry, the hydrogen atom can be depicted exclusively at locant #1, rather than "floating" between the two nitrogen atoms. In other words, there is no need for an ambiguous locant designator **a** — as shall be the case for some of the molecules to be described in Chapter 9.

[§] As with phenanthridine above, the orientation and locant numbering follow the IUPAC standard, rather than that of its parent, phenanthrene. Thus the alternate name should be 1,5-diazaphenanthrene, rather than the 1,7 name that results when one uses the locant numbering that IUPAC assigned. Note there is no ambiguity, since there is no standard orientation that can vary according to historical development or whim, in the proposed system. The canonical name will follow the shorter perimeter path between the two nitrogen atoms and will select as the direction the shorter path to a ring fusion; that is, counterclockwise starting from the upper nitrogen in the standard picture.

[**] Instead of the locant numbering of its parent, anthracene, IUPAC reverts to the standard locant numbering for this compound; thus the two nitrogen atoms of this molecule are at locant numbers 5 and 10. In presenting the alternate name, rather using the locants depicted, the standard locant numbers (9 and 10) are used

[††] Same comments as for phenazine. Also, the arsenic atom is given priority over the nitrogen and thus has the lower locant number.

[‡‡] Alternately, the two systems can be combined to give as an accepted IUPAC name 2-Azathiophene.

45.	Isoxazole	1-Oxa-2-azacyclopentadiene	$O\beta N\beta(\underline{C}\beta)_3$
46.	Furazan	1-Oxa-2,5-diazacyclopentadiene	$O\beta N\beta(\underline{C}\beta)_2N\beta$
47.	Phenoxazine	9-Oxa-10-azaanthracene	$O\beta C\beta(\underline{C}\beta)_4C\beta N\beta C\beta(\underline{C}\beta)_4C\beta{:}^{(3\text{-}13,17\text{-}27)}\beta$

Table 4:
IUPAC Rule B-2.12 Trivial and semi-trivial names

1. Isochroman

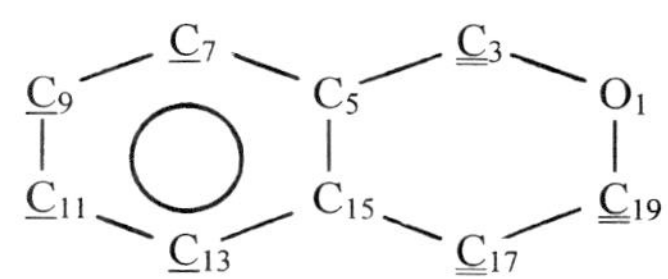

$O1\underline{C}1C\beta(\underline{C}\beta)_4C1(\underline{C}1)_2{:}^{(5\text{-}15)}(\beta)$

2. Chroman

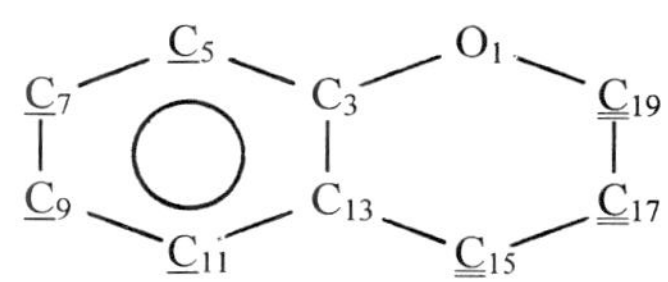

$O1C\beta(\underline{C}\beta)_4C1(\underline{C}1)_3{:}^{(3\text{-}13)}(\beta)$

3. Pyrrolidine

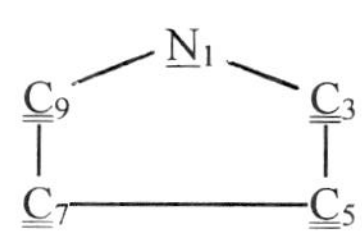

$N1(\underline{C}1)_4$

4. Pyrroline

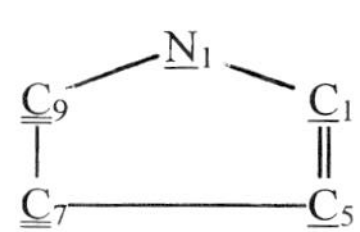

$N1\underline{C}2\underline{C}1(\underline{C}1)_2$

2-Pyrroline shown
also 1-Pyrroline and
3-Pyrroline exist

5. Imidazolidine

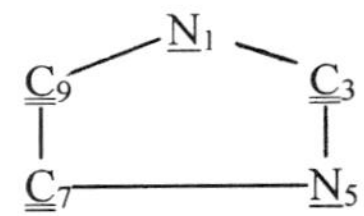

N1C1N1(C1)$_2$

6. Imidazoline

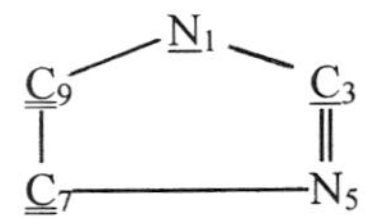

N2C1N1(C1)$_2$

7. Pyrazolidine

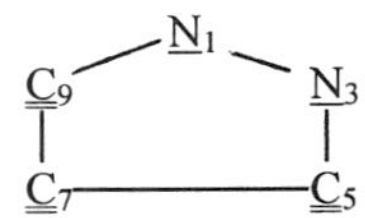

(N1)$_2$(C1)$_3$

8. Pyrazoline

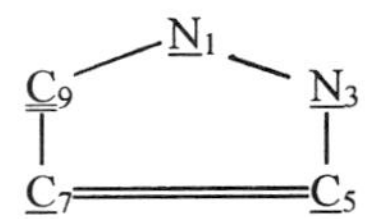

(N1)$_2$C2C1C1

3-Pyrazoline shown also 1-Pyrazoline and 2-Pyrazoline exist

9. Piperidine

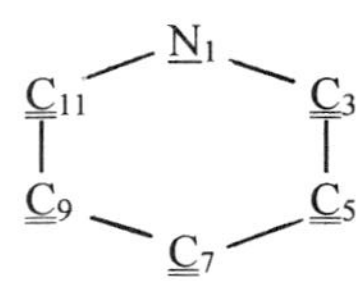

N1(C1)$_5$

10. Piperazine

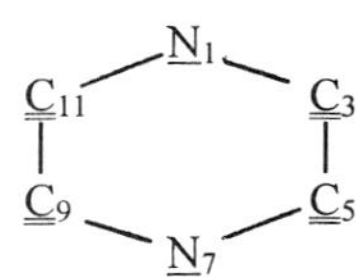

[N1(C1)$_2$]$_3$

11.	Indoline	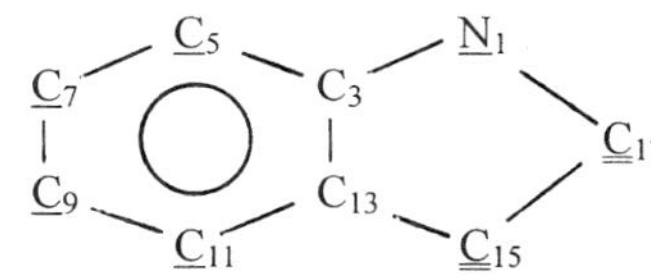	$N1C\beta(\underline{C}\beta)_4C1(\underline{C}1)_2{:}^{(3\text{-}13)}(\beta)$
12.	Isoindoline	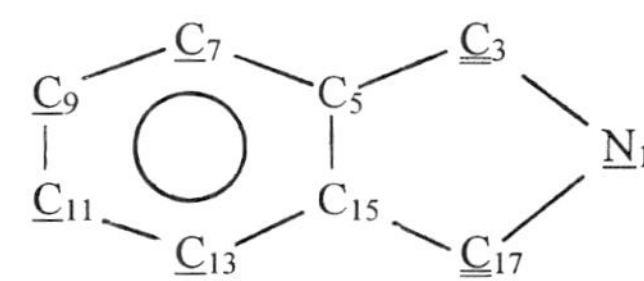	$N1\underline{C}1C\beta(\underline{C}\beta)_4C1\underline{C}1{:}^{(5\text{-}15)}(\beta)$
13.	Quinuclidine	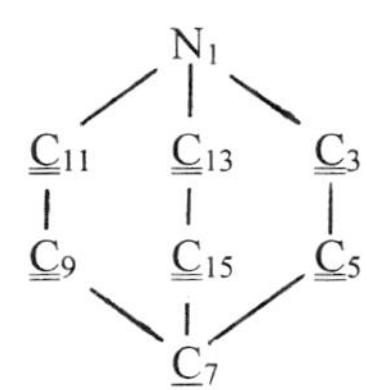	$N1(\underline{C}1)_2\underline{C}1(\underline{C}\,1)_2{:}^{(1\text{-}7)}(1\underline{C}^{(=13)}1\underline{C}^{(=15)}1)$
14.	Morpholine	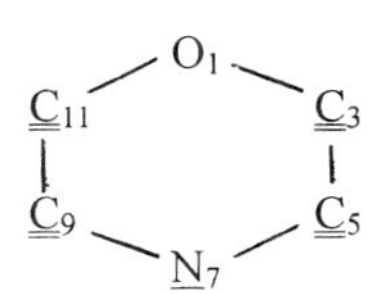	$O1(\underline{C}1)_2\underline{N}1(\underline{C}1)_2$

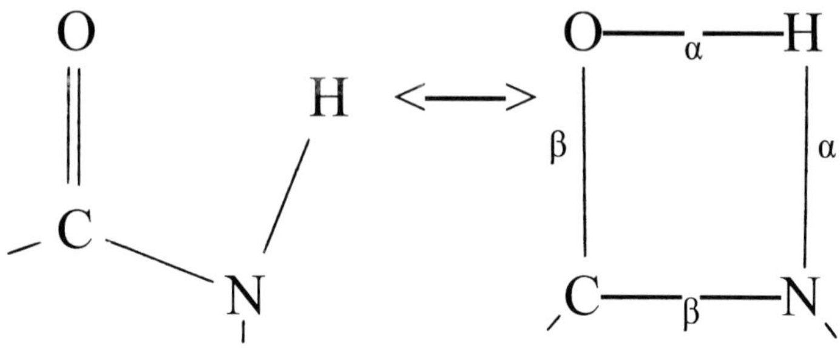

Fig. 15: General Formula for "Biochemistry"

(18) The introduction of α and β bonds into the nomenclature illustrates an important synergy when a molecule contains a carbonyl and an adjacent amine (or imine) group. Instead of the traditional representation of the hydrogen atom being singly-bonded to the nitrogen atom, the oxygen atom doubly-bonded to the carbon atom and there being no bond between the oxygen and hydrogen, this group may be viewed as an intra-molecular four member ring containing 2 α and 2 β bonds (Figure 15); namely, half of the pi bond of the double bond in the carbonyl group forms an alpha bond from the oxygen to the hydrogen; thereby leaving a beta bond to the carbon. The other half of this double bond forms a beta bond between the carbon and nitrogen withdrawing half of the bond between the hydrogen and nitrogen atoms leaving an alpha bond between these two atoms. As shall be further developed in Chapter 9, this four member "ring" OβCβNαHα is the heuristic that separates "biochemistry" from "organic chemistry".

(19) The benzyne molecule (Figure 16) illustrates a similar type of question regarding aromaticity. Here, instead of the traditional picture having a directly isolatable triple bond, which induces adjacent single bonds and thus the remaining assignment of two double bonds, it should be acknowledged that the remaining aromaticity has NOT been suppressed. Consequently, instead of:

$$C3C1(\underline{C}2\underline{C}1)_2 \qquad (27)$$

a more appropriate name would be:

$C\gamma C\beta(\underline{C}\beta)_4$ (28)

(20) In a similar manner, the first metallabenzyne to be prepared [37]

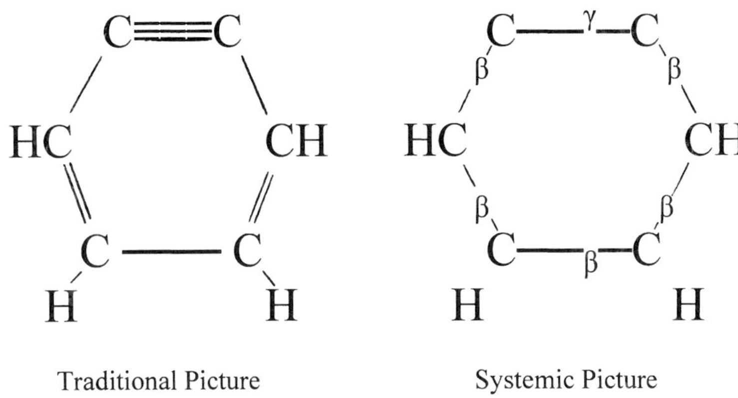

Fig. 16: Benzyne

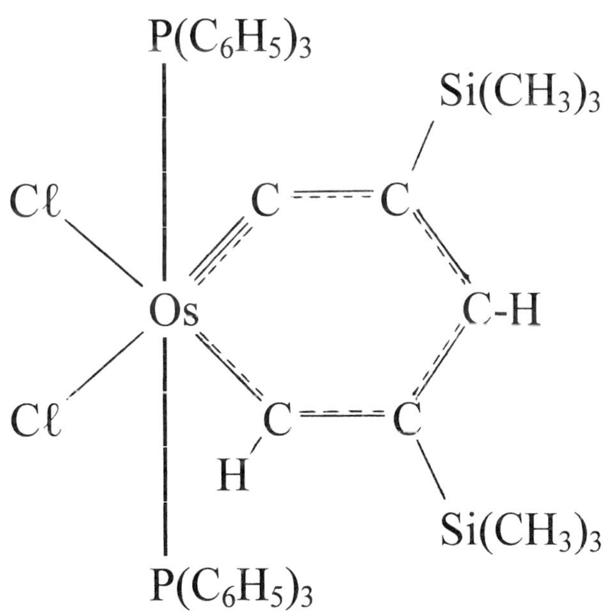

Fig. 17: An osmium metallabenzyne

(Figure 17) would be nomenclated as:

$$\mathrm{Os}\ \gamma(\mathrm{C}\beta)_5\mathrm{:}^{(1,1)}(1\mathrm{C}\ell);^{(1,1)}[1\mathrm{P}1\mathrm{Ph}\mathrm{:}^{(3,3)}(1\mathrm{Ph})];^{(5,9)}(1\mathrm{Si}1\underline{\underline{\mathrm{C}}}1\mathrm{H})\mathrm{:}^{(3,3)}(1\underline{\underline{\mathrm{C}}}1\mathrm{H});^{(7,11)}(1\mathrm{H}) \qquad (29)$$

(21) One advantage, or disadvantage, of the proposed, systemic nomenclature is the need for highly detailed knowledge of the entire bonding matrix. In the natural product, heme, shown in Figure 18, the reason for using ℵ bonds for the four iron-nitrogen bonds, instead of, as represented in textbooks [38-40], using single bonds is illustrated.

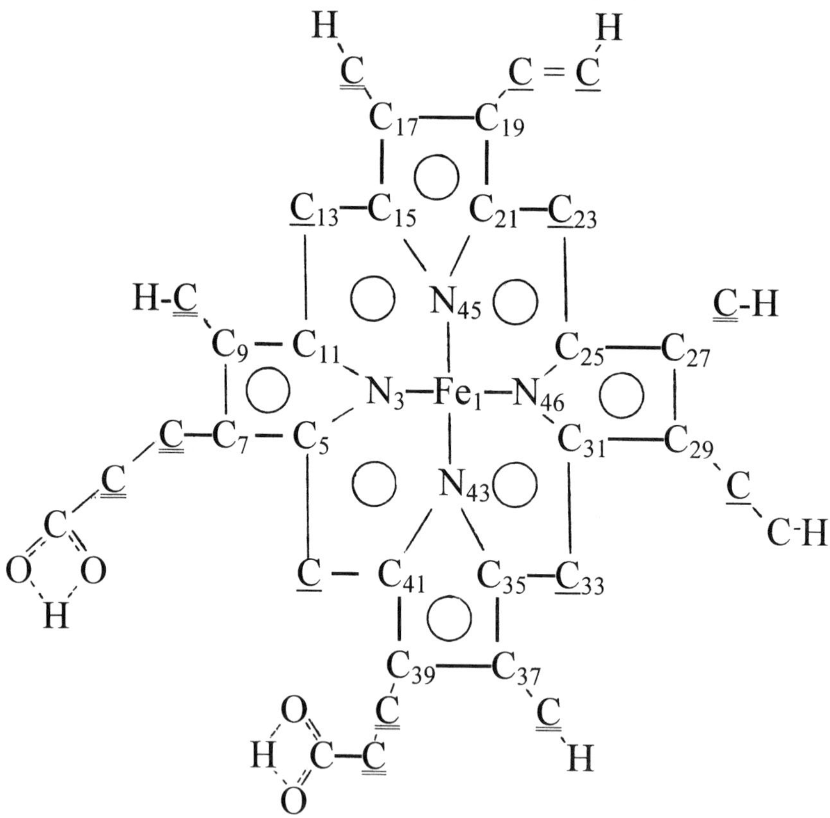

Fig. 18: Heme

By this choice, the use of Robinson circles clearly indicates the aromatic, even if not truly double bond, character throughout the entire recticular [41] kernel of the molecule. The systemic canonical name for heme thus becomes:

$$\mathrm{Fe}\aleph\mathrm{N}\beta(\mathrm{C}\beta)_{19}\mathrm{N}\aleph{:}^{(1\text{-}15)}(\aleph\mathrm{N}^{(=45)}\beta);^{(1\text{-}25)}(\aleph\mathrm{N}^{(=46)}\beta);\,^{(5\text{-}41)}(\beta\underline{\mathrm{C}}\beta);\,^{(3\text{-}11,21\text{-}45,31\text{-}46,35\text{-}43)}(\beta);^{(7,39)}[1(\underline{\underline{\mathrm{C}}}1)_2](\mathrm{C}\beta\mathrm{O}\alpha\mathrm{H}\alpha\mathrm{O}\beta);^{(19,29)}(1\underline{\mathrm{C}}2\underline{\mathrm{C}}1\mathrm{H});\,^{(9,17,27,37)}(1\underline{\underline{\mathrm{C}}}1\mathrm{H});^{(13,23,33)}(1\mathrm{H}) \qquad (30)$$

(22) One of the main reasons why nodal nomenclature [42] was created was that the IUPAC organic nomenclature did not allow for assigning unique canonical names to members of the class of molecules referred to as "cyclophanes"; a problem that would not have arisen had beta bonds been available. This is not to devalue the many virtues of nodal nomenclature, rather merely to show how the proposed system nomenclates these compounds without difficulty. For example, consider [2,3]-ortho,para-cyclophane (Figure 19). Before presenting the nomenclature assignment for this molecule, it is instructive to examine both the current traditional IUPAC "organic"

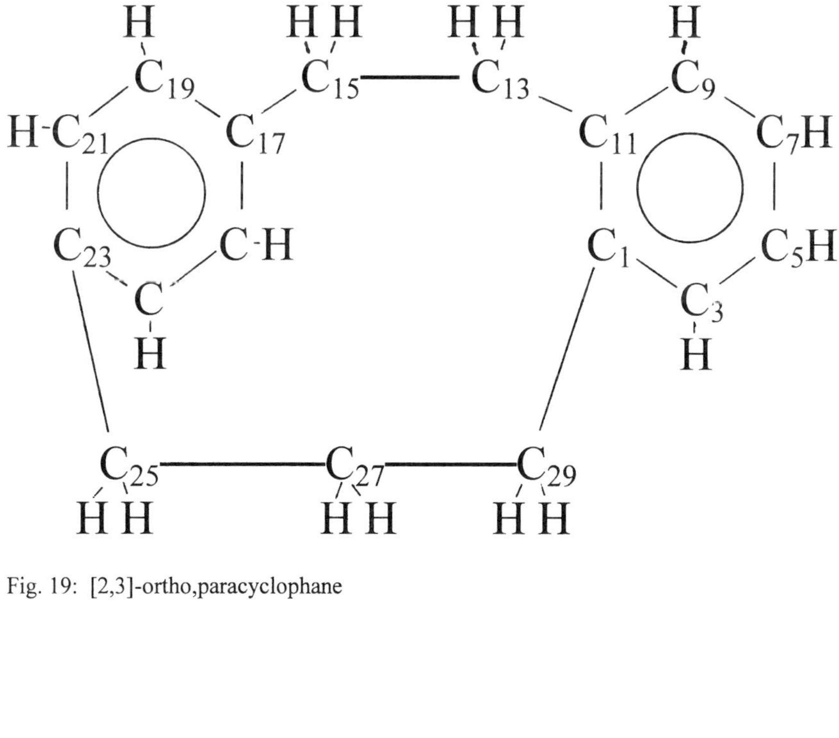

Fig. 19: [2,3]-ortho,paracyclophane

name (the one in which an entire extra layer of specialized names; viz., cyclophanes has not been created), as well as the name that "inorganic" chemists would apply to a "similar"[*] compound. The standardized name according to organic chemistry rules[†] is: 1,1':2,4'-dimethylenetrimethylenedibenzene. On the other hand, "inorganic" chemists would see two "core" modules (in this case benzene rings)[‡] connected by two bridging groups, which they designate by the prefix μ. Thus, their name must include a *bis*benzene core, a μ-ethylene bridge, a μ-propylene bridge and a set of locant numbers detailing how the bridges are connected to the core. This process is indicated by the symbol η (alphabetically eta, but often referred to as "hapto" from the Greek word to fasten). The resulting IUPAC "inorganic" name is: μ-ethylene-μ-propylene-[1,2-η;1,4-η] bisbenzene. Instead of either of these names, the name for this molecule is formulated starting from the longest non-intersecting contiguous closed path that can be assigned, which is 15 atoms long. Additionally, the lowest locant numbering is achieved by starting from the union of the ortho ring with the longer chain and progressing through the shorter chain first. The systemic name thus produced is:

$$C\beta(\underline{C}\beta)_4C1(\underline{\underline{C}}1)_2C\beta(\underline{C}\beta)_2C1(\underline{\underline{C}}1)_3:^{(17\text{-}23)}((\beta\underline{C})_2\beta);^{(1\text{-}11)}(\beta) \qquad (31)$$

Two notes of significance at this time are:

(a) In exactly the same manner as for pyradizine above, inclusion of the β symbol in the nomenclature allowed successive individual bonds to be listed without incorrectly ascribing either a single or a double bond to any place where there is a "natural" ambiguity. This is in contradistinction to the evasion of the problem by the use of the term "benzene" as a module in both "organic" and "inorganic" IUPAC nomenclature.

(b) Had one of the non-bridging carbon atoms in the para ring been a heteroatom, the rest of the locant number assignment would remain the same (inasmuch as the identity of nodes is low on the priority list); however, which of the two two-atom segments

---

[*] See Footnote on bottom of page 20.

[†] Because IUPAC's ordering system is based on alphabetization, the prefix "di" is included before "tri".

[‡] This is in contradistinction to the more familiar instances in which the core module of organo-metallic compounds is a single metal atom.

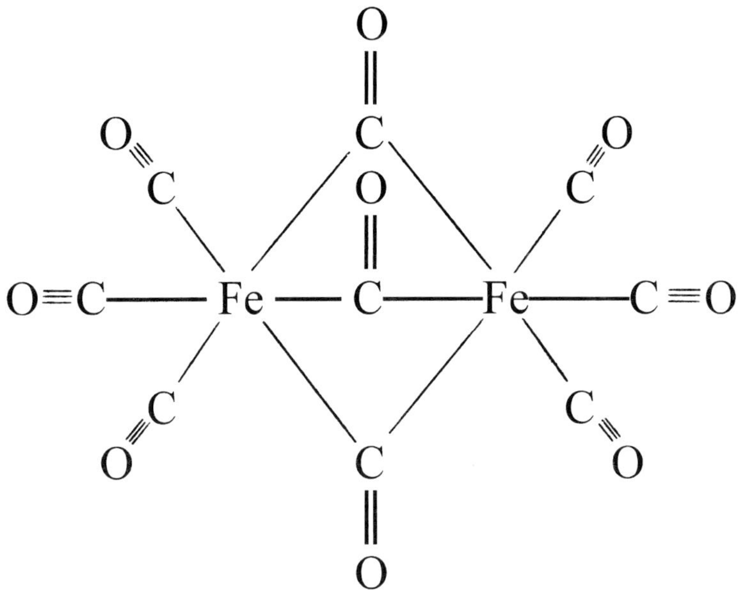

Fig. 20: Tri-μ-carbonyl-bis(tricarbonyliron)

is to be regarded as part of the principal cycle and which the bridge depends on the atomic number of the heteroatom; namely, a nitrogen atom would be on this principal cycle, but a boron atom would NOT. Conversely, the presence of a heteroatom on an ortho- or meta- connected ring would have no effect, inasmuch as the longer chain would always be selected as part of the principal cycle.

(23) Having introduced the most common prefix of IUPAC inorganic nomenclature, (μ) in the previous example, another "typical" inorganic coordination compound, $Fe_2(CO)_9$, is presented as Figure 20. The names assigned to this compound are:

(a) in IUPAC inorganic nomenclature [43]: tri-μ-carbonyl-bis(tricarbonyliron);

(b) its "pseudo-organic" name, assuming one wished to name this

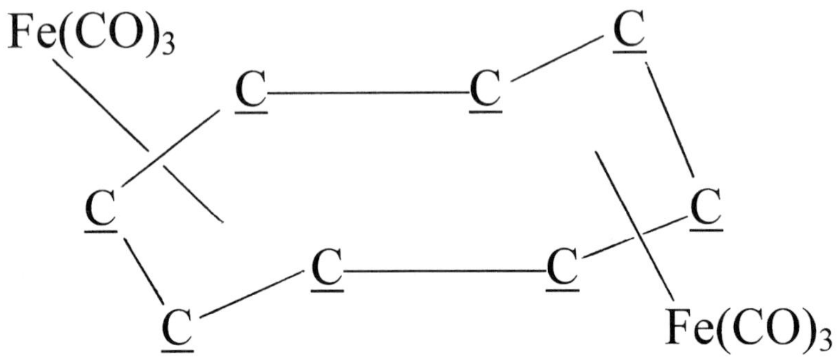

Fig. 21: A cyclooctatetraene-tricarbonyliron compound

compound using IUPAC organic nomenclature *: 1,3-Bistricarbonyl-2,4-bisoxa-bicyclo[1.1.1]carbonyl-buta-1,3-ferretane†

(c) the name that would be assigned to it in the proposed system:

$$(Fe\aleph C\aleph)_2{:}^{(1\text{-}5)}(\aleph C\aleph{:}^3 2O);^{(1,1,1,5,5,5)}(\aleph C3O);^{(3,7)}(2O) \qquad (32)$$

Note that this last name does not use any complicated set of add-on prefixes, such as μ, η, etc. (inorganic) or esoteric suffixes, such as the extended Hantsch-Widman system (organic) [44]. Moreover, in Chapter 3, a further simplification will be introduced using "cylindrical" nomenclature.

(24) In Figure 21, attention is directed to a molecule for which identical modules are bound to non-overlapping halves of a ring. Cahn and Dermer [45], name this compound: trans-μ-(1-4-η:5-8-η-cyclo-octatetraene)- bis-(tricarbonyliron). Godwin [46], on the other hand, recognizes that there is the possibility for connections to only some of the atoms and thus elects to include all relevant locant numbers; namely: μ-(1,2,3,4-η-:5,6,7,8-η-cycloocta-1,3,5,7-tetraene)- bis-(tricarbonyliron). In the nomenclature being developed in this

* Pseudo since the iron atoms each have a coordination of six -- a coordination that is not evident for "organic" molecules.

† Hantsch-Widman nomenclature uses the suffix **etane** in ferretane to indicate the presence of a four member ring containing no nitrogen atoms.

treatise, each of the iron atoms in this molecule should be viewed as being aleph bonded to the appropriate four carbon atoms of the ring. Consequently, instead of the eight atom carbon ring being the primary nomenclature focus with the iron carbonyl groups of secondary interest, consider the 10 atoms long cycle, illustrated in Figure 22,

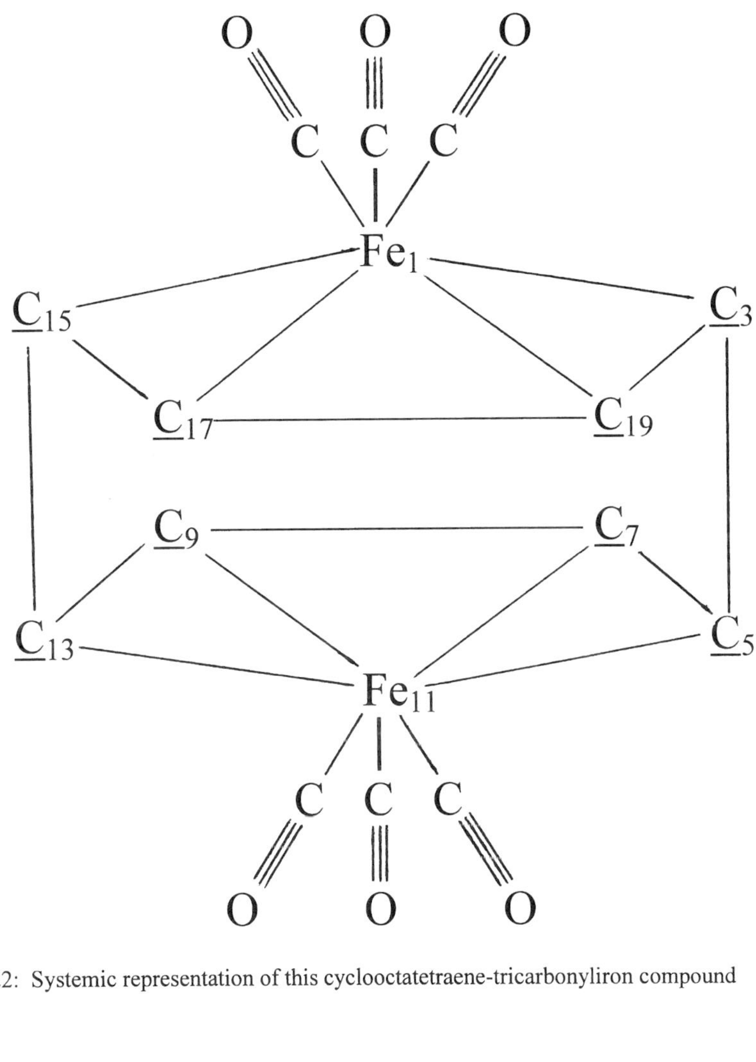

Fig. 22: Systemic representation of this cyclooctatetraene-tricarbonyliron compound

which includes the two iron atoms and is nomenclated* as:

$$[\mathrm{Fe}\aleph(\underline{\mathrm{C}}\beta)_3\underline{\mathrm{C}}\aleph]_2{:}^{(3\text{-}19,9\text{-}13)}(\beta);^{(1\text{-}15,1\text{-}17,5\text{-}11,7\text{-}11)}(\aleph);^{(1,1,1,11,11,11)}(1\mathrm{C}3\mathrm{O}) \qquad (33)$$

(25) Although this feature of focusing on the ten member ring including both iron atoms, instead of just an eight member ring without the iron atoms is an extremely convenient one, there shall arise scenarios in which this is not always possible. One such example occurs when all aleph-bonded cycles involving the metal atom are shorter than some other ring in the molecule that does not contain the metal atom. In such cases the metal atom has to be viewed as being on a bridge between two longer paths forming the principal cycle. In Figure 23, for example, the molecule which IUPAC calls: tetracarbonyl(η-1,5-cyclooctadiene) molybdenum and which CAS augments with an oxidation number of zero: tetracarbonyl(η-1,5-cyclo-octadiene) molybdenum(0), would have as its systemic name:

$$[\mathrm{C}2\underline{\underline{\mathrm{C}}}1(\underline{\underline{\mathrm{C}}}1)_2]_2{:}^{(1\text{-}11)}(\aleph\mathrm{Mo}^{(=17)}\aleph);^{(17,17,17,17)}(1\mathrm{C}3\mathrm{O}) \qquad (34)$$

Because the naming algorithm selected counts all atoms in a ring as equal for the purpose of selecting the principal cycle, the molybdenum atom has been relegated to being on a bridge, despite that it is the focus (coordinating atom) in most, if not all, inorganic and organo-metallic nomenclature systems. An alternate name, that is

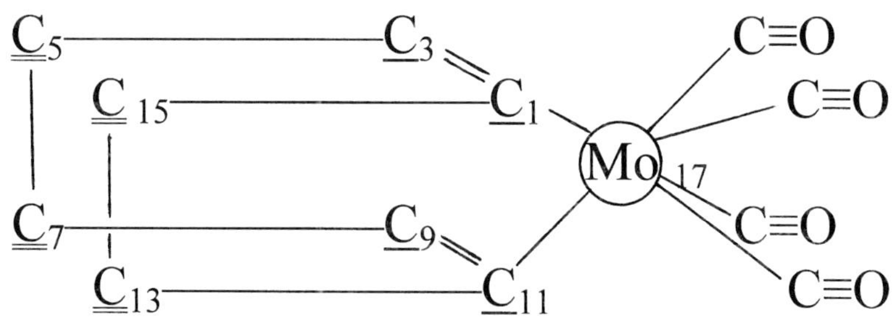

Fig. 23: Tetracarbonyl (η-1,5-cyclooctadiene) molybdenum

*Because β has priority over ℵ at locant 6, (33) is, as written, rather than as: $\mathrm{Fe}\aleph\underline{\mathrm{C}}\beta\underline{\mathrm{C}}\aleph\mathrm{Fe}\aleph(\underline{\mathrm{C}}\beta)_5\underline{\mathrm{C}}\aleph{:}^{(3\text{-}19,5\text{-}9)}(\beta);\ ^{(1\text{-}15,1\text{-}17,7\text{-}11,7\text{-}13)}(\aleph);\ ^{(1,1,1,7,7,7)}(1\mathrm{C}3\mathrm{O})$, which would be the canonical name based only on atom size.

closer to traditional inorganic nomenclature, will be created using spherical nomenclature in Chapter 6.

(26) Instead of the concept of "bridging groups" expressed by the symbol μ and the subscripts that are appended to it in IUPAC inorganic nomenclature, in order to be able to indicate precisely which atoms are being bridged by which other atoms, there is a straight-forward sequential listing of the atoms and bonds in the proposed system. For example, Figure 24, which has as its IUPAC name: [47] Tri-μ_2-hydrido-μ_3-hydrido-biscyclopentdienyllutetium, would be nomenclated as:

$$(\mathrm{Lu}\alpha\mathrm{H}\alpha)_3{:}^{(1\text{-}5)}(\alpha\mathrm{H}^{(=13)}\alpha;^{(9\text{-}13)}\alpha;^{(1,1,5,5,9,9)}(1\mathrm{C}(\beta\underline{\mathrm{C}})_4\beta) \quad (35)$$

Alternately, as is the practice in IUPAC nomenclature, selected abbreviations for frequently used groups of atoms that act as a single

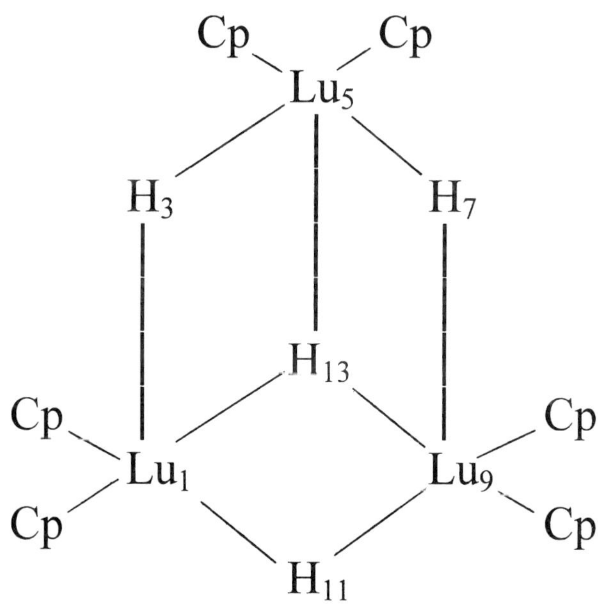

Fig. 24: Systemic representation for molecule that IUPAC calls Tri-μ_2-hydrido-μ_3-hydrido-biscyclopentdienyllutetium

unit may be introduced†. In this instance Cp for the cyclopentadienyl module, C(β$\underline{C}$)$_4$β. Thus (35) may be written as:

$$(\mathrm{Lu}\alpha\mathrm{H}\alpha)_3{:}^{(1\text{-}5)}(\alpha\mathrm{H}^{(=13)}\alpha;^{(9\text{-}13)}\alpha;\ ^{(1,1,5,5,9,9)}(1\mathrm{Cp}). \qquad (36)$$

In this name it is clearly spelled out that all of the bridges are alpha bridges. Furthermore, from the name one can determine that three of the hydrogen atoms are bonded to two lutetium atoms; i.e., bond order equals one-half, while the last of the hydrogen atoms is bonded to all three lutetium atoms; i.e., bond order equals only one-third. Which of

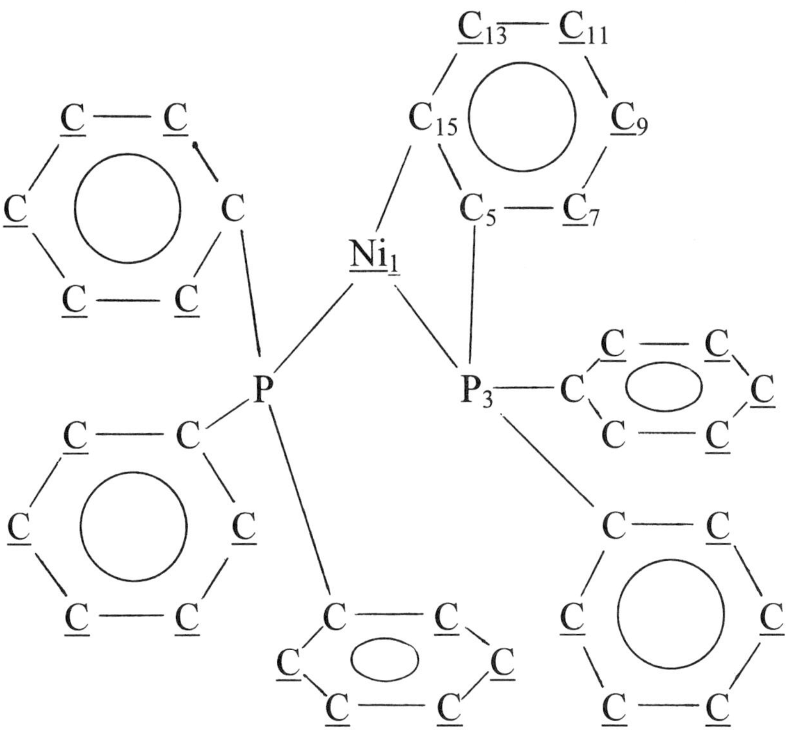

Fig. 25: A Molecule that Uses the Kappa Convention in IUPAC Nomenclature

† Other modules often abbreviated are Ph for the phenyl group Cβ($\underline{C}$β)$_5$ and tBu for the tertiary butyl group C1$\underline{\underline{C}}$1H:$^{(3,3)}$(1$\underline{\underline{C}}$1H). Additionally, ortho, meta and para C_6H_4 groups can be abbreviated as oPh, mPh and pPh respectively.

these hydrogen atoms is only one-third bonded is constantly interchanging in a pseudo-rotational manner. To illustrate this, Figure 24 was deliberately drawn with M. C. Escher* in mind; i.e., with the perspective distorted, so that one can not tell whether the hydrogen atom with subscript 11 or the one with subscript 13 is in front of the plane of the three lutetium atoms. Moreover, a similar picture emerges with any of the other three hydrogen atoms selected as the one with connectivity three. An additional comment about the systemic nomenclature is that, once again, as with the cyclophanes in Figure 19, greater simplicity and uniformity are introduced by not needing a special symbol, such as η, to identify the points of contact.

(27) As well as finessing the use of μ and η in "inorganic" nomenclature, the proposed nomenclature discards the entire kappa convention [48]†. The molecule illustrated in Figure 25, which has the IUPAC name: [2(-diphenylphosphino-κP)-phenyl-κC^1]hydrido (triphenylphosphine-κP)nickel(II), using the phenyl abbreviation described in the footnote on page 55, would be named:

$$\underline{\mathrm{Ni}}\mathrm{1P1C\beta(\underline{C}\beta)_4C1{:}^{(1)}1P1Ph{:}^{(3,3)}(1Ph);^{(3,3)}(1Ph);^{(5\text{-}15)}(\beta)} \qquad (37)$$

(28) Not only are some inorganic compounds named using the kappa convention, others augment this kappa convention with a "priming" convention [49]. All of these are, similarly, made obsolete by the proposed system. For example, the ion illustrated in Figure 25, which IUPAC names as: aqua[(1,2-ethanediyldinitrilo-κ^2N,N') (tetraacetato-κ^3O,O",O"")]cobaltate (1-) would be named as:

$$\mathrm{[Co1O1C1\underline{\underline{C}}1N1(C1)_2N1\underline{\underline{C}}1C1O1{:}^{(1\text{-}9)}(1O1C1\underline{\underline{C}}1){:}^{(5)}(2O);^{(1\text{-}9,1\text{-}15)}(1);}$$
$$\mathrm{^{(1)}(1O1H){:}^{(3)}(1H);^{(5,19)}(2O);^{(15)}(1\underline{\underline{C}}1C\beta O\alpha){:}^{(5)}(\beta O\alpha)]^-} \qquad (38)$$

A similar, but even more complicated, IUPAC name (requiring a triply double primed oxygen atom) is formed for that ion in which a fourth CH_2COO group replaces one of the Co-N bonds. Instead of

---

* The Dutch artist M. C. Escher was famous for his pictures distorting perspective so as to create illusions that defy physical reality, such as water that flows uphill, staircases that go nowhere, one hand drawing the other hand, etc.

† Per Section I-10.6.2.2. of [48]: "In the nomenclature of polydentate chelate complexes, single ligating atom attachments of a polyatomic ligand to a coordinating centre are indicated by the italic element symbol preceded by the Greek letter kappa, κ."

the IUPAC name: aqua[(1,2-ethanediyldinitrilo-κN)(tetraacetato-κ^4O,O",O"",O""")]cobaltate (1-), the systemic nomenclature for this compound would be:

$$\mathrm{Co1O1C1\underline{C}1N1(C1)_2N1\underline{C}1C1O1{:}^{(1\text{-}9,1\text{-}15)}(1O1C1\underline{C}1){:}^{(5)}(2O);^{(1\text{-}9)}(1);\; ^{(1)}(1O1H){:}^{\overline{(3)}}(1H);^{(5,19)}(2O);^{(15)}(1\underline{C}1C\beta O\alpha){:}^{(5)}(\beta O\alpha)]^-} \quad (39)$$

In all cases, no use has been made of the prime symbol, nor is there any need to delve into why IUPAC chooses to use double primes with the oxygen atoms in these names vs. their use of single primes with the examples they showed using nitrogen atoms [50].

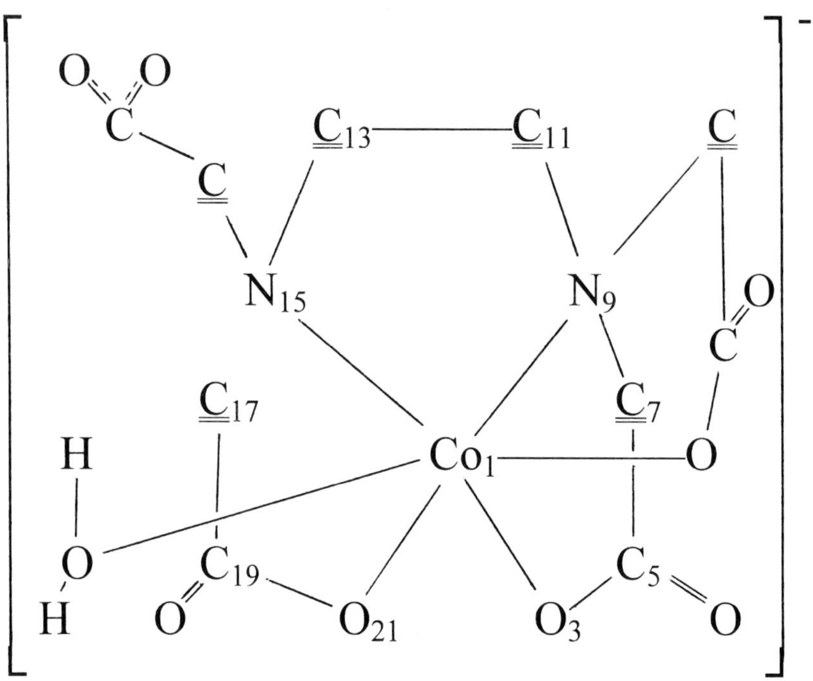

Fig. 26: An Ion that Needs Both the Kappa and the Priming Conventions in IUPAC Nomenclature

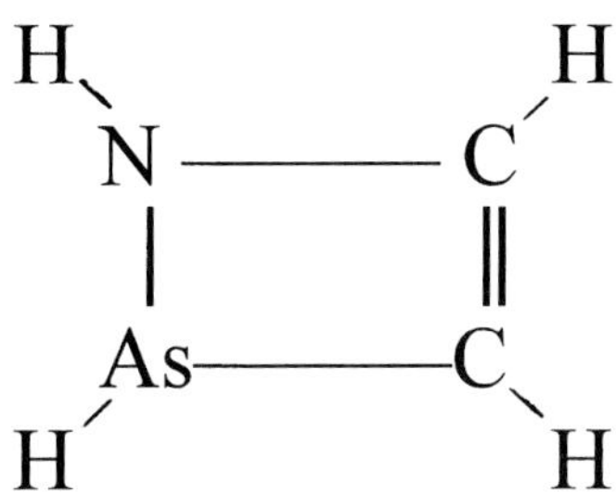

Fig. 27: IUPAC Name: Δ^3-1,2-Azarsetine

(29) Another example of the prevailing use of "unnecessary"* affixes in IUPAC inorganic nomenclature occurs in the extended Hantzsch-Widman system. Here a Greek capital delta (Δ) is inserted to denote a double bond in a heterocyclic ring. For example, the four-member ring which IUPAC would name as: Δ^3-1,2-Azarsetine (Figure 27) in the proposed system is simply:

As1N1C2C1 (40)

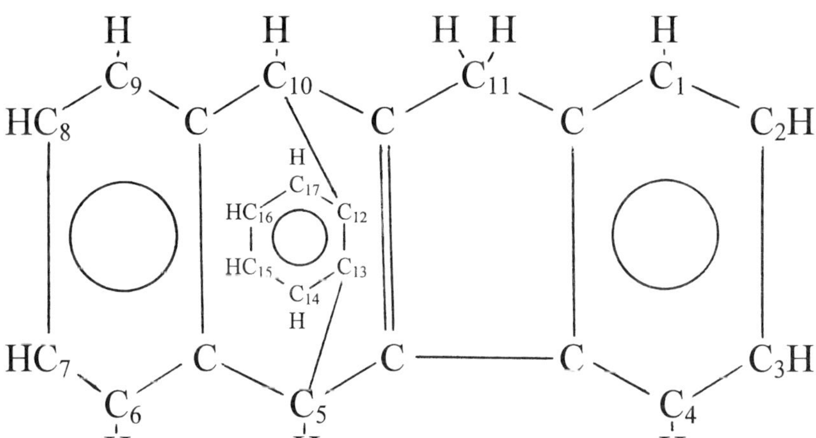

Fig. 28: Molecule IUPAC names: 5,10-o-Benzeno-5,10-dihydro-benzo[b]fluorene

*In a follow-on report, it will be shown that affixes are useful in any domain that can not be uniquely described by graph theory and/or measure theory. In the case of chemistry, this occurs for enantiomers.

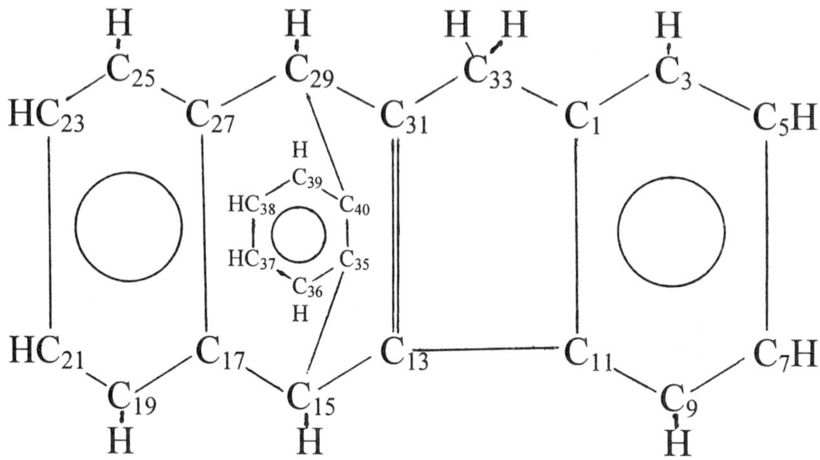

Fig. 29: Five ring system with systemic name: $C\beta(\underline{C}\beta)_4(C1)_3C\beta(\underline{C}\beta)_4(C1)_3\underline{\underline{C}}1$: $^{(15\text{-}29)}(1C^{(=35)}\beta[\underline{C}^{(=36\rightarrow 39)}\beta]_4C^{(=40)}1);^{(13\text{-}31)}2;^{(1\text{-}11,17\text{-}27,35\text{-}40)}\beta$

(30) Similar to the complexity in the IUPAC naming of inorganic compounds which requires the postulation of affixes and esoteric conventions, attention is again focused on the corresponding lack of simplicity and logic in IUPAC's organic rules; in particular, the above described long tables of "standard" forms to be memorized and then modified, such as what Tables 1, 2 and 3 above corrected. This chapter ends with a compound that is sufficiently tedious in the IUPAC system that even expert books on nomenclature get mired in the details and sometimes publish inconsistent names. Figure 29 on page 87 of Cahn and Dermer's monograph [51], is presented herein as Figure 28. The molecule whose name was published as: 10,11-Dihydro-5,10-o-benzeno-5H- benzo[b]fluorene, has also been included in another Dermer monograph (with Fletcher and Fox) [52] — this time with the correct locant numbers: 5,10-Dihydro-5,10-o-benzeno-11H-benzo[b] fluorene. Two further remarks about the IUPAC fiat concerning this name are the selection of the locant number for the "extra" H and the promotion of the "Hydro" (Dihydro in this case) to priority over the lower establishing the ordering sequence). In other words, without IUPAC's selective alphabetic considerations, only the "Benzeno" but not the o-, m- and p- are candidates for alphabetizing, consistency should have dictated an IUPAC name of: 5,10-o-Benzeno-5,10-dihydrobenzo [b]fluorene.

The proposed systemic name (Figure 29), on the other hand, although longer, relies on neither a memorized table of "standard" names, nor any other conventions:

$$C\beta(\underline{C}\beta)_4(C1)_3C\beta(\underline{C}\beta)_4(C1)_3\underline{\underline{C}}1:^{(15\text{-}29)}(1C^{(=35)}(\beta C^{(=36\text{-}>39)}{}_4\beta C^{(=40)}1);{}^{(13\text{-}31)}(2);{}^{(1\text{-}11.17\text{-}27,5\text{-}40)}(\beta) \quad (41)$$

consequently, there is less chance of either error or ambiguity.

Additional further usage of these standardized bond with intermediate bond orders shall be introduced in later sections. One especially important use of ℵ bonds will be described in Section 6 for “organic” ring assemblies involving four and eight member rings.

REFERENCES

[1] Pauling, L., The Nature of the Chemical Bond and the Structure of Molecules and Crystals, 3-rd Ed., Cornell University Press, Ithaca, New York, 1960, 6.

[2] D.F. Shriver, P. Atkins, C.H. Langford, Inorganic Chemistry, 2-nd Ed., W.H. Freeman, New York, 1994, 74.

[3] A.F. Wells, Structural Inorganic Chemistry, 3-rd Ed., Clarendon Press, Oxford, U.K., 1962, 60.

[4] T.W. Armitt and R. Robinson, J.Chem.Soc.,127 (1925) 1604.

[5] Ibid #2, 343.

[6] F. Harary, Graph Theory, Addison-Wesley, Reading, Ma., 1969, 26.

[7] International Union of Pure and Applied Chemistry, Nomenclature of Inorganic Chemistry, 2-nd Ed., Definitive Rules, 1970, Butterworths, London, 27.

[8] International Union of Pure and Applied Chemistry, Nomenclature of Organic Chemistry: Section A, Pergamon Press: Oxford, U.K., 1979.

[9] Ibid #3.

[10] A. Streitwieser and C.H. Heathcock, Introduction to Organic Chemistry, 2-nd Ed.; Macmillan, New York, 1981, 498.

[11] J.J. Novoa, P. Lafuente, R.E. Del Sesto and J.S., Miller, J.S., Angew.Chem.Int. Ed.,40 (2001) 2540.

[12] S.B. Elk, MATCH 31 (1994) 89.

[13] Ibid #2, 42.

[14] R. Steudel, Chemistry of the Non-Metals, de Gruyter Textbook, (Engl. Ed. by F.C. Nachol & J.J. Zuckerman), New York, 1977, 167.

[15] T.W.G. Solomons, Organic Chemistry, 5-th Ed.; Wiley, New York, 1992, 151.

[16] R.T. Morrison and R.N. Boyd, Organic Chemistry, 5-th Ed.; Allyn and Bacon, Inc., Boston, 1987, 1174.

[17] Ibid #10, 1039.

[18] F.L. Taylor, Ind.Eng.Chem., 40 (1948) 734.

[19] A.L. Goodsdon, J.Chem. Inf.Comput.Sci., 20 (1980) 172.
[20] S.B. Elk, THEOCHEM, 453 (1998) 29.
[21] S.B. Elk, MATCH, 31 (1994) 89.
[22] S.B. Elk, MATCH, 8 (1980) 121.
[23] S.B. Elk, MATCH, 17 (1985) 255.
[24] S.B. Elk, Graph Theory Notes N.Y., 27 (1994) 16.
[25] S.B. Elk, J.Chem.Inf.Comput.Sci., 34 (1994) 942.
[26] S.B. Elk, MATCH, 23 (1988) 9.
[27] Ibid #8, 19
[28] E. Clar, The Aromatic Sextet, William Clowes & Sons, Ltd., London, 1972, 103.
[29] Ibid #8, 23-24.
[30] S.B. Elk, MATCH, 13 (1982) 239.
[31] A.M. Patterson, J.Am.Chem.Soc. 47 (1925) 543.
[32] Ibid #8, 24.
[33] Ibid #22.
[34] Ibid #4.
[35] Cotton, F.A., Wilkinson, G., Murillo, C.A., Bochmann, M., **Advanced Inorganic Chemistry**, 6. Ed., Wiley, New York, 1999, p.351
[36] Solomons. T.W.G., ***Organic Chemistry***, 5. Ed., Wiley, New York, 1992, p. 828.
[37] T.B. Wen, Z.Y. Zhou and,J. Gouchen, Angew.Chem. Int.Ed.,40 (2001) 1951.
[38] Ibid # 10, 976.
[39] Ibid #15, 1133.
[40] Ibid #16, 1367.
[41] Ibid #18.
[42] N. Lozac'h, A.L. Goodson and W.H. Powell, Angew.Chem.Int.Ed.Engl. 18 (1979) 887.
[43] R.S. Cahn and O.C. Dermer, **Introduction to Chemical Nomenclature**, 5-th Ed., Butterworths, London, U.K., 1979, 32.
[44] Ibid, 95 (Table 5-6).
[45] Ibid #43, 31.
[46] E.W. Godly, "Chemical Nomenclature", Editor: Thurlow, K.J.; Kluwer Academic Publishers, Dordrecht, The Netherlands, 18.
[47] Ibid #2, 1120-21 Formula 19-II.
[48] G.J. Leigh, Nomenclature of Inorganic Chemistry Recommendations 1990, Blackwell Scientific Publications, London, 174.
[49] Ibid, 180.
[50] Ibid, 177 & 178.
[51] Ibid #43, 87.
[52] J.H. Fletcher, O.C. Dermer and R.B. Fox, "Nomenclature of Organic Compounds - Principles and Practice", American Chemical Society's Advances in Chemistry Series 126, Washington, D.C., 1974, 38.

Chapter 3

Other significant differences from existing systems

CHAPTER ABSTRACT:
Because the objective of this treatise is to form a uniform nomenclature that spans all of chemistry, many of the starting premises that are the foundation of the disparate subdivisions of chemistry have been re-examined. These premises were mainly of a historical, rather than a rational, origin, and, in many instances have been shown to be inconsistent. In their place, choices have been made with a global, rather than a local, and with an analytic, rather than a synthetic, approach in mind. Beginning with the selection in Chapter 1 of a "primary path", the nomenclature being developed eliminates:

(1) the need for both SSSR (smallest set of smallest rings) in "organic" and chelation and dentation in "inorganic" chemistry, along with all of the consistency problems that these concepts created;

(2) the use of a premier language; all ordering is based strictly on atomic number/weight;

(3) subordinating any atom (e.g., hydrogen) in describing molecular structure.

There is no need for a complicated system of artificial word endings to designate different functional groups. This is done by the particular juxtaposition of the atoms and bonds which expresses that function. All of the attributes associated with a functional group are described in terms of just the atoms and bonds involved. Moreover, by expanding the set of bonds that may be used, the nomenclature more closely approximates the actual geometry of the moiety being named. Significant differences from existent systems include the ability to distinguish between aliphatic and aromatic compounds in "organic chemistry", as well as to nomenclate compounds that have "unusual" bonding patterns, such as the zero order bond in the propellanes, etc. In addition, comparable to coordinate systems in geometry, an alternative "cylindrical" nomenclature has been developed, which, in selected instances, produces a simpler name than the "Cartesian" name.

In addition to differences described in Chapters 1 and 2 between the proposed system of nomenclature versus the present, internationally-accepted standard IUPAC system, such as:

(1) Consecutive odd integers, instead of all integers, are assigned as the locant numbers for atoms on the principal chain or ring. Even locant numbers are reserved for bonds,

(2) Terminal hydrogen atoms are significant and are included in selecting primary chains,

(3) The perspective for assigning canonical names is strictly an analytic one, never a synthetic one [1],

as well as some important piecemeal innovations and improvements to this standard that nodal nomenclature [2] had introduced: *

(4) There are no "parts of the structure that are to be excluded from the portion of the structure for which general nodal nomenclature is to be applied",

(5) No simplifying "replacement" modules are introduced, such as shrinking a ring to a node

further observations about the systemic nomenclature scheme being developed include:

(6) Whenever there exists more than a single graph theoretical cycle in the graph of a molecule, a different choice is made as to which cycle is most important and thus should be given priority [3].

(7) There is no need for a premier language to be used in ordering or indexing; instead all ordering of atoms are based on atomic number (/weights for isotopes).

(8) There is no need for any artificial word ending, such as ane, ene, yne, ol, al, etc. in IUPAC organic nomenclature, ide, ate, ite, ic vs. ous acid, etc. in IUPAC inorganic nomenclature, as well as nodane in nodal nomenclature.

A detailed description of these proposals follows:

(6) The method for selecting the principal cycle eliminates dependence on concepts that not only are antiquated, but also not expandable to all of the historical subdivisions of chemistry. The distinct, and uncoordinated, protocols in use today in "organic" vs. "inorganic" chemistry nomenclature are replaced by a common, graph theory based, superstructure. In doing so,

---

*One of the main motivations for the development of nodal nomenclature was to be able to compensate for inadequacies that prevented the assignment of a consistent set of canonical names to selected, then recently formulated, molecules — especially the cyclophanes.

all reliance on the smallest set of smallest rings (SSSR) in "organic chemistry", as well as the entire concept of chelation in "inorganic chemistry" is purged.

Before examining the different perspectives that have historically influenced the choice of descriptors in the various subdivisions of chemistry, especially organic vs. inorganic, note that, in the study of geometry, one typically views figures in terms of either the space that they occupy, called the "content-defined" definition, or the union of the segments that enclose this content, called the "boundary-defined" definition[4]. Moreover, for figures having the heuristic called "simple" described in Chapter 2, one can readily inter-convert between content and boundary-defined figures without ambiguity. This expectation of interconvertibilty, however, is a liability when describing the covalent bonding of atoms that form a graph theoretical cycle in the various branches of chemistry. This is a consequence of the fact that the best geometrical description of selected molecules is one of skew polygons — which are **not** "simple" in the geometric sense of the "term".

(6a) Organic: As noted in the above mentioned earlier study [4], for multi-cycle aliphatic compounds, and thus for its nomenclature, the IUPAC system of "organic chemistry" taxonomy is based on geometrically covering a one-dimensional (edge) set in a three-dimensional (embedding) space — an idea that is in contrast to the traditional geometry picture in which the boundary of a simple figure is always one dimension less than its content. This is not a problem with aromatic compounds, which have **that** attribute of geometric simplicity of being two-dimensional in a two-dimensional embedding space. However, because the faces, rather than the edges, are the primary geometric figure to be covered, a different geometrical difficulty arises; namely, describing aromatic compounds in terms limited to an integer bond edge set may be just as distorting as it is to describe aliphatic compounds in terms of rings; i.e., face sets. It is, therefore, not surprising that the bond set available in traditional organic chemistry is inadequate. On the other hand, by expanding the allowable bond set, a means has been proposed wherein these differences can be finessed — even if they can not be made to disappear.

Because the nomenclating of aliphatic compounds is based on the above described covering set, any use of SSSR [5] in the nomenclature is irrelevant. In its place, one might expect the optimum theoretical goal for nomenclature to be the naming of a canonical Eulerian [6-7] cycle which spans the edge set. Unfortunately, the existence of such a spanning cycle is rare. Consequently, a reasonable expectation is that the goal has been shifted to one of finding that particular Eulerian cycle which includes a maximum number of atoms in the molecule and to then augment this cycle with a minimum set of bridges that spans the uncovered edges. This goal **would** be the quest **had** the context in which both the words "maximum" and "minimum" were used been unambiguous.

While it is true that in this context, the word "maximum" depends only on number (namely, the number of atoms) and is, therefore, unambiguous; **HOWEVER,** the word "minimum" refers to a set of bridges. Not only is the number of bridges important, also one must consider certain properties of each individual bridge. This is analogous to the comparison of triangles in Chapter 1 wherein shape had to be considered. In other words, there is, in most instances, more than one way in which the desired set of bridges can be delineated.

For example, consider the five ring aromatic compound whose both IUPAC and common name is "perylene"*. In Figure 1, three different criteria that might be used for allocating locant numbering when assigning a

---

*Per IUPAC Rule A-21.2 [8] perylene (see page 80 above) is number 23 in a list of 35 reference fused polycyclic hydrocarbons. Also, as noted in Chapter 2, this list is just one of several lists to be memorized Furthermore, because these lists do not begin to cover all of the possibilities; a complex system of what is called "ortho-fused" and "ortho- and peri-fused" [9,10] additions to these names had to be promulgated. Moreover, a list of alternate names and of exceptions was also included. It was for precisely this reason that a more logical system, that did not rely on memorization of long tables for naming both polycyclic aromatic compounds of ring size six [11] (therein called "polybenzenes") and of the larger general class of "arenas" [12] was devised. These new techniques included both adding new rings to a smaller base structure (called a "synthetic name") and viewing a given molecule as an entity formed from the longest chain of specified modules augmented by the fusion of other modules at prescribed locations (called an "analytic name"). As noted in [11], although there is a heuristic advantage in being able to build up from smaller modules, consistency can not be maintained as modules are affixed at all of the topologically possible locations; consequently, despite the appeal for both indexing and cataloging, any system of nomenclature that is to be expandable when new moieties are created **must** be an analytical one, rather than a synthetic one. Moreover, a synthetic nomenclature is likely "to require a tedious series of frequent reorientations with concomitant recalculations" every time an additional ring is addended to aggreagtion. Some of the other avenues that this author followed included: (a) using a triangular shaped hexagonal tessellation envelope which could then be assigned a unique binary name, as well as a more heuristic "cluster name" [13]; (b) performing an isometric pseudoconversion of polybenzenes into acyclic polyenynes [14]; creating a metric distance ordering on a hexagonal grid using selected sectors of the plane so as to approximate the ordering created using Patterson's nomenclature rules [15]; as well as the extensions using the Matula system that were mentioned on Chapter 1. Meanwhile, other researchers who similarly made a major contribution to the evolving nomenclature of this important class whose writings were frequently consulted in formulating this section

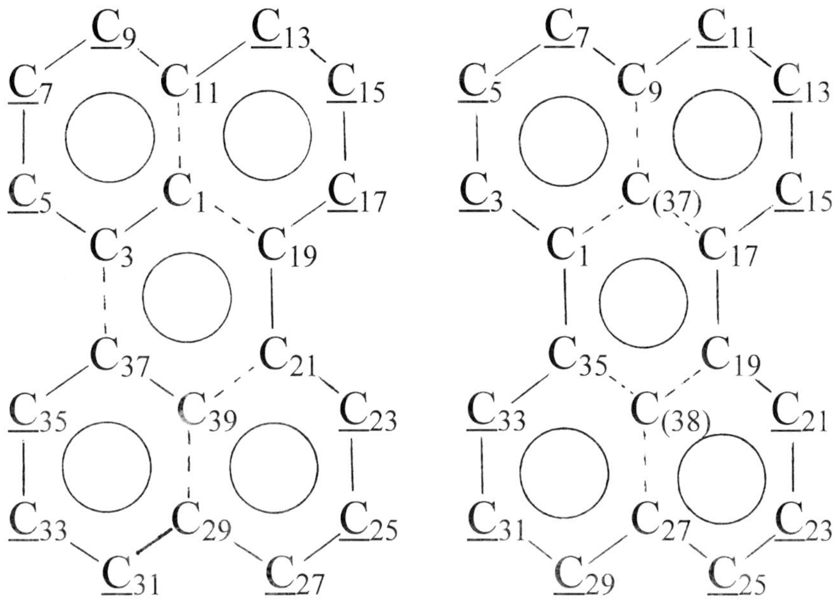

a. An optimal Hamiltonian path (length = 20)

b. Largest Eulerian cycle with branched bridges

Fig. 1: Selected Locant Numberings of Perylene (page 1 of 2)

canonical name to this molecule are listed. The reason why more than one candidate for canonical name had to be examined is that there does not exist an Eulerian cycle, (or even a Eulerian path) through all 24 edges. Additionally, there is not a Hamiltonian cycle (length = 20) through all twenty of the carbon atoms. There are, however, Hamiltonian paths through all 20 carbon atoms; i.e., path length = 19. The one which produces the lowest locant numbers starts at one triple point and ends at the other (part a). This would be nomenclated as:

$$(C\beta)_{19}C{:}^{(1\text{-}11,1\text{-}19,3\text{-}37,21\text{-}39,29\text{-}39)}(\beta);^{(5,7,9,13,15,17,23,25,27,31,33,35)}(1H) \quad (1)$$

The largest Eulerian cycle, which does not cover all of the carbon atoms (part b), has length = 18. To use this cycle as the basis for nomenclature, it

---

include studies by: Balaban [16], Bonchev [17], Cyvin [18], Dias [19], Herndon [20], Trinajstic [21]; to name but a few.

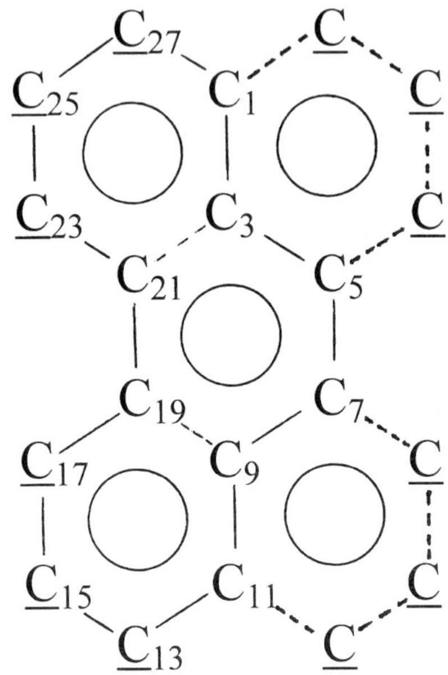

c. Largest Eulerian cycle with unbranched bridges

Fig. 1: Selected Locant Numberings of Perylene (page 2 of 2)

must be augmented with both a primary and a secondary bridge in order to cover the entire edge set; i.e., there is the need to assign locant numbers to vertices beyond the principal cycle. On the other hand, the largest Eulerian cycle which can be augmented using **only** primary (unbranched) bridges (part c) is of length 14. These are named respectively as:

$$\{[C\beta(\underline{C}\beta)_3]_2C\beta\}_2{:}^{(1\text{-}9)}(\beta C^{(=37)}\beta);^{(19\text{-}27)}(\beta C^{(=38)}\beta);^{(17\text{-}37,35\text{-}38)}(\beta) \quad (2)$$

and

$$(C\beta)_6(\underline{C}\beta)_3(C\beta)_2(\underline{C}\beta)_3{:}^{(1\text{-}5,7\text{-}11)}[(\beta\underline{C})_3\beta];^{(3\text{-}21,9\text{-}19)}(\beta) \quad (3)$$

Consequently, from a pragmatic perspective, a trade-off has to be made between the length of the longest chain and the complexity of the bridging.

The above discussion overlooks the fact that **there does exist** a system in which a single number can unambiguously describe a set of bridges. A **very major** problem with the Matula prime number-based system [22] is its esoteric nature. Matula numbers, and even worse the Elk-Matula numbers [23-27], are tedious to work with. Coupling this with the fact that, as well as being unfamiliar to chemists, they are unknown by most mathematicians, it is **very unlikely** that they could be promoted enough to gain wide usage. Instead, chemists have opted to tabulate a sequence of numbers to describe the bridging. Unfortunately, there is no agreement as to what qualifies as the desired "minimum" set. For pragmatic reasons, all presently familiar systematic nomenclature schemes ascribe priority to unbranched longer chains over shorter branched ones. By such a protocol, fewer parameters are needed in order to guarantee that there is the desired one-to-one relationship between name and structure. IUPAC [28] tries to evade the issue by assigning prefixes to "commonly used" groups, such as "iso", "neo", "sec", etc. Nodal nomenclature, on the other hand, allocates integers as the length designator for each chain. However, in the process, one frequently has to assign locant numbers beyond those in the primary chain; thereby greatly increasing the complexity of the nomenclature. Re-examining Figure 12 in Chapter 1, one can find five "main line" terms, with a locant descriptor for all but the first (for which it is unnecessary as this is the primary chain). By the postulates of the nodal system, every secondary and higher chain is viewed as starting from the connecting node, rather than allowing it to be part of a potentially longer chain. As illustrated in that figure, the chain connected to carbon atom with locant number 7 is treated as though it was only five atoms in length; the "sixth" carbon in that chain was relegated to being a tertiary chain (which required a new locant number, rather than being a continuing part of the secondary chain).

Furthermore, it should be noted that when the space being spanned is both uniform and two-dimensional, use of Eulerian cycles often becomes inefficient. In its place, a better system, using Hamiltonian cycles, can be formulated. Unlike both IUPAC and nodal nomenclature, the focus of the proposed systemic nomenclature is the longest contiguous path in a principal ring (when there is one) or chain — when the molecule is acyclic, irrespective of **any** considerations of local geometry.* This nomenclature considers smaller rings inside of larger rings **merely** as bridges in the principal ring structure. Figure 2 is an example of such a molecule. In the graph of this molecule, the principal path is one that ignores smaller rings, whether they be aliphatic or aromatic, and focuses on the largest graph

---

* An important reason for the historical reliance on local geometry was eliminated by the postulation of β bonds, especially for the class of cyclophanes.

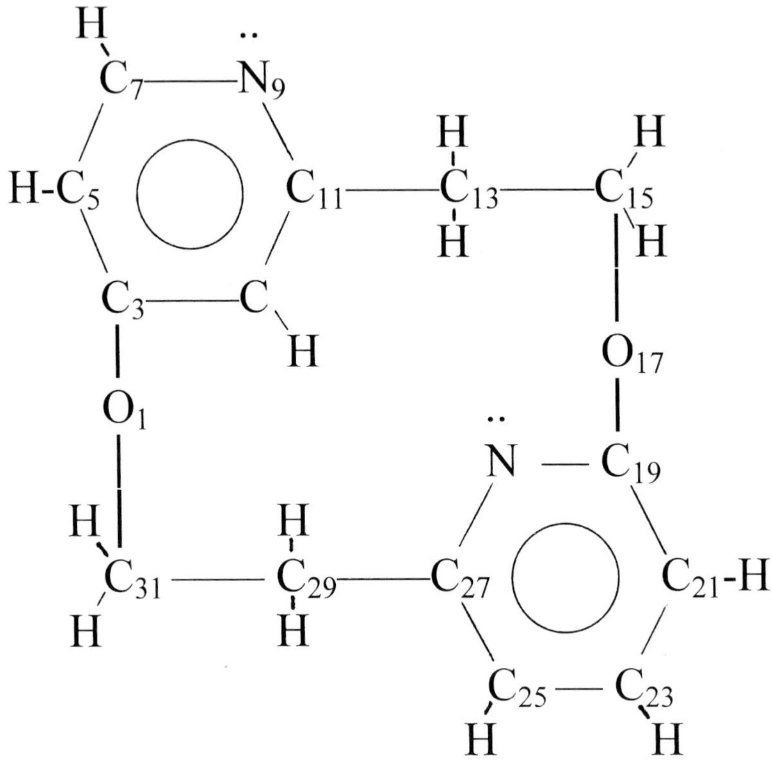

Fig. 2: Canonically naming a compound which has rings within larger rings

theoretical cycle — which for this molecule is 16 atoms long. Similarly, the locant numbering starts at one of the highest atomic number elements (oxygen in this case) and traces in the direction which maximizes the next different either bond multiplicity or atomic number. This translates to locant #1 being the left oxygen atom as drawn in Figure 2 and the direction of increasing locants is clockwise (1Cβ) vs. 1C1). Similarly, had one chosen the other oxygen atom to be locant #1, the first difference (in the clockwise paths) occurs at location #9 (N has preference over C). Thus the name assigned to this molecule is:*

* Note that, unlike in nodal nomenclature, for this molecule, there is no need to assign locant numbers to either the nitrogen atom or the carbon atom in the two respective bridges. Such an assignment would be necessary only when there are either non-hydrogen ligands or else "secondary" bridges projecting from atoms on a "primary"

$$\mathrm{O1C\beta(\underline{C}\beta)_2N\beta C1(\underline{\underline{C}}1)_2O1C\beta(\underline{C}\beta)_3C1(\underline{\underline{C}}1)_2{:}^{(3\text{-}11)}(\beta\underline{C}\beta);^{(19\text{-}27)}(\beta N\beta)} \qquad (4)$$

By such a choice, the entire concept of SSSR (smallest set of smallest rings), including modifications to that subject such as ESER (essential set of essential rings) [29], may be viewed as a "fix" that **doesn't** work, in much the same manner as the geocentric system of astronomy that reached its zenith under Ptolemy was discarded when Copernicus formulated the heliocentric system.* Moreover, excluding the subset of strictly coplanar molecules for which there is a pragmatic (even if not theoretical) utility, one need not be concerned with the problems of the type described in [5]. Observe that the molecule depicted in Figure 2 can be projected onto a planar surface without there being any evident interaction of non-neighboring atoms. This was what was referred to in Chapter 1 as "an intrinsically two-dimensional molecule in a three dimensional embedding space", and even though the use of Taylor's term "reticular" [30] would not apply due to his considering the "bridged aspect of the molecule" more

---

bridge. For example, had there been a methyl group, rather than a hydrogen atom attached to the, so far unnumbered, carbon atom of the 3-11 bridge, one would now need to assign a locant number to this carbon in order to be able to unambiguously locate the methyl group. Similarly, when two atoms not on the principal ring are connected by a bridge containing at least one additional atom, then further locant numbering is required. Although in theory this is only a small gain, pragmatically it is very advantageous in that the need to apply a sequencing rule for successive locants beyond those in the principal ring/chain has been delayed, even though it has not been eliminated.

* If enough modifications are permitted, **any** system can be made to fit a given data set. For example, a system with 100 data points can ***always*** be ***exactly*** covered with a 99-th degree polynomial. However, there is ***no*** scientific merit in doing so. The system of astronomy developed between 127 and 151 A.D. by Claudius Ptolemaeus was the unchallenged descriptor of the heavens for 1300 years. In what is now referred to as "the Ptolemaic system" the supremacy of the circle was unquestioned. Consequently, in order to adequately describe planetary motion, a description of the orbit of a celestial body was first modified from a simple circle with the earth at the center to one in which the earth was moved eccentrically. However, as more accurate measurements of the planetary orbits were developed, this initial "improvement" was seen to be insufficient. This "minor inconvenience" was remedied by including a combination of two circles in a cycloidal arrangement in which a second circle, called the "epicycle" rolled along a primary circle, called the "deferent". However, what "worked" for the sun was inadequate to describe the motion of the moon, etc. And so it went — with corrections to corrections to corrections, but always keeping the circle front and center. The needed paradigm shift did not occur until Copernicus in 1543 replaced the circle with an ellipse and set the sun at one of the foci of this ellipse; thereby relegating the earth to being just another planet.

important than its fusion properties, most chemists would view this molecule as the fusion of a 12 member ring with two six member rings.

(6b) Inorganic: Historically, attention was focused on a single atom in a molecule, to the exclusion of all others. This is seen in the names assigned to the common inorganic acids (and anions). Although there are multiple oxygen atoms and only one sulfur or phosphorus or nitrogen atom, that latter atom was the one used to determine the name of this acid (or ion). The number of "supporting" oxygen atoms was relegated to being an inconstant suffix that relied on an assumed maximum number of oxygen atoms that could "be held in thrall" by this principal atom.* This idea has been carried over today to continuing the focus on a single atom in a chain or cycle. In order to achieve the desired unification of nomenclature across all of chemistry, this practice is discontinued and all atoms, including hydrogen, are treated equally. The basic concepts of chelation (one atom grasping a chain of lesser important atoms) and dentation, (coordination of closed cycles emanating from a single atom) are discarded, with the focus now being directed to a global vs. a local geometry of the moiety. The very word "chelation" seems to be deliberately ambiguous. This author was left with the question: Is it more than just coincidence that, in its latest (1990) recommendations, IUPAC [31] gives a formal definition for *coordination*

---

* In the earliest known (and still the most familiar) oxy acids, there are four oxygen atoms attached to a central sulfur or phosphorus atom; however, there are only three oxygen atoms attached to a central nitrogen or chlorine atom. These were considered to be the "normal" acids and were assigned names ending with the suffix "-ic"; namely, sulfuric, phosphoric, nitric and chloric acid respectively. Note that there is no uniformity as to whether the whole element name (sulfur) or a part of the name was to be retained (it is nitric, not "nitrogenic" acid.) Next, lower numbers of support atoms (ous acids = 1 less than "normal" oxygen atom, hypo...ous acids = 2 less than "normal" oxygen atoms, etc.) were discovered; along with a higher number of oxygen atoms for chloric acid ($HClO_4$) which was dubbed "perchloric acid. This was next followed by other acids having a higher number of attached oxygen atoms, some of which were named with just the prefix "per-", others with the prefix "peroxy-", still others with a prefix "superoxy-", etc. The names assigned to these "newer" acids, for the most part, do not follow any consistent or logical naming progression; rather there has evolved an uncoordinated collage of names such as: peroxymonosulfuric acid for H_2SO_5 , pyrosulphuric acid for $H_2S_2O_7$, peroxydisulfuric acid for $H_2S_2O_8$, etc. One exception to this randomness is seen in the numeric prefixes for selected ions of phosphate, also sulfate, selenate, etc.; namely, O-PO_3 is phosphate, O-$(PO_3)_2$ is diphosphate, O-$(PO_3)_3$ is triphosphate, etc. In the systemic nomenclature being proposed, the linear chain of these oxy-ions has systemic name: $O[(\beta P\beta O){:}^{(3,3)}(\beta O)]_n^{(n+2)-}$, while the monocycle has one less oxygen and thus the systemic name of: $[P\beta O\beta){:}^{(1,1)}(\beta O)]_n^{(n)-}$. Likewise the sulfur linear analog is systemically named: $O[(\beta S\beta O){:}^{(3,3)}(\beta O)]_n^{(2)-}$ and the silicon analog: $O[(\beta Si\beta O){:}^{(3,3)}(\beta O)]_n^{(2n+2)-}$, etc.

number ("the number of sigma bonds between ligands and central atom"), but evades giving a similar denotation for *chelation*? [32][*] Note that in the system being developed, this is not a matter of concern. Similarly, it is unimportant that, by using IUPAC's definitions, the term "monodentate chelation" is an oxymoron. Neither *dentation* nor *chelation* have a place in the proposed system. Meanwhile, note that the smallest chelation illustrated in IUPAC's set of recommendations is didentate. By contrast, monodentate coordination presented no problem to this committee and was discussed on the next page.

(7) Concerning the lack of a premier language, note that every part of the name associated with an aggregation of atoms, such as C_2H_5, depends only on the atoms therein. This is in contradistinction to the use of words that would be alphabetized differently in different languages. One such example is *Äthyl* in German vs. *Ethyl* in English. Likewise, intra-language inconsistencies of ordering are eliminated. For example, alphabetically *iso* comes before *n*, comes before *sec*, comes before *tert*, etc. Structurally, however, since the atomic number of carbon is greater than of hydrogen, a more rational ordering would be: tert > sec > iso >n. This idea is also applicable to indexing functional groups such as the halogens; namely, by considering size as the important parameter: I > Br > Cl > F when this is the only difference between compounds that are to be ordered, or alternately by considering speed of reactivity F > Cl > Br > I. From a chemistry perspective, there is nothing to recommend the use of alphabetical ordering.

(8) In lieu of any artificial word endings, all references to functional groups have been completely separated from the names assigned to principal chains. Consequently, there is no use for any of the morphemic suffixes that are prevalent today. Additionally, because neither hydrogen, nor any other atom has been subordinated, one should not consider the five atom molecule of methane $\mathrm{H1\underline{\underline{C}}1H}$ as being smaller than the four atom molecules of ethyne H1C3C1H or formaldehyde $\mathrm{O2\underline{C}1H}$, etc.

Having enumerated these three additional salient differences, attention is next directed to the influence of dimension across the entire spectrum of chemistry; in particular, focus is directed to some molecules that are intrinsically three dimensional. As a first example, consider the graph shown in Figure 3, along with some representative molecules that are

---

[*] "Chelation involves coordination of more than one sigma - electron pair donor group from the same ligand to the same central atom". Note that in this **quoted** description the choice of the word "involves" appears to be deliberately more nebulous than what one expects from a formal definition of a term.

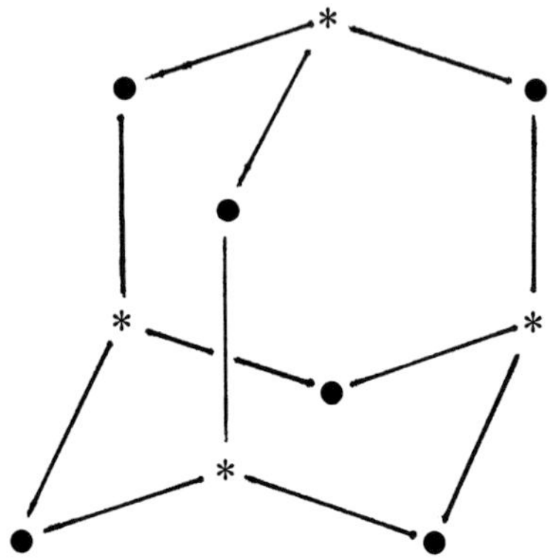

Fig. 3: An intrinsically three dimensional module formed from two different submodules (part 1 of 5)

important in different branches of chemistry:
(a) In the domain of "organic chemistry" (parts 2 and 3 of Figure 3): two molecules closely associated with three dimensionality are examined. Before doing this, however, it should be noted that the simplest figure that mathematically can be created in three dimensional space is the tetrahedron – a quantity topologically referred to as a "3-simplex", where the 3 denotes the dimension and the word "simplex" indicates that this is the simplest figure that can be formed in the specified space [33].* (If one now ignores the • atoms and focuses on the atoms designated by the asterisk symbols in Figure 3, one has a model of such a 3-simplex). Chemically, however a compound with carbon atoms at the vertices of a tetrahedron would be highly strained; so much so that formulation of such a molecule continues to remain high on the wish list of synthetic organic chemists [34]. On the other hand, by the insertion of another module (atom) in the middle of each of the six carbon to carbon bonds of such a theoretical tetrahedrane, the angle strain is relieved and some common molecules can be formed. One such insertion of both mathematical and practical interest is that of a methylene group $\underline{\underline{C}}$ (part 2 of Figure 3). This compound, which has a valence of four on all ten of the carbon atoms, is known by the common name: adamantane. When multiple copies of this module are combined so that each hydrogen atom is replaced by a carbon of another adamantane module, the diamond

---

* Note that a triangle is a 2-simplex, a line segment a 1-simplex and a point a 0-simplex.

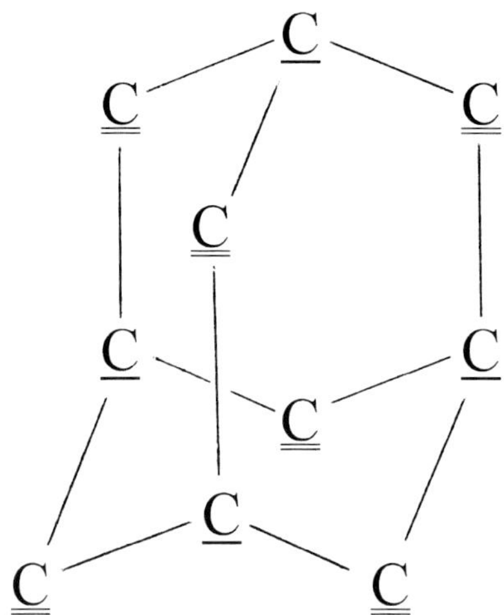

Adamantane

Fig. 3: An intrinsically three dimensional module formed from two different submodules (part 2 of 5)

crystal is formed. This important form of carbon, which will described and nomenclated in detail in Chapter 8.

The IUPAC name for adamantane: Tricyclo[3.3.1.1^{3,7}]decane bears some critical examination at this point. This name, especially the "tricyclo" part of the name is, at best, confusing inasmuch as (see [35]) there are four, rather than three, important "hexagons" in this molecule. Moreover, were it not for IUPAC fiat, the focus might have been on either the larger set of hexagons that could be formed, or, better yet, on the larger (eight member) rings. By IUPAC's basing its system of nomenclature on SSSR, all of the octagonal rings in adamantane have been relegated to a status of insignificance. To the contrary, in determining what is important in establishing the proposed system of nomenclature, cognizance is taken of the fact that for aliphatic compounds all of the "hexagonal" faces are not only skew hexagons, but also that no planar subset of four of the six vertices is constantly maintained. For this reason a greater consistency is achieved by placing emphasis on the largest (in this case octagonal) ring for all such molecules, rather than on any set of less significant hexagonal ones. Consequently, the systemic name for adamantane is:

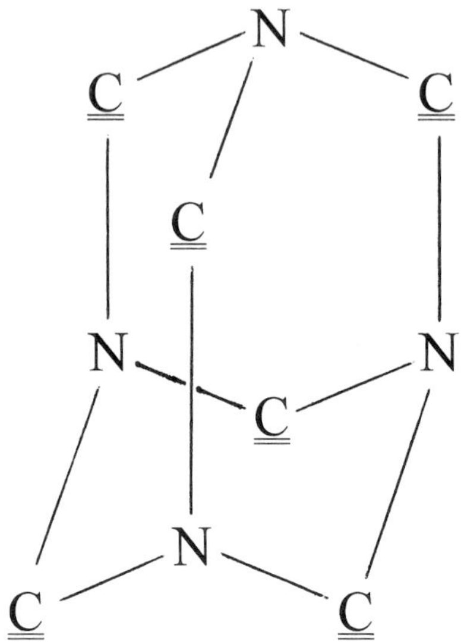

Adamanzane

Fig. 3: An intrinsically three dimensional module formed from two different submodules (part 3 of 5)

$$(\underline{C}1\underline{\underline{C}}1)_4{:}^{(1\text{-}9,5\text{-}13)}(1\underline{\underline{C}}1) \qquad (5)$$

Similarly, starting with nitrogen atoms at the starred locations of the tetrahedron and again inserting methylene groups produces the molecule known by the common name adamanzane. This is equivalent to replacing each of the $\underline{C}$ modules in (5) by a module containing the single nitrogen atom; thereby producing the systemic name:

$$(N1\underline{\underline{C}}1)_4{:}^{(1\text{-}9,5\text{-}13)}(1\underline{\underline{C}}1) \qquad (6)$$

At this time, attention is directed to the similarity between aliphatic compounds in "organic" chemistry and corresponding compounds in "inorganic" chemistry — a similarity that, in many instances, is far greater than between members of the classes of aliphatic vs. aromatic compounds in the traditional domain of "organic" chemistry. For this purpose, consider the same bonding arrangement in the two phosphorus-oxides that are formed by replacing the $\underline{\underline{C}}$ groups with O in both compounds and the $\underline{C}$ pairs

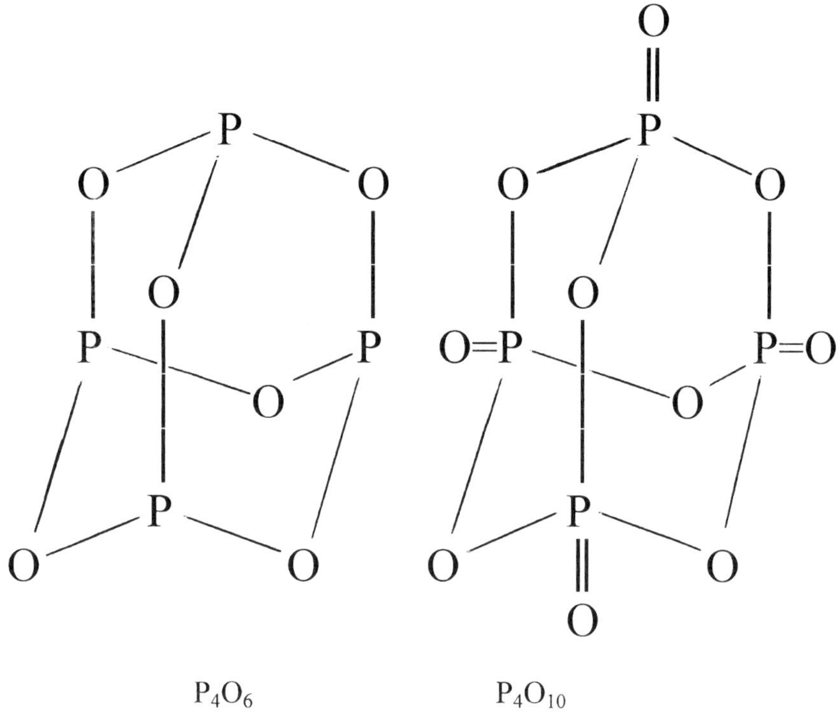

Fig. 3: An intrinsically three dimensional module formed from two different submodules (parts 4 and 5 of 5)

with P for P_4O_6 (part 4) and with P2O for P_4O_{10} (part 5). These are named, respectively, as:

$$(P1O1)_4{:}\ ^{(1\text{-}9,5\text{-}13)}(1O1) \qquad (7)$$

and

$$(P1O1)_4{:}\ ^{(1\text{-}9,5\text{-}13)}(1O1);\ ^{(1,5,9,13)}(2O) \qquad (8)$$

Additionally, this bonding arrangement is applicable to the cage in which hydrogen sodide ($H^+ Na^-$) was recently encapsulated [36]. Here the six • symbols in part 1 of Figure 3 have been replaced by $1(\underline{\underline{C}}1)_3$ groups. The authors of this study gave this compound the ad hoc name: 3^6 adamanzane. However, there is a readily determinable systemic name:

$$[(N1\underline{\underline{C}}1_3)]_4{:}^{(1\text{-}17,9\text{-}25)}[1(\underline{\underline{C}}1)_3] \qquad (9)$$

Note that the price paid for this "across the field" standardization is the ability to readily compare molecular structures. Such a price is paid by every analytic, versus synthetic, nomenclature [37]. For example, of the four mathematically possible diamantanes, the point fusion of two adamantane modules (part a of Figure 4), would be nomenclated as:

$$C1\underline{C}1(\underline{\underline{C}}1\underline{C}1)_3C^{(17=1)}1\underline{C}1(\underline{\underline{C}}1\underline{C}1)_3{:}^{(3\text{-}11,7\text{-}15,\ 19\text{-}27,23\text{-}31)}(1\underline{\underline{C}}1) \qquad (10)$$

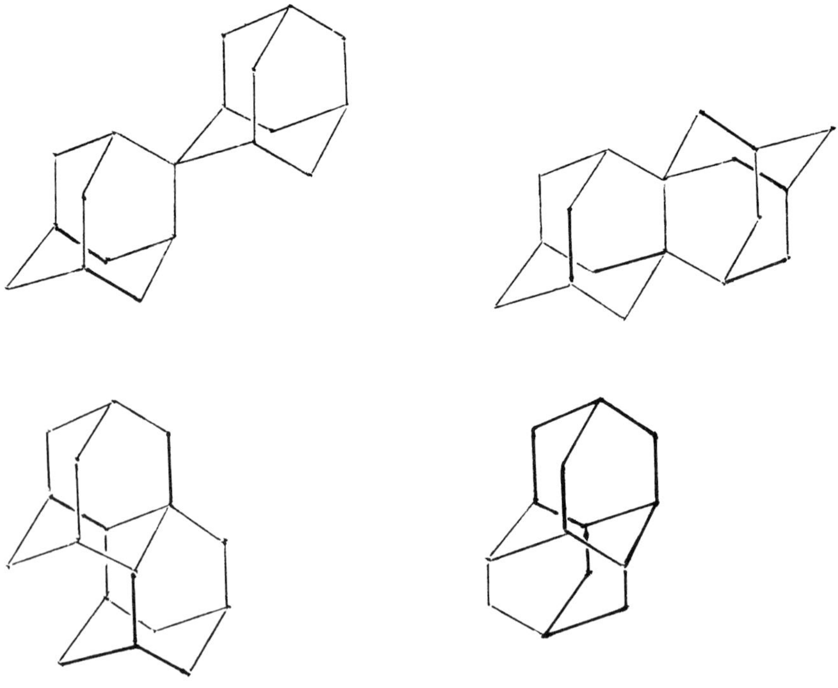

Fig. 4: Fusions of two adamantane modules: (a) vertex; (b) edge; (c) angle; (d) plane

Upon comparing this name to the name assigned to adamantane [see (5)], one notes that there is insufficient overlap in the two names for the proposed nomenclature to be useful in QSAR (quantitative structure activity relationships) studies. Without translating the name into a structural formula and then making the comparison of connectivities, none of the systematic nomenclatures (IUPAC, nodal, or the proposed one) alone allow for determining how close two structure really are to each other. Moreover, the degree of similarity gets progressively less as one examines the canonical names of edge fused adamantane modules (part b of Figure 4):

$$\mathrm{C1(\underline{C}1\underline{\underline{C}}1)_3C1(\underline{C}1\underline{\underline{C}}1)_3{:}^{(1\text{-}21,3\text{-}11,7\text{-}15,17\text{-}25)}(1\underline{\underline{C}}1);^{(1\text{-}15)}(1)} \qquad (11)$$

This name, while still bearing some resemblance to (5) in that certain combinations of atom-bond sequences and of locant sequences are repeated, has a familial relation that is nowhere near as clear. By the time that one reaches the face fusion (part d of Figure 4), nearly all of the "local" similarity in the canonical name

$$\mathrm{(\underline{C}1)_4(\underline{\underline{C}}1\underline{C}1)_2(\underline{C}1\underline{\underline{C}}1)_2{:}^{(3\text{-}11,5\text{-}21)}(1\underline{\underline{C}}1)} \qquad (12)$$

has been obscured.

At this point, it should be noted that the module of significance for the class of compounds referred to as polymantanes is adamantane. Unlike the fusion of benzene modules to form the polybenzenes, the number and variety of polymantanes increases much more rapidly. Figure 4, which illustrates the four different diamantanes is the second tier of this class of compounds.

As further illustration of the importance of the global, in contrast to the local, perspective in the proposed systemic nomenclature some topological isomers are next examined: The first two examples chosen are the "linear" Moebius strip [38] $C_{42}H_{72}O_{18}$ (Figure 5) and its untwisted isomer (Figure 6) -- a molecule which is the geometrical ideal of a cylinder. Although the only difference between these two molecules is the location of two bonds, there is a major difference in their chemical properties and in their canonical names. This is seen in the selection of the largest non-redundant graph theoretical cycle; namely, a cycle through all 60 of the carbon and oxygen atoms is viable for Figure 5, but the longest such cycle is only 42 atoms long for Figure 6. Systemic names for these two isomers are:

$$\mathrm{\{[O1(\underline{\underline{C}}1)_2]_2O1\underline{\underline{C}}1C1\underline{\underline{C}}1\}_6{:}^{(17\text{-}77,37\text{-}97,57\text{-}117)}(2)} \qquad (13)$$

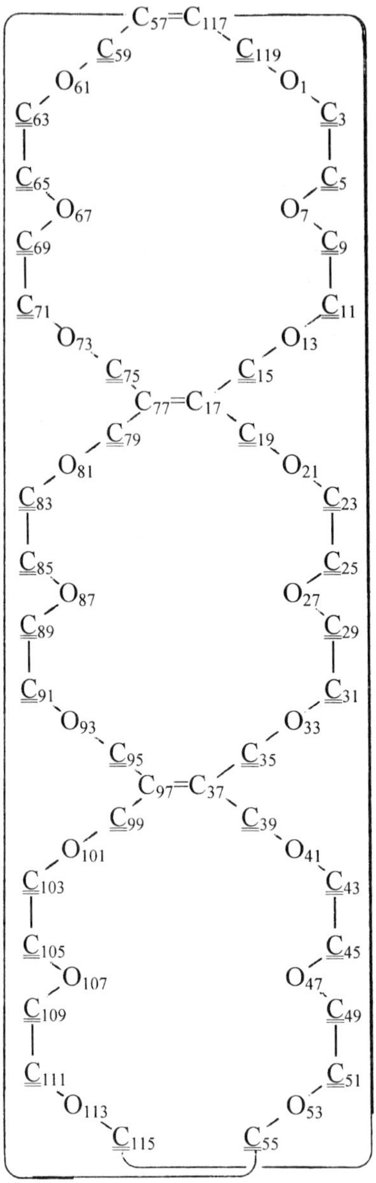

Fig. 5: Walba's Moebiane

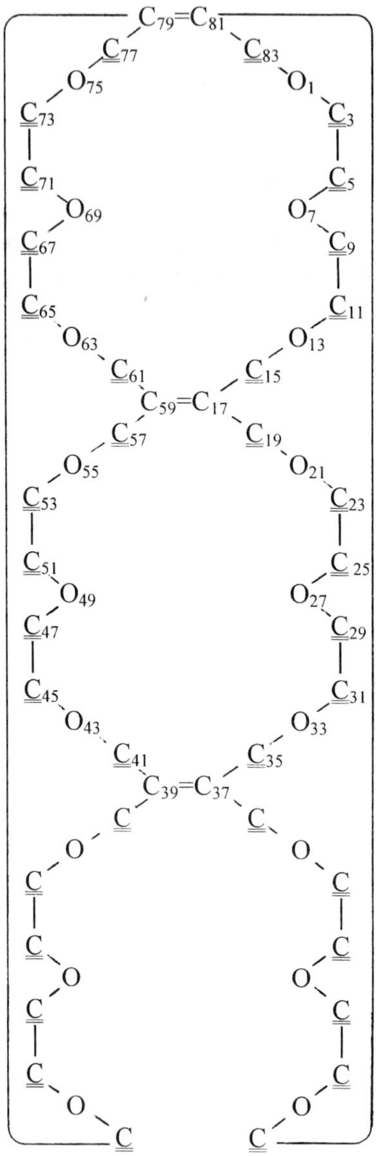

Fig. 6: Untwisted Cylinder
Isomeric with Moebiane

and

$$\{[O1(\underline{\underline{C}}1)_2]_2O1(\underline{\underline{C}}1C1\underline{\underline{C}}1[O1(\underline{\underline{C}}1)_2]_2O1\underline{\underline{C}}1C2C1\underline{\underline{C}}1\}_2{:}^{(37\text{-}81,39\text{-}79)}[1\underline{\underline{C}}1(O1(\underline{\underline{C}}1)_2)_2]O1\underline{\underline{C}}1;^{(17\text{-}59,37\text{-}39,79\text{-}81)}(2) \quad (14)$$

respectively. Later in this chapter, a twelve ring planar Moebius structure [39] shall be nomenclated, as well as a six ring aggregation that may be topologically viewed as the degenerate case of the Moebius strip in which a "twist" along a longitudinal axis has been pressed into a co-planar aggregation.

Instead attention is next directed to polycyclic compounds for which there is a different formal bond order than that which simplistic graph theory would imply. Unlike the subject matter introduced in Chapter 2, here traditional covalent bonds, in contradistinction to consideration of hydrogen bonds, are an integral part of the nomenclature. As a first example, note that, unlike the larger propellanes, the bond order is equal to zero between the bridgehead carbon atoms in [1,1,1]-propellane [40] (Figure 7). This "anomoly" is accounted for in the nomenclature by the name*

$$(C1\underline{\underline{C}}1)_2{:}^{(1\text{-}5)}(1\underline{\underline{C}}1);^{(1\text{-}5)}0 \quad (15)$$

If one were to follow the IUPAC practice, which considers this molecule as

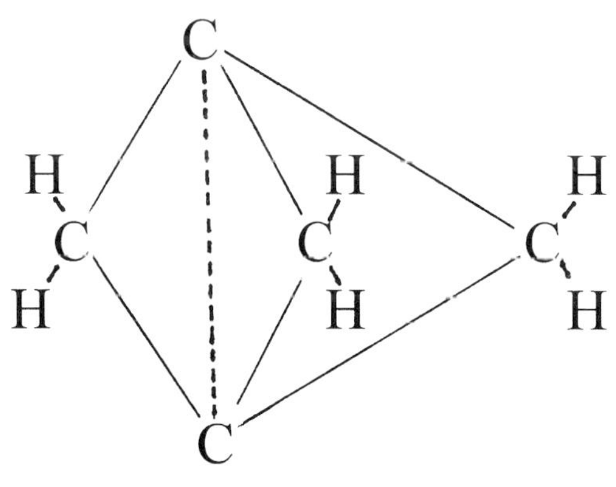

Fig. 7: [1.1.1]-Propellane — Actually observed (zero bond)

---

*The coefficient zero indicates that there is neither a bond nor an element between locants 1 and 5.

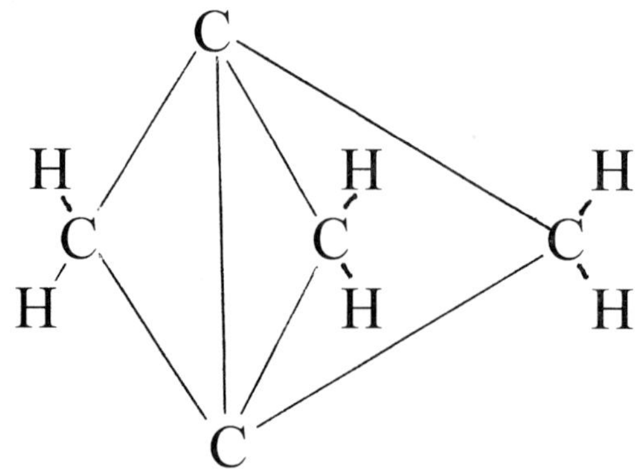

Fig. 8: [1.1.1]-Propellane — Prototypical analogy (with fictitious bond)

[1.1.1.0]-Tricyclopentane, and thus includes a non-existent bond (Figure 8), then the systemic name of this compound would have been:

$$(C1\underline{\underline{C}}1)_2{:}^{(1\text{-}5)}(1\underline{\underline{C}}1);^{(1\text{-}5)}1 \qquad (16)$$

Note that, even though it is not *absolutely* required, the inclusion of the zero bridge in (15) is **recommended** -- In this way the reader will not be misled into thinking that there was a typographical error. On the other hand, for most, if not all, larger propellanes*, this bond is existent and these compounds would have names comparable to (16), rather than (15). In other words, the proposed systemic nomenclature accurately describes the actual chemistry that has been measured, instead of some fictitious "logical extrapolation" from "similar" compounds.

Additionally, for comparison purposes, included herein are systematic names for the mathematically simplest ring overlap compound (Figure 9):

$$(\underline{C}1\underline{\underline{C}}1)_2{:}^{(1\text{-}5)}(1\underline{\underline{C}}1) \qquad (17)$$

and for the mathematically simplest paddlane (Figure 10):

$$(\underline{C}1\underline{\underline{C}}1)_2{:}^{(1\text{-}5,1\text{-}5)}(1\underline{\underline{C}}1) \qquad (18)$$

---

*See, for example, section 6B of Greenberg and Liebman's monograph [41] describing a large number of such "pathologic" molecules.

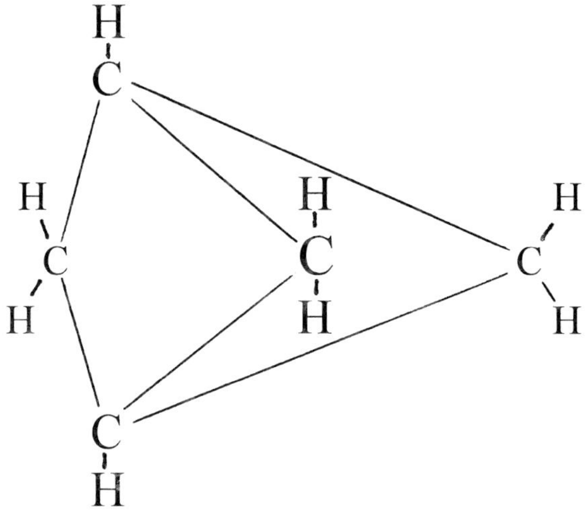

Fig. 3-9: $H1C(1\underline{\underline{C}}1)_3C1H$ — Mathematically Simplest Overlap Compound

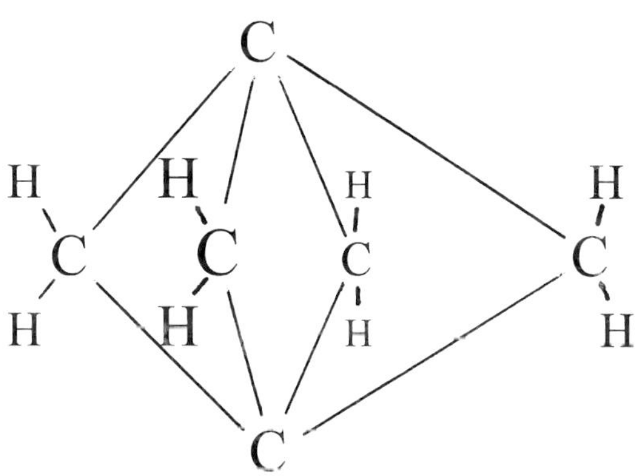

Fig. 3-10: $C(1\underline{\underline{C}}1)_4C$ — Mathematically Simplest Paddlane

These three classes of "organic" compounds (propellanes, overlap compounds, and paddlanes), as well as the corresponding "inorganic" ones, such as the cryptands, provide an introduction to a more general class of "cylindrical"[*] (vs. intrinsically planar) compounds. Because the entire

[*] A more accurate description would be "pseudo-cylindrical", inasmuch as the connecting chains between bridgehead atoms are not stationary. Instead, they should be viewed as

structure of the systemic nomenclature being developed has, up to this point, been based on an alternation of bonds and angles, a sequence of two successive bonds or of two successive atoms would never be encountered. Therefore, a new, special interpretation *could* be introduced for two bonds together without an intervening atom and also for two atoms together without an intervening bond. For a special limitation of the first of these codings, the assignment chosen is that of a symmetric graph theoretical cycle between two "anchor" atoms. Now, in exactly the same manner as certain geometrical figures are more simply described using cylindrical, vs. Cartesian (also called "rectangular") coordinates, similarly, "cylindrical" names (distinct from the Cartesian names that have been used up to this point) can be formulated for molecules. As with coordinate systems in mathematics, one usually uses other coordinate systems besides the Cartesian system **only** when there is an appropriate type of symmetry that results in a simpler name using this system. For example, the cylindrical name for a circle of radius 3 and center at the origin requires a quadratic equation in two variables ($x^2 + y^2 = 9$) using Cartesian coordinates; however, there exists a linear equation using only one variable ($r = 3$) in a cylindrical coordinate system. As with the description of figures using the different types of coordinates, only some cylindrical names will be more desirable than the Cartesian names developed earlier. For example, the heuristic concept of simplicity is increased by assigning cylindrical names to Figures 9 and 10:

$$\mathrm{H1C(1\underline{\underline{C}}1)_3C1H} \qquad (19)$$

and

$$\mathrm{C(1\underline{\underline{C}}1)_4C} \qquad (20)$$

respectively*. Likewise, for the "inorganic" cryptand shown in Figure 11, its cylindrical name:

---

loose interdigiting "jump ropes" whose envelope is between a cylinder and an ellipsoid of revolution.

* Although the protocol with respect to parentheses established in Section 1 would seem to allow (19) to be written as $\mathrm{H1C1\underline{\underline{C}}11\underline{\underline{C}}11\underline{\underline{C}}1C1H}$, and, similarly, for (20), a different interpretation is given to this code. Namely, the parentheses, in such cases, are regarded as an integral part of the cylindrical name, rather than an abbreviating factor, and thus the above expansion is not a valid one. This aspect of the nomenclature will be clarified in Chapter 6.

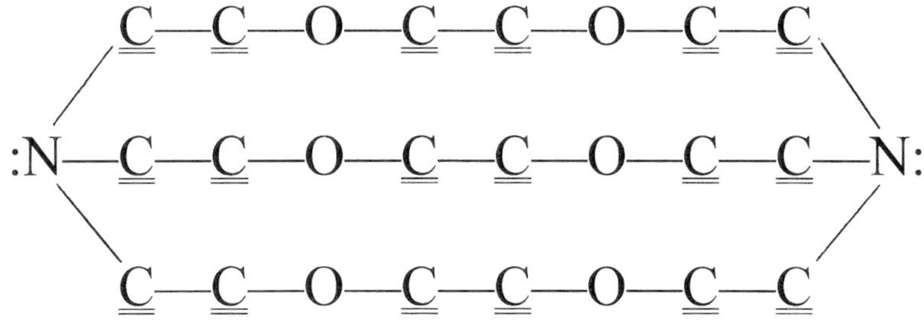

Fig. 11: A Symmetric Cryptand

$$N\{1[(\underline{C}1)_2O1]_2(\underline{C}1)_2\}_3N \qquad (21)$$

is a major simplification over its Cartesian name. That this not always the case is seen by examining a typical propellane, with chain lengths m, n and p. As in analytic geometry, when m, n and p are not all equal (Figure 12), for this molecule there is no simplifying symmetry. Consequently, the use of cylindrical nomenclature may be counterproductive. For example, consider the "semi-symmetric" cryptand illustrated in Figure 13. By replacing one of the six oxygen atoms with a sulfur atom in Figure 11 the cylindrical name of this compound would be[*]:

$$N\{1(\underline{C}1)_2S1(\underline{C}1)_2O1(\underline{C}1)_2\},\{1[(\underline{C}1)_2O1]_2(\underline{C}1)_2\}_2N \qquad (22)$$

Now, some of the advantage of cylindrical naming has been lost over the Cartesian method of naming the largest cycle and an augmenting bridge, which would have produced as the canonical Cartesian name:

$$S1(\underline{C}1)_2O1(\underline{C}1)_2N1[(\underline{C}1)_2O1]_2(\underline{C}1)_2:^{(13\text{-}31)}\{1[(\underline{C}1)_2O1]_2(\underline{C}1)_2\} \qquad (23)$$

Moreover, if a second oxygen atom at locant #19 were replaced by a sulfur atom, there is now no useful symmetry for the cylindrical name:

$$N\{1(\underline{C}1)_2S1(\underline{C}1)_2O1(\underline{C}1)_2,1(\underline{C}1)_2O1(\underline{C}1)_2S1(\underline{C}1)_2,1(\underline{C}1)_2[O1(\underline{C}1)_2]_2\}N \quad (24)$$

but there is for the Cartesian one:

---

[*] Note the use of the comma to separate disjoint chains between the anchor atoms, in exactly the same way as had been done in the superscripts of Table 2 (See Chapter 1).

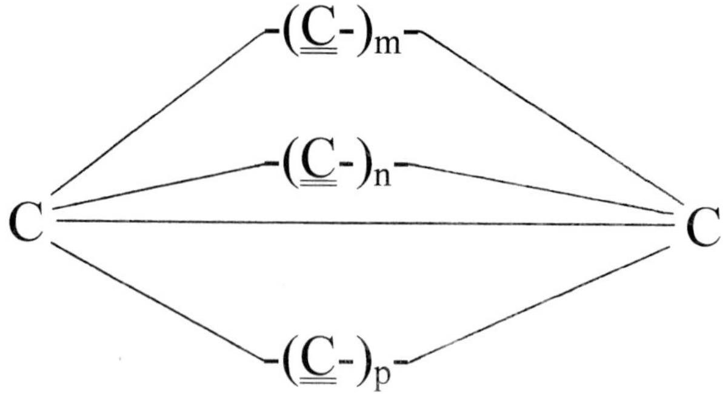

Fig. 12: A typical propellane

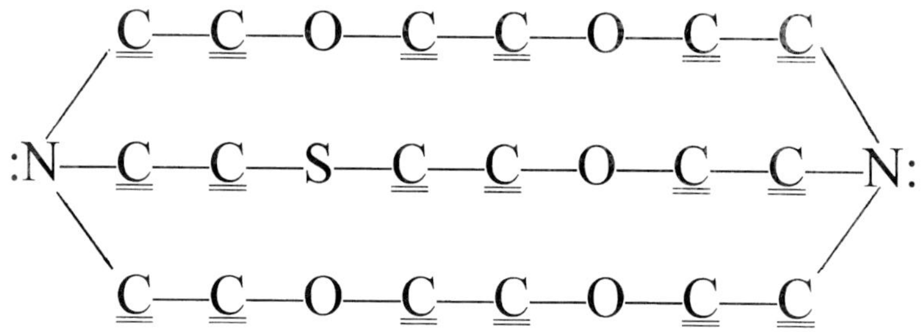

Fig. 13: A Semi-Symmetric Cryptand

$[(S1(\underline{\underline{C}}1)_2)_2N1(\underline{\underline{C}}1)_2O1(\underline{\underline{C}}1)_2]_2:^{(7\text{-}25)}(1[(\underline{\underline{C}}1)_2O1]_2(\underline{\underline{C}}1)_2$ (25)

Additionally, note that one of the molecules described in an earlier chapter (See Figure 17 in Chapter 2) would have a somewhat simpler name using the cylindrical format:

$Fe(\aleph C\aleph:^{(3)}2O)_3Fe;^{(1,1,1,5,5,5)}(\aleph C3O)$ (26)

vs. $(Fe\aleph C\aleph)_2:^{(1\text{-}5)}(\aleph C\aleph:^{3}2O);^{(1,1,1,5,5,5)}(\aleph C3O);^{(3,7)}(2O)$ (Ch. 2- 32)

Next, observe that in all of the cylindrical molecules nomenclated so far, the chains between the bridgehead atoms are "independent"; i.e., there is no atom common to two or more chains between the bridgehead atoms. On the other hand, had there been such "cross-linkage" in three dimensional space then the "jump rope effect" between the individual chains would be

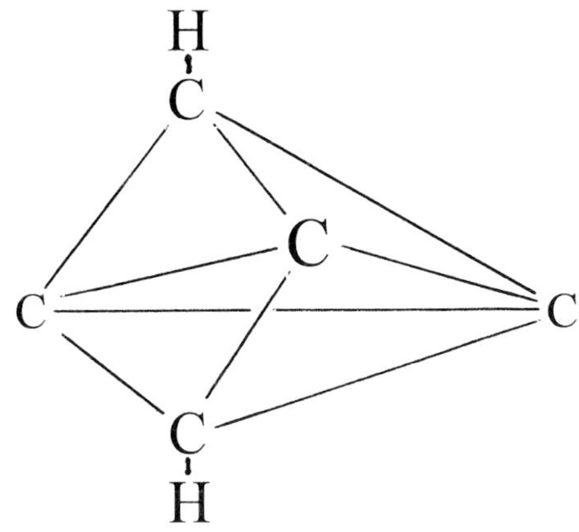

Fig. 14: The mathematically simplest trigonal bipyramid

impossible. The mathematically simplest such figure is the trigonal bipyramid C_5H_2 (Figure 14). Using the bridge method first described, this moiety would be nomenclated as:

$$H1(C1)_4C1H:^{(3\text{-}7,3\text{-}9,5\text{-}9,5\text{-}11,7\text{-}11)}(1) \qquad (27)$$

Alternately, one could use the cylindrical model method and assign names to each of the chains between the bridgehead atoms (usually designated by unprime, prime and double prime if exactly three chains and alphabetically when more than three chains):

$$H1C(1C1)_3C1H:^{(5\text{-}5',5\text{-}5'',5'\text{-}5'')}(1) \qquad (28)$$

At this point it should be noted that there may not be an immediately obvious counterpart of the cylindrical method for many compounds, as there may not exist two unique bridgehead atoms. Nevertheless, there does exist cylindrical names for them. To find one such name, theoretically, all that is needed is to select **any** two atoms as the (quote) bridgehead (unquote) atoms and all atoms on chains between them as the "primary" bridge. Any additional atoms not on this bridge may be further designated as lying on a "secondary", “tertiary”, etc. bridge. A canonical cylindrical name arises when there are two significant bridgehead atoms, nearly all the remaining

Table 4: Names for "alkanes" in the shape of the Platonic solids

Tetrahedrane	
IUPAC	Tricyclo$[1,1,0,0^{2,4}]$butane
Systemic Cartesian	$(\underline{C}1)_4{:}^{(1-5,3-7)}(1)$
Systemic cylindrical	$H1C(1\underline{C}1)_2C1H{:}^{(3-7,5-9)}(1)$
Hexahedrane	
IUPAC[*]	Pentacyclo$[2,2,1^{1,3},1^{4,6},0^{2,8},0^{5,7}]$octane
Systemic Cartesian	$(\underline{C}1)_8{:}^{(1-7,3-13,5-11,9-15)}(1)$
Systemic cylindrical	$H1C[1(\underline{C}1)_2]_3C1H{:}^{(5-17,7-11,13-15)}(1)$
Octahedrane	
IUPAC[†]	No IUPAC name assigned to this structure
Systemic Cartesian	$(\underline{C}1)_6{:}^{(1-5,1-7,3-9,3-11,5-9,7-11)}(1)$
Systemic cylindrical	$C(1C1)_4C{:}^{(5-9,5-13,9-11,11-13)}(1)$
Icosahedrane	
IUPAC[‡]	No IUPAC name assigned to this structure
Systemic Cartesian	$(X1)_{12}{:}^{(1-5,1-7,1-9,3-15,3-17,3-23,5-13,,5-15,7-11,7-13,9-21,9-23,11-19,11-21,13-19,15-19,17-21,17-23)}(1)$
Systemic cylindrical	$X[1(X1)_2]_5X{:}^{(5-11,5-23,5-25,7-11,7-13,7-25,11-15,13-15,13-17,15-19,17-19,17-21,19-23,21-23,21-25)}(1)$
Dodecahedrane	
IUPAC[§]	No IUPAC name found for this compound; however a Chem Abstracts name [44] was[**]
Systemic Cartesian	$(\underline{C}1)_{20}{:}^{(1-9,3-23,5-19,7-15,11-37,13-33,17-31,21-29,25-39,27-35)}(1)$

---

[*]The desired designation of locant numbers is the one that produces the lowest sequence of superscripts. Note that in order to have a Hamiltonian cycle, adjacent to locant #1 (all vertices are equal so any one may be chosen) must be the next numbered vertex, 3, and the last numbered one, 15. There are now two choices for vertex number 5, and having chosen one, two choices for vertex 7. Thus by following four potential paths the canonical name for cubane is established. A similar protocol produces the canonical names for the three larger polyhedra. For tetrahedrane all permutations of assignment are identical and produce the same sequence (1-5,3-7) for the two edges that are not part of the spanning Hamiltonian cycle.

[†]An important commentary on IUPAC's lack of a name for these larger Platonic solids is in order. Namely, IUPAC assigns names only when there is a consensus as to what is appropriate. Its mandate is not to be an innovator, but to codify, and thus standardize, what is the accepted form.

[‡]See above footnote for octahedrane.

[§]See above footnote for octahedrane.

[**]Using the inverted order prevalent for abstracting, the Chemical Abstracts Service (CAS) name originally assigned [42] was: 5,2,1,6,3,4-[2,3]Butanylidenedipentapeno[2,1,6-cde:2'1'6'-gha] pentalene,hexadecahydro-. Although IUPAC does not always follow CAS's lead, if they did the name would be nearly the same; namely, the hexadecahydro would be first, rather than last.

Systemic cylindrical $H1C[1(\underline{C}1)_4]_3C1H{:}^{(5\text{-}21,11\text{-}23,15\text{-}29)}[1(\underline{C}1)_2];_{(7\text{-}37,9\text{-}33,17\text{-}31,19\text{-}41,25\text{-}39,27\text{-}35)}(1)$

atoms lie on chains connecting these two bridgehead atoms and the bridgehead atoms are a maximum graph theoretical distance apart from each other. The IUPAC name, the Cartesian name, and the cylindrical name for the alkane that is modeled by each of the Platonic solids are given in Table 4. Additionally, the systemic Cartesian names for the conjugated alkenes — all of which are theoretically possible, even though only the dodecahedrene has been produced — is created by replacing all of the single bonds by β bonds and deleting the hydrogen atoms; i.e., the systemic Cartesian name listed in Table 4 would be modified by replacing the ($\underline{C}$1) by (Cβ) and the (1) by (β). (Meanwhile, Figures 15 through 19 picture these solids, along with the locant numbering that is relevant for assigning canonical Cartesian names.) Next, because all of the atoms in each solid are equivalent, randomly select as the bridgehead atoms two that are at opposite corners of the solid; i.e., have the largest GTD between them. This is unambiguous for all but the tetrahedron, for which any two vertices may be chosen. For the hexahedron, choose vertices 1 and 11 (Figure 16); for the octahedron vertices 1 and 9 (Figure 17), for the dodecahedron vertices 1 and 31 (Figure 18) and for the icosahedron vertices 1 and 19 (Figure 19). Having done so, next assign locant numbers along the principal chain. The corresponding figures that shall be used to produce the cylindrical nomenclature for the associated "alkanes" are presented as Figures 20 through 24. Note that, for convenience of representation, these figures depict the hydrogen suppressed

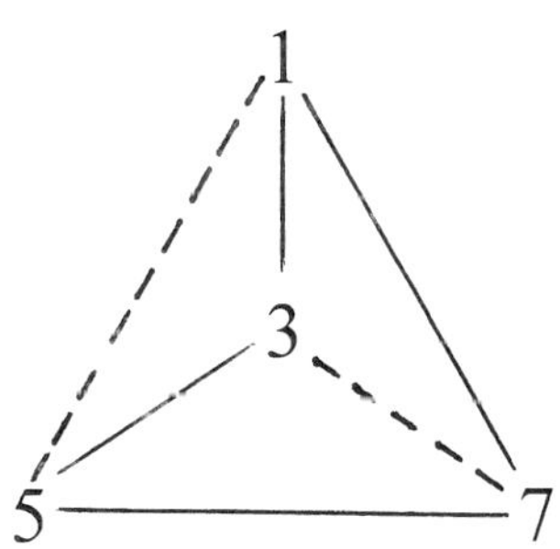

Figure 15: Tetrahedrane (Cartesian locant numbering)

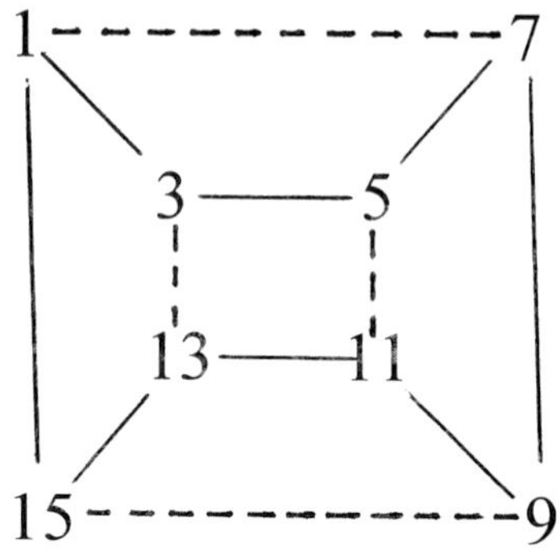

Fig. 16: Hexahedrane (Cubane) (Cartesian locant numbering)

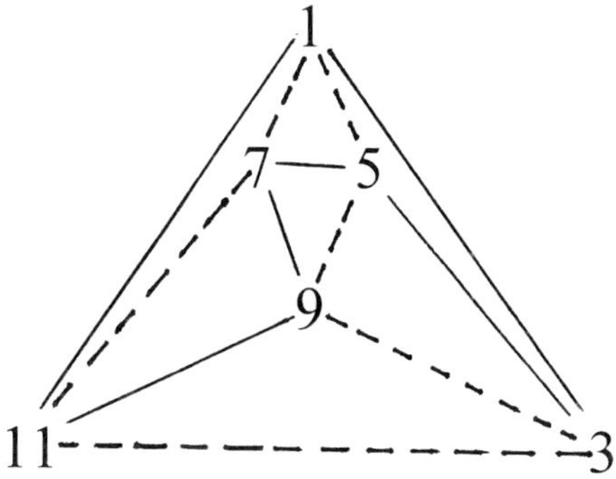

Fig. 17: Octahedrane (Cartesian locant numbering)

model; consequently, the longest contiguous path for each will start with a hydrogen atom. Meanwhile, Figures 20 through 24 list the locant numbers only of the carbon atoms. For example, in Fig. 21, the bridgehead atoms of cubane are locants #3 and 9*. Continue the process of assigning locant numbers by covering as many edges as possible using non-redundant paths between the two bridgehead atoms. In doing so, however, only 9 of the 12 edges of cubane are covered. To cover the remaining three edges one must supplement this set of paths with as many as necessary (3 in this case)

---

* Locants #1 and 11 of the principal chain are hydrogen atoms.

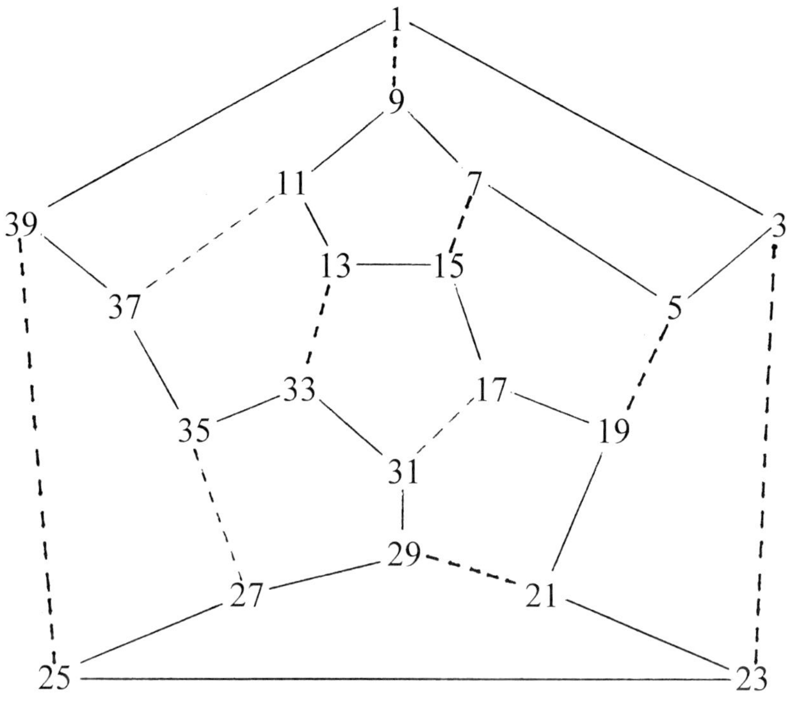

Fig. 18: Dodecahedrane (Cartesian locant numbering)

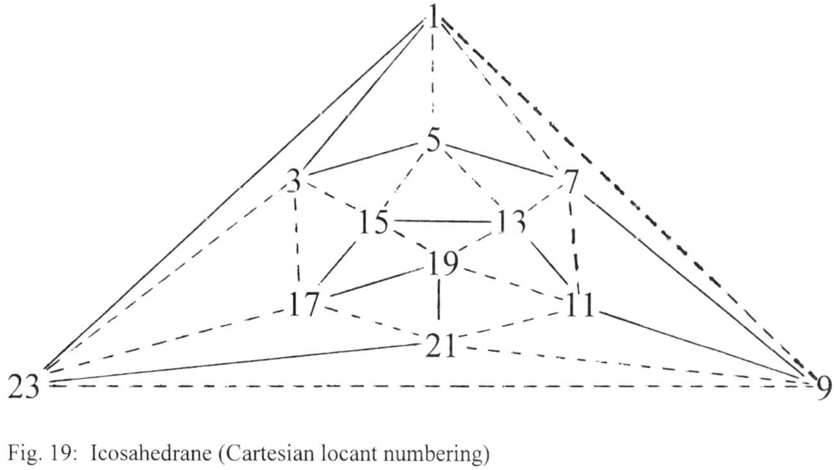

Fig. 19: Icosahedrane (Cartesian locant numbering)

"cross-links" (dash lines). The canonical cylindrical name that may now be assigned to cubane is:

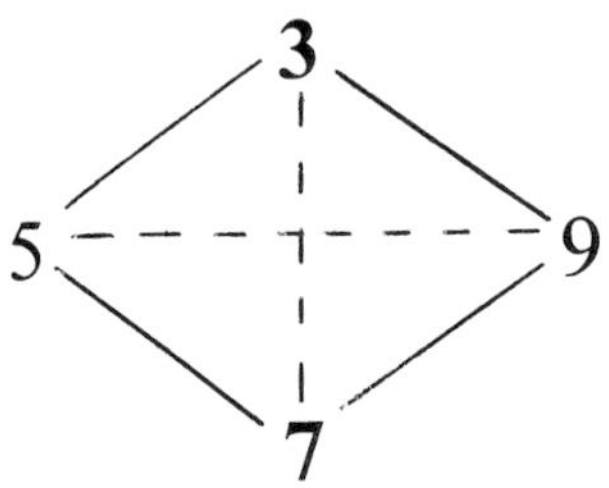

Fig. 20: Tetrahedrane (Cylindrical)

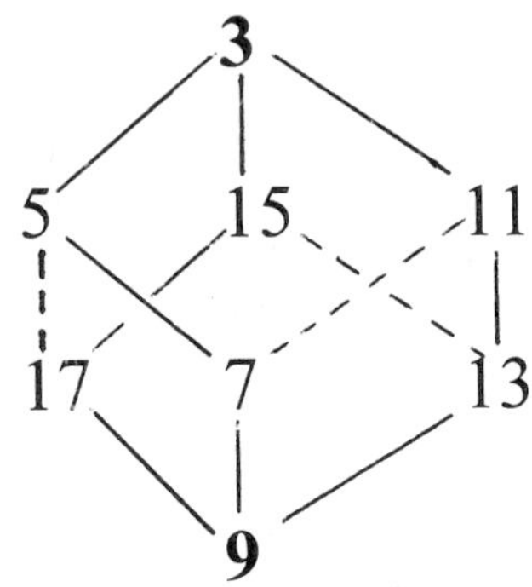

Fig. 21: Hexahedrane (Cylindrical)

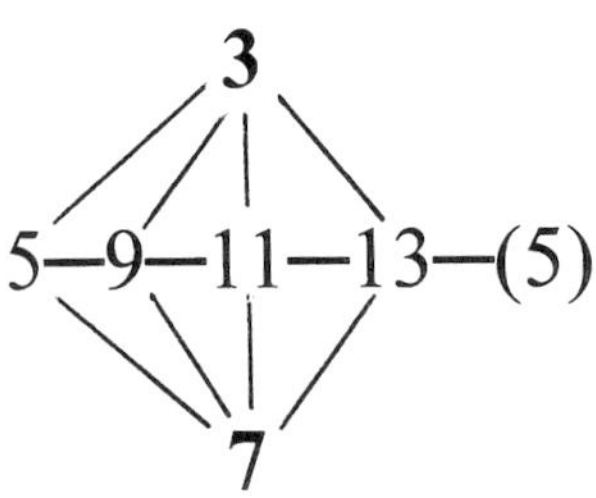

Fig. 22: Octahedrane (Cylindrical)

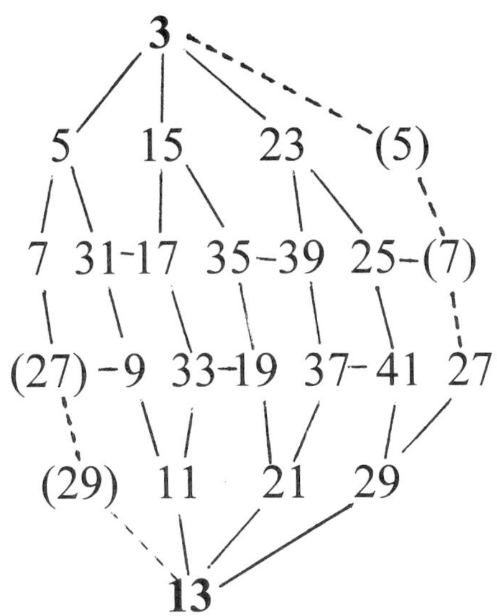

Fig. 23: Dodecahedrane (Cylindrical)

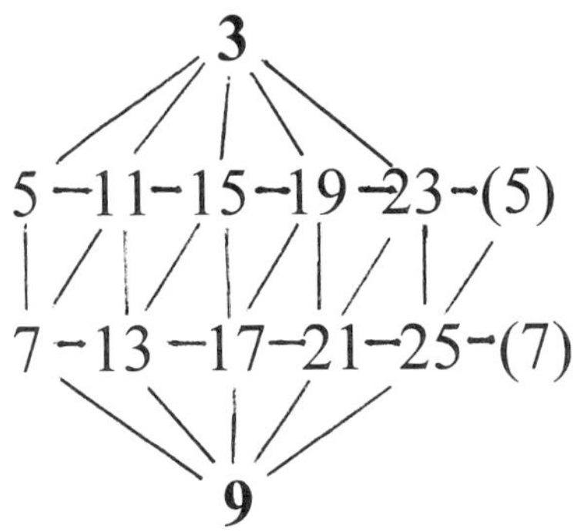

Fig. 24: Icosahedrane (Cylindrical)

$$H1C(1\underline{C}1\underline{C}1)_3C1H:^{(5\text{-}13,7\text{-}15,11\text{-}17)}(1) \qquad (29)$$

Upon comparing these two techniques, note that sometimes the rectangular method and sometimes the cylindrical method yield a simpler name. In general, the presence of cylindrical symmetry is the attribute necessary to make the use of cylindrical nomenclature of pragmatic value.

Historically it should be noted that the set of Platonic solids (the five regular polyhedra having all of its faces and all of its trihedral angles congruent) is of interest primarily because of aesthetics. Molecules which are associated with these particular three dimensional shapes are of two major types:

(1) a central atom is surrounded by n identical ligands; namely, this is a subset of the class of stars described on page 16 of Chapter 1. It will best be nomenclated in Chapter 6, after the introduction of the third important variety of nomenclature — spherical nomenclature.

and

(2) a connected cage of n identical atoms. For all such cages, the first part of the canonical name will be: $(X1)_n$, where X can be any atom

For both of these sets, there exists a static stable equilibrium for only these five specific values of n. For all other values of n the relevant geometry and how it affect the various canonical names (Cartesian, cylindrical, etc.) that shall be assigned to such moieties will be described in later parts of this book. Meanwhile, note that cages for all values of n are possible (both regular and not-regular) and that each such cage is named using the Greek prefix for the number of faces followed by the suffix –hedron. Consequently, the common name given to two theoretically possible sets of molecules is that same prefix and the suffixes -hedrane for a saturated molecule and –hedrene for a conjugated molecule.*

At this point it is important to note that as well as wanting to include in the nomenclature the presence of no bond, when one was expected; namely, what had been called in Chapter 2 a zero bond, (which indicated a lack of attachment at the indicated point), also one might want to include the presence of repulsion between selected atoms in a molecule. One such example is the aromatic compound illustrated in Figure 25, which has IUPAC name: 9,18-diphenyltetrabenz[a,c,h,j]anthracene. This molecule which starts with the same seven rings as does the typical coplanar aromatic molecule illustrated in Figure 13 in Chapter 2, but has two additional rings, has a strong interaction between each of the phenyl rings and the nearest

*The only Platonic polyhedrene that is presently known is dodecahedrene [43]

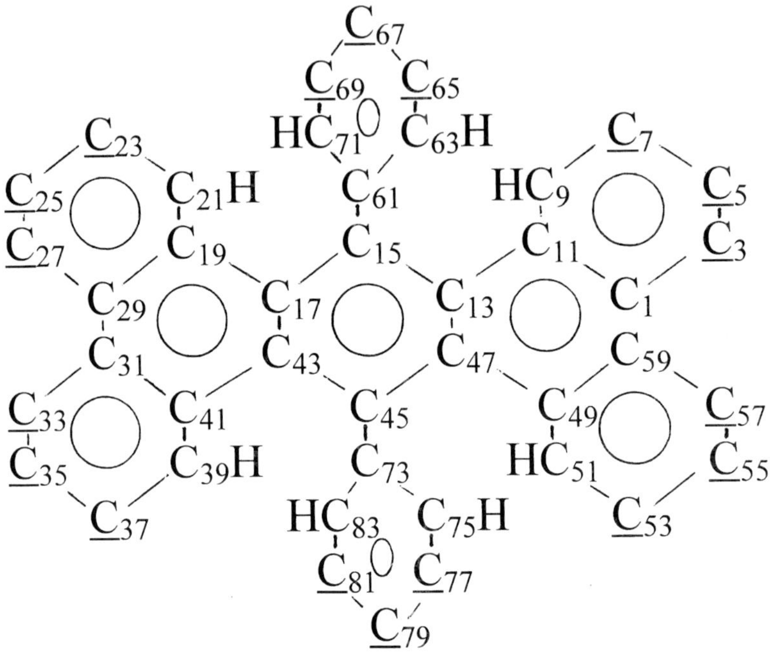

Fig. 25: Nomenclating a strained aromatic compound in which Coulomb repulsion, rather than attraction, between parts is the dominant feature

hydrogen atoms on the "core" (unbridged) aggregation that warps this otherwise planar molecule. Note that this IUPAC name gives no indication of there being a problem. Moreover, were it not for the hydrogen atoms (which are suppressed in the IUPAC name) there would not be one. Therefore, it is recommended that instead of the extrapolation of the IUPAC name with the direct system name:

$$[C\beta(\underline{C}\beta)_4(C\beta)_5(\underline{C}\beta)_4(C\beta)]_2\text{:}\,^{(1\text{-}11,13\text{-}47,17\text{-}43,19\text{-}29,31\text{-}41,49\text{-}59)}(\beta);\,^{(15,45)}1[Ph] \qquad (30)$$

the Ph abbreviation be expanded to the full atom-bond combination, in order that locant numbers on the phenyl rings be spelled out and that a new symbol which indicates negative bonding (i.e., repulsion) between selected atoms be incorporated; e.g., Φ. With this understanding the systemic name of this molecule becomes:

$$[C\beta(\underline{C}\beta)_4(C\beta)_5(\underline{C}\beta)_4(C\beta)]_2{:}\,^{(1\text{-}11,13\text{-}47,17\text{-}43,19\text{-}29,31\text{-}41,49\text{-}59)}(\beta);\,^{(15,45)}1[C\beta(\underline{C}\beta)_5];\,^{(9\text{-}63,21\text{-}71,39\text{-}83,51\text{-}75)}(\Phi) \quad (31)$$

Returning to the previous topic of which of the various nomenclature types are most efficient for a given molecule, observe one advantage of using cylindrical vs. Cartesian, names is that it is often easier to find a convenient starting point for large, highly symmetric molecules; namely, arbitrarily select one atom as one of the bridgeheads and then that atom which is a maximum graph theoretical distance (GTD) from it will be the other bridgehead. For example, attempting to find the rectangular name for buckminsterfullerene is daunting inasmuch as finding one, no less, all of the Hamiltonian cycles that might exist is extremely tedious. On the other hand, as shown in Figure 26 (a reproduction of Figure 5 in [44]), labeling any carbon as the starting point "0", there is exactly one maximally distant carbon (GTD = 9). There is now a much smaller set of possible paths to follow in order to create the desired locant numbering and the canonical name (Figure 27). In its expanded (completely detailed) form, this compound would be named:

$$C[(\beta C)_8\beta]_3C{:}\,^{(3\text{-}7)}(\beta C^{=53}\beta C^{=55}\beta C^{=57}\beta);\,^{(5\text{-}21)}(\beta C^{=59}\beta);\,^{(9\text{-}59)}(\beta C^{=61}\beta C^{=63}\beta);\,^{(11\text{-}57)}(\beta C^{=65}\beta);\,^{(13\text{-}61)}(\beta C^{=67}\beta C^{=69}\beta);\,^{(15\text{-}65)}(\beta C^{=71}\beta C^{=73}\beta);\,^{(17\text{-}67)}(\beta C^{=75}\beta);\,^{(23\text{-}37)}(\beta C^{=77}\beta C^{=79}\beta);\,^{(25\text{-}63)}(\beta C^{=81}\beta);\,^{(27\text{-}77)}(\beta C^{=83}\beta);\,^{(29\text{-}81)}(\beta C^{=85}\beta C^{=87}\beta);\,^{(31\text{-}83)}(\beta C^{=89}\beta C^{=91}\beta);\,^{(33\text{-}85)}(\beta C^{=93}\beta);\,^{(35\text{-}89)}(\beta C^{=95}\beta C^{=97}\beta);\,^{(39\text{-}53)}(\beta C^{=99}\beta);\,^{(41\text{-}79)}(\beta C^{=101}\beta);\,^{(43\text{-}99)}(\beta C^{=103}\beta C^{=105}\beta);\,^{(45\text{-}101)}(\beta C^{=107}\beta C^{=109}\beta);\,^{(47\text{-}103)}(\beta C^{=111}\beta);\,^{(49\text{-}95)}(\beta);\,^{(51\text{-}111)}(\beta C^{=113}\beta C^{=115}\beta);\,^{(55\text{-}73)}(\beta C^{=117}\beta C^{=119}\beta);\,^{(69\text{-}87,71\text{-}113,75\text{-}93,91\text{-}109,97\text{-}107,105\text{-}117,115\text{-}119)}(\beta) \quad (32)$$

This name, however, *could* be somewhat simplified if one agrees to follow the sequential locant assignment scheme of nodal nomenclature [45]. In this way similar bridges may be listed together with the resultant name being:

$$C[\beta(C\beta)_8]_3C{:}\,^{(3\text{-}7)}(\beta(C\beta)_3);\,^{(9\text{-}59,13\text{-}61,15\text{-}65,23\text{-}37,29\text{-}81,31\text{-}83,35\ 89,43\text{-}99,45\text{-}101,51\text{-}11,55\text{-}7)}(\beta(C\beta)_2);\,^{(5\text{-}21,11\text{-}57,17\text{-}67,25\text{-}63,27\text{-}77,33\text{-}85,\ 39\text{-}53,41\text{-}79,47\text{-}103)}(\beta C\beta);\,^{(49\text{-}95,69\text{-}87,71\text{-}113,75\text{-}93,91\text{-}109,97\text{-}107,105\text{-}117,115\text{-}119)}(\beta) \quad (33)$$

This gain in shortening (Formula 32 to Formula 33) is, however, offset by requiring extreme attention to detail in naming the vertices sequentially and is recommended only for computer, rather than human, nomenclating.

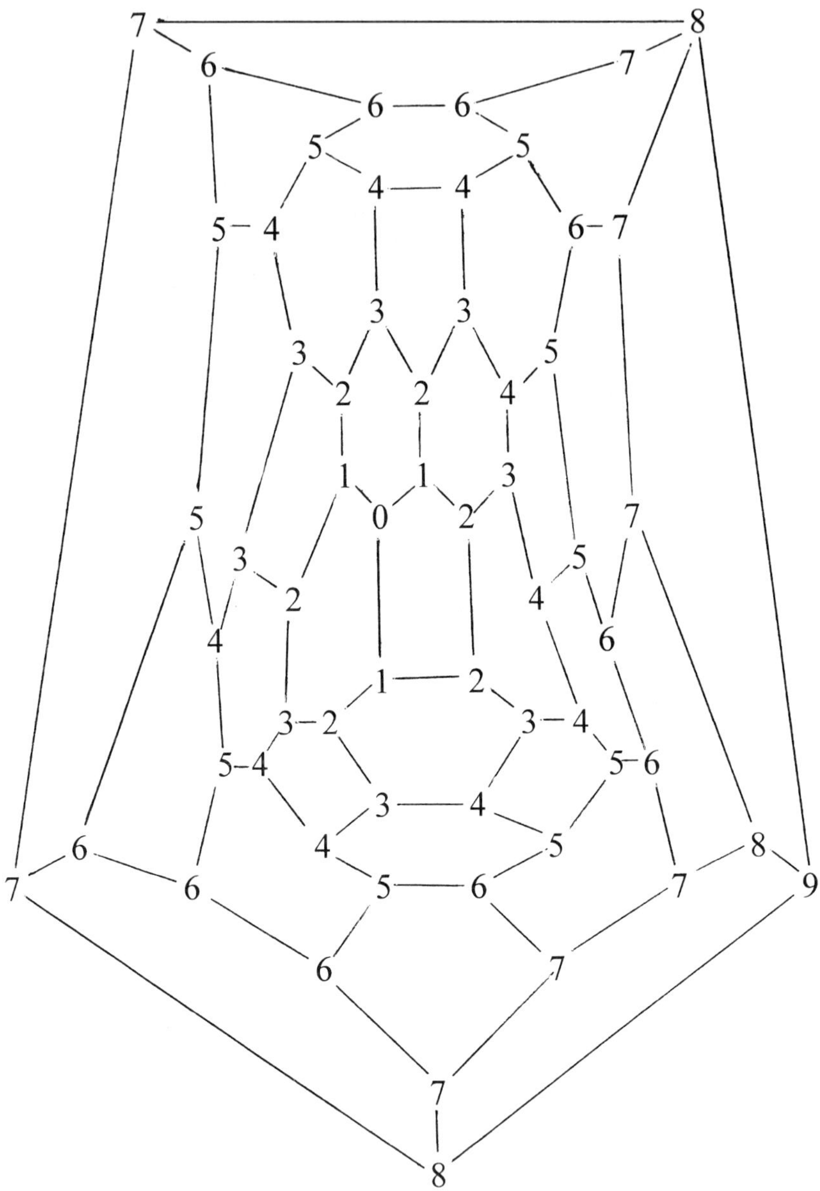

Fig. 26: Graph Theoretical Distances (GTD) From Reference Node (0) in C_{60}

Not only are Figures 26 and 27 useful for nomenclating C_{60} (buckminsterfullerene), they are also the ones to be used in labeling the

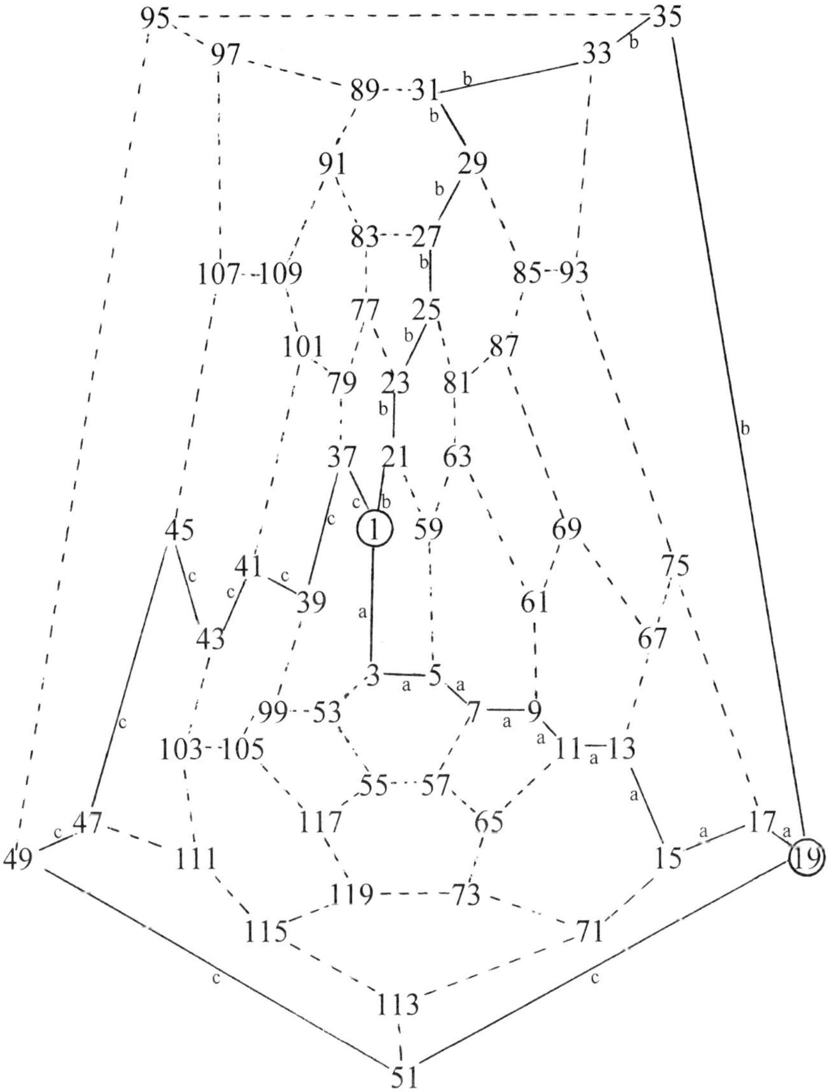

Fig. 27: Locant numbering for the cylindrical naming of C_{60}

saturated compound $C_{60}H_{60}$ (buckminsterfullerane). For this compound, it superficially appears that all one has to do is to replace each of the β bonds with a single bond and each of the carbon atoms with a carbon-hydrogen combination $\underline{C}$; i.e., that formulas (32) and (33) would become respectively:

$$\begin{gathered}\underline{C}[(1\underline{C})_8 1]_3\underline{C}\colon{}^{(3\text{-}7)}(1\underline{C}^{=53}1\underline{C}^{=55}1\underline{C}^{=57}1);{}^{(5\text{-}21)}(1\underline{C}^{=59}1);{}^{(9\text{-}59)}(1\underline{C}^{=61}1\underline{C}^{=63}1);\\ {}^{(11\text{-}57)}(1\underline{C}^{=65}1);{}^{(13\text{-}61)}(1\underline{C}^{=67}1\underline{C}^{=69}1);{}^{(15\text{-}65)}(1\underline{C}^{=71}1\underline{C}^{=73}1);{}^{(17\text{-}67)}(1\underline{C}^{=75}1);\\ {}^{(23\text{-}37)}(1\underline{C}^{=77}1\underline{C}^{=79}1);{}^{(25\text{-}63)}(1\underline{C}^{=81}1);{}^{(27\text{-}77)}(1\underline{C}^{=83}1);{}^{(29\text{-}81)}(1\underline{C}^{=85}1\underline{C}^{=87}1);\\ {}^{(31\text{-}83)}(1\underline{C}^{=89}1\underline{C}^{=91}1);{}^{(33\text{-}85)}(1\underline{C}^{=93}1);{}^{(35\text{-}89)}(1\underline{C}^{=95}1\underline{C}^{=97}1);{}^{(39\text{-}53)}(1\underline{C}^{=99}1);\\ {}^{(41\text{-}79)}(1\underline{C}^{=101}1);{}^{(43\text{-}99)}(1\underline{C}^{=103}1\underline{C}^{=105}1);{}^{(45\text{-}101)}(1\underline{C}^{=107}1\underline{C}^{=109}1);{}^{(47\text{-}103)}(1\underline{C}^{=111}1);\\ {}^{(49\text{-}95)}(1);{}^{(51\text{-}111)}(1\underline{C}^{=113}1\underline{C}^{=115}1);{}^{(55\text{-}73)}(1\underline{C}^{=117}1\underline{C}^{=119}1);\\ {}^{(69\text{-}87,71\text{-}113,75\text{-}93,91\text{-}109,97\text{-}107,105\text{-}117,115\text{-}119)}(1)\end{gathered} \tag{34}$$

and

$$\begin{gathered}\underline{C}[1(\underline{C}1)_8]_3\underline{C}\colon{}^{(3\text{-}7)}(1(\underline{C}1)_3);{}^{(9\text{-}59,13\text{-}61,15\text{-}65,23\text{-}37,29\text{-}81,31\text{-}83,35\text{-}89,43\text{-}99,45\text{-}101,51\text{-}11,55\text{-}7)}\\ (1(\underline{C}1)_2);{}^{(5\text{-}21,11\text{-}57,17\text{-}67,25\text{-}63,27\text{-}77,33\text{-}85,\ 39\text{-}53,41\text{-}79,47\text{-}103)}(1\underline{C}1);{}^{(49\text{-}95,69\text{-}87,71\text{-}113,75\text{-}93,}\\ {}^{91\text{-}109,97\text{-}107,105\text{-}117,115\text{-}119)}(1)\end{gathered} \tag{35}$$

This, however, is an oversimplification. Although fullerenes divide three-dimensional space into "inside", "boundary" and "outside" sets and that all chemical properties of interest are constrained to the boundary set, the same can not be said for fulleranes. In particular, as Saunders [46] demonstrated, the 60 hydrogen atoms will not all lie outside the boundary, as the simplified model would suggest. To the contrary, exactly as cyclobutane is NOT coplanar, despite that by having 90° angles between the carbon atoms would seem to minimize strain between successive carbon atoms, the presence of repulsion between the hydrogen atoms yields a lower energy minimum when the molecule is warped out of the plane. In fact, in cyclobutane this non-planar distortion is up to 36° [47]. Similarly, had the neighboring hydrogen atoms in fullerane been all exterior, this would lead to a highly strained molecule having 90 unfavorably eclipsed CCCC dihedral angles and 120 CCC planar angles, rather than the energetically desired tetrahedral angles. To keep these hexagonal faces coplanar, while having all of the hydrogen atoms outside requires a steric energy of 836.1 kcal/mol. This excess energy is reduced by over 53 kcal/mol when just one of the hydrogen atoms is interior to the "cage" because there are now three fewer eclipsing interactions as well as many of the strained 120° angles are reduced to closer to the desired 109°28.5'. Furthermore, Saunders, upon running an MM3 optimization program concluded that the lowest steric energy would occur when ten hydrogen atoms are interior and the remaining 50 exterior. To try to canonically name individual "topological" isomers with selected hydrogen atoms interior is beyond our capacity inasmuch as the number of such isomers is 2^{60} divided by the average number of identical copies; i.e., in the range of trillions to quadrillions.

As well as new properties of these molecules that have come to light with the discovery and nomenclating of the higher fullerenes and fulleranes, an extension that is unfathomable using traditional nomenclature, but is readily explained using beta bonds is next described when some of the carbon atoms are replaced by metal atoms in various of the smaller fullerenes. These molecules, which historically are not in the domain of

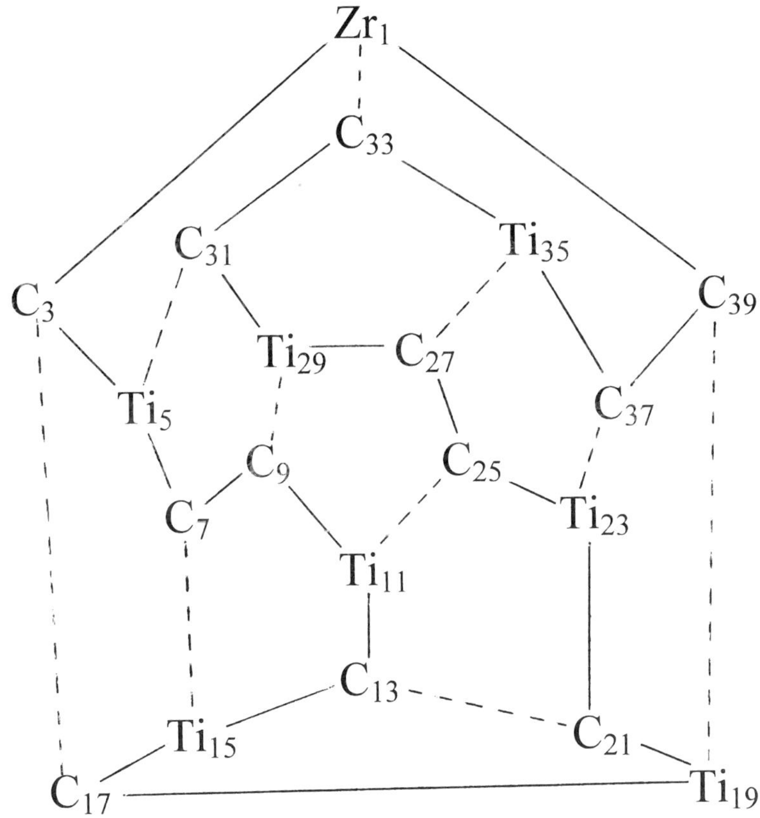

Fig. 28: A Titanium-Zirconium "Metcar"

organic chemistry, but rather are grouped into their own "fiefdom" (organometallic compounds) and are referred to as "metallocarbohedrenes", which is usually abbreviated to “metcars”. Figure 28 illustrates a molecule in which eight of the carbon atoms in the dodecahedrene structure have been replaced with atoms from Columns 4 and 5 of the periodic table (in particular — from Column 4: titanium, zirconium and hafnium, and from Column 5: vanadium and niobium) [48]. The Cartesian systemic name for this Ti_7ZrC_{12} compound appears to be:

$$Zr1C1[Ti1(C1)_2]_2\,(Ti1C1)_3[Ti(C1)_2]_2{:}^{(1\text{-}33,3\text{-}17,5\text{-}31,7\text{-}15,9\text{-}29,11\text{-}25,13\text{-}21,19\text{-}39,23\text{-}37,27\text{-}35)}(1) \qquad (36)$$

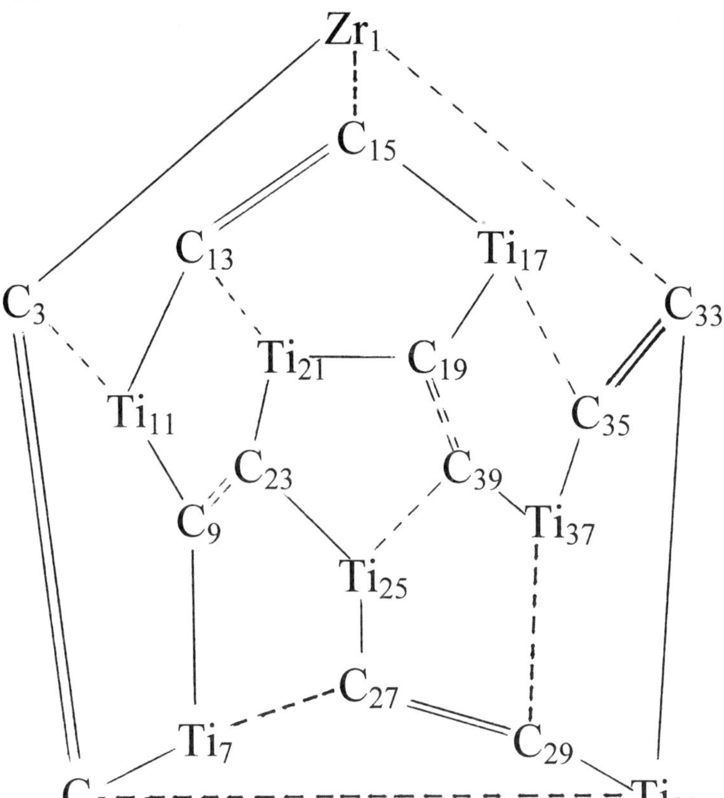

Fig. 29: An allocation of double vs. single bonds that forms a Hamiltonian path in a titanium-zirconium metcar

This name would be correct had each of the carbon atoms only a valence of three. To the contrary, if the molecule involved was a metallocarborAne and thus had as its molecular formula, $Ti_7ZrC_{12}H_{12}$, the systemic name would be:

$$\mathrm{Zr1\underline{C}1[Ti1(\underline{C}1)_2]_2\,(Ti1\underline{C}1)_3[Ti(\underline{C}1)_2]_2:^{(1\text{-}33,3\text{-}17,5\text{-}31,7\text{-}15,9\text{-}29,11\text{-}25,13\text{-}21,19\text{-}39,23\text{-}37,27\text{-}35)}(1)} \qquad (31)$$

Nevertheless, since the given molecular formula does not contain hydrogen atoms, thereby indicating that this is a metallocarborEne, a different locant assignment might appear to be desirable in order to accommodate the double bond prior to the higher atomic number element (Ti) from the carbon atom designated as locant #3 (Figure 29). Such a revised assignment of locant numbers is not necessary when one recognizes all of the bonds as being either β or ℵ bonds, rather than double vs. single bonds. Thus it is immaterial that, when trying to create a Hamiltonian cycle, there is not a consistent set of double vs. single bonds (as would be the requirement of a traditional nomenclature). The best that one can do is to describe a Hamiltonian path through the twenty vertices. For such a path the systemic, but not the canonical, name is:

$$\mathrm{Zr1(C2C1Ti1C1Ti1)_2(C1Ti1C2)_2C1Ti1C:^{(9\text{-}23,19\text{-}39,\ )}(2);\ _{(1\text{-}15,1\text{-}33,3\text{-}11,5\text{-}31,7\text{-}27,13\text{-}21,17\text{-}35,25\text{-}39,29\text{-}37)}(1)} \qquad (34)$$

For this metcar the appropriate canonical name is obtained from (32) by replacing all of the single bonds with β bonds between the carbon atom pairs and ℵ bonds between each carbon-metal pair. Consequently, the proposed Cartesian canonical name is:

$$\mathrm{Zr\aleph C\aleph Ti\aleph C\beta C(\aleph Ti\aleph C)_3(\aleph Ti\aleph C\beta C)_3\aleph:^{(3\text{-}17,13\text{-}21)}(\beta);\ _{(1\text{-}33,5\text{-}31,7\text{-}15,9\text{-}29,11\text{-}25,19\text{-}39,23\text{-}37,27\text{-}35)}(\aleph)} \qquad (35)$$

Turning now to forming the canonical cylindrical name for this molecule (Figure 30), one of the poles is the zirconium atom (again labeled as locant #1, while the other is the diametrically opposite titanium atom (locant # 11 in Figure 28 and locant #25 in Figure 29). The three paths from locants 1 thru 11 uniquely label twelve of the vertices (3 thru 27). The remaining six nodes lie on three sets of two each, labeled from locants 3 to 7 (29 and 31), from 5 to 9 (33 and 35) and from 13 to 27 (37 and 39). The six edges still to be covered include one β and five ℵ bonds. This produces the canonical cylindrical name of:

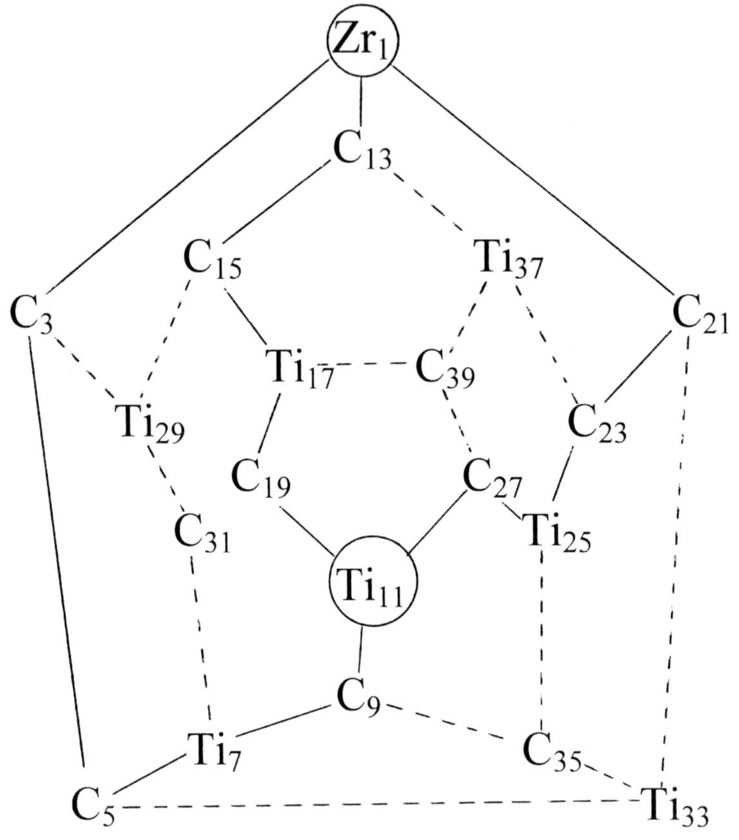

Fig. 30: Canonical locant numbering for assigning a cylindrical name to a titanium-zirconium metallocarborane

$$Zr(\aleph C\beta C\aleph Ti\aleph C\aleph)_3Ti{:}^{(3\text{-}7,5\text{-}9,13\text{-}17)}(\aleph Ti\aleph C\aleph);^{(19\text{-}31)}(\beta);^{(15\text{-}29,21\text{-}33,23\text{-}37,25\text{-}35,27\text{-}39)}(\aleph) \quad (36)$$

which is only slightly shorter than the Cartesian name.

Not only can a canonical name be assigned to "simple" molecules such as dodecahedrene (with 20 atoms), it is also "relatively easy" to canonically name high symmetry moieties, such as the 5880 atom ion with empirical formula $C_{2900}H_{2300}N_{60}P_{120}S_{60}O_{200}F_{180}Pt_{60}$ that self-assemblies into a supermolecular dodecahedron [49] where each of the twenty carbon atoms

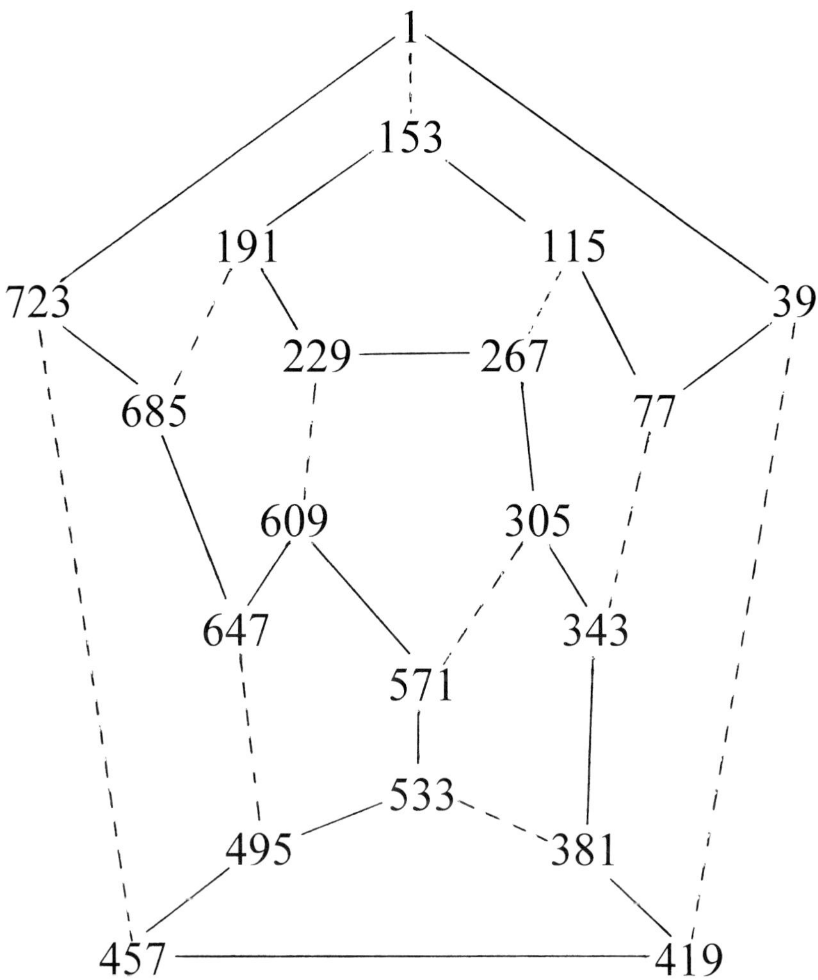

Fig. 31: Locant numbering of initial atom in each of the modules of a self-assembled platinum cation

has been replaced by a module IUPAC calls: tri(4'-pyridyl)methanol. The systemic name for this module is:

$$N\beta(\underline{C}\beta)_2(C1)_2C\beta(\underline{C}\beta)_2N:^{(1\text{-}7,11\text{-}17)}(\beta(\underline{C}\beta)_2;^{(9)}\ (1O1H);^{(9)}\ (1C\beta(\underline{C}\beta)_2N\beta(\underline{C}\beta)_2 \qquad (37)$$

Also, each of the thirty carbon-carbon bonds forming the dodecahedron are replaced by a diphenyldiplatinum complex where each of the platinum atoms has two phosphorus triphenyl ligands (which remain as part of the self-assembling molecule) and a trifluoromethylsulfoxo ligand (that separates as 60 $OSO_2CF_3^-$ ions). The systemic name for this reacting complex is:

$$\mathrm{F1C1S1O1Pt1C\beta(\underline{C}\beta)_2\,(C\;1)_2(\underline{C}\beta)_2C1Pt1O1S1C1F{:}^{(11\text{-}17,19\text{-}25)}(\beta(C\beta)_2);\;{}^{(9,9,27,27)}(1P1PhH{:}^{(3,3)}PhH)} \qquad (38)$$

Next note that in the forming the name for the entire cation, because the highest atomic number atom is in that module which corresponds to the edge of the dodecahedron, rather than being associated with the vertex, choose the longest path starting from a platinum atom in the "edge" module and, inasmuch as all of the bonds in this principal cycle are singular, progress to the adjoining nitrogen atom of the "vertex" module. This produces a locant numbering as shown in Fig. 31. Consequently, although sequences of atoms and bonds may be recognizable, the pattern of atoms in this super-molecule is not identical to that of the dodecahedron. Instead the name for the cation is:

$$\mathrm{\{[Pt1N\beta(\underline{C}\beta)_2C\beta(\underline{C}\beta)N1Pt1C\beta(\underline{C}\beta)_2(C1)_2(\underline{C}\beta)_2C1]_{20}{:}^{(1\text{-}153,39\text{-}419,77\text{-}343,115\text{-}267,191\text{-}723,229\text{-}609,305\text{-}571,381\text{-}533,457\text{-}723,495\text{-}647)}[1C\beta(\underline{C}\beta)_2(C1)_2(\underline{C}\beta)_2C1{:}^{(3\text{-}9,11\text{-}17)}(\beta(\underline{C}\beta)_2)];\;{}^{(3\text{-}9,\,13\text{-}19,23\text{-}29,31\text{-}37)}(\beta(\underline{C}\beta)_2);{}^{(1,1)}[1P1PhH{:}^{(3,3)}(1PhH{:}^{(3,3)}(1PhH)];{}^{(11)}(1O);\;{}^{(11)}(1C\beta(\underline{C}\beta)_2N\beta(\underline{C}\beta)_2)\}^{60+}} \qquad (39)$$

One important chemistry fallout from the creation of cylindrical names shall arise when the two anchor atoms normally do not have a coordination greater than two -- such as is the situation with oxygen. In Chapter 6, after having developed properties of the boranes and the metallocenes, cylindrical nomenclature shall be applied to some supramolecular clusters.

Returning now to the mathematical problem of spanning a set of points (atoms) connected by a set of edges (bonds) in a three dimensional Euclidean space. Note that, although a Hamiltonian spanning cycle [50] will, in general, be the basis for formulating the optimal rectangular name for a fisular [51] compound, the twin problems of existence of such a cycle for some molecules and of uniqueness for others are encountered:

(1) Occurrence of non-existence was shown earlier for the untwisted isomer of Moebiane (Fig. 6);

(2) Occurrence of more than one distinct Hamiltonian cycle is often encountered in high symmetry polyhedra [52] that have high incidence vertices. For example, a different Hamiltonian cycle from the one through the 12 incidence = 5 vertices of an icosahedron that was portrayed in Figure 24 is illustrated in Figure 31. For this latter picture the more traditional Schlegel projection of the icosahedron is used, rather than one based on GTDs described in [53]. Several others are also available. Consequently, in order to be able to find the canonical name, one must examine the locant numberings of all possible Hamiltonian cycles -- a potentially tedious task that grows in magnitude as the coordination increases. On the other hand, because there is a single atom (or, for the graph of some molecules, at most a pair of atoms) diametrically opposed to any selected inital atom (atoms 1 and 19 in Fig. 19 or atoms 1 and 7 in Figure 32), and, inasmuch as there is five-fold symmetry along five paths (labeled by letter a thru e) between these "poles", the paths followed, which determine the locant numbering, are interchangeable. Consequently, the cylindrical system is independent of considerations of Hamiltonian cycles and its usage minimizes the problem of

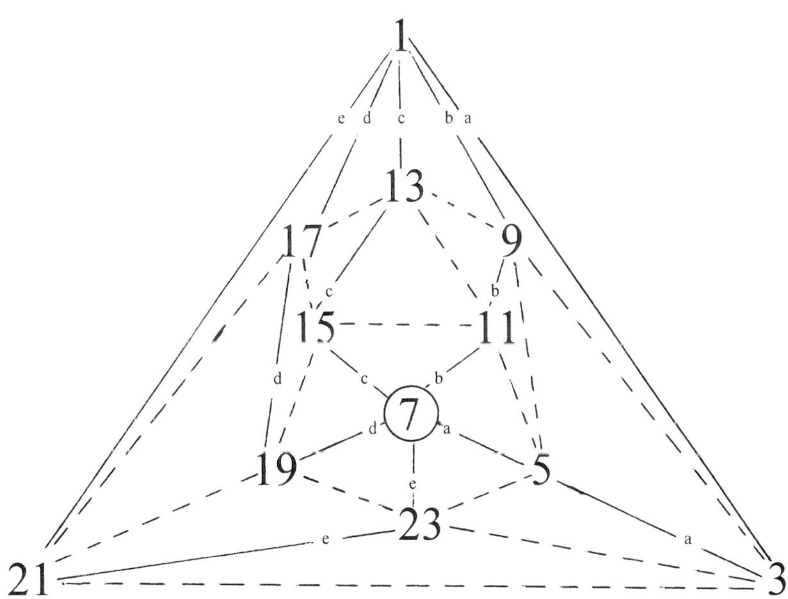

Fig. 32: Another cylindrical name for an icosahedral molecule

uniqueness. In this system, the name of this moiety is:

$$X(1X1X1)_5X{:}^{(3\text{-}9,3\text{-}21,3\text{-}23,5\text{-}9,5\text{-}11,5\text{-}23,9\text{-}13,11\text{-}13,11\text{-}15,13\text{-}17,15\text{-}17,15\text{-}19,17\text{-}21,19\text{-}21,19\text{-}23)}(1) \tag{40}$$

An important observation of the above two naming schemes for fisular compounds is that there has been no usage of, nor reliance or reference to, the smallest set of smallest rings [54], or any other inappropriate planar measure of an essentially linear set. Consequently, the question raised in an earlier study [55] as to whether the nomenclature for

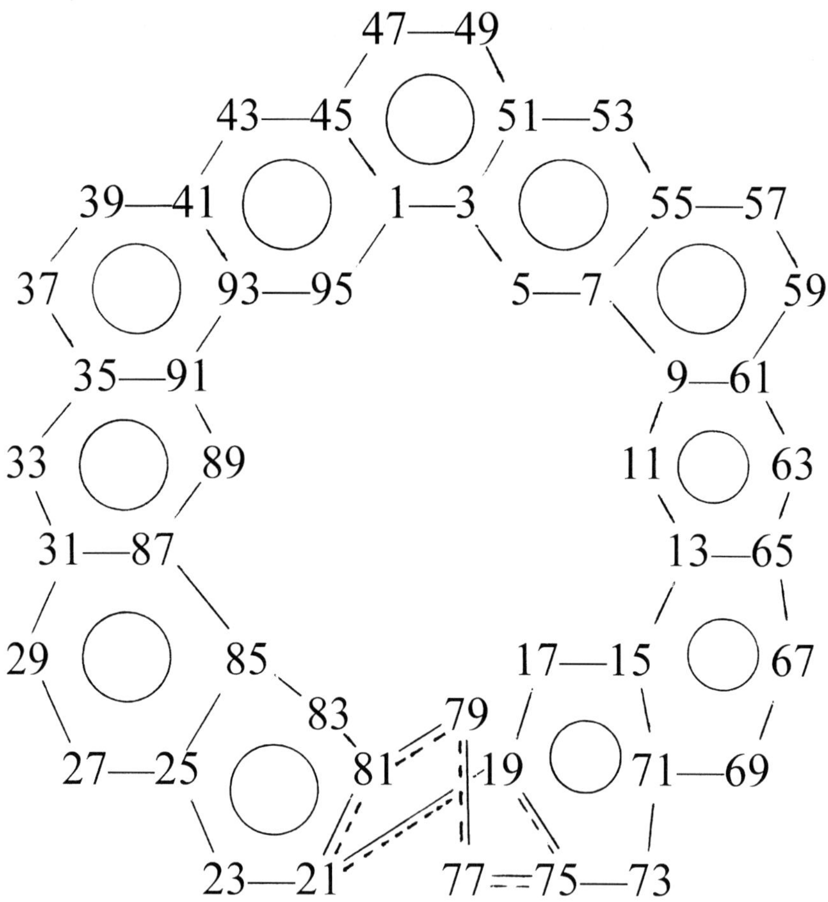

Fig. 33: Planar Moebiane

cubane should be based on the six rings that physically comprise the boundary of a cube, or the five squares (less the boundary) that form the basis of SSSR, or the minimum spanning set (of four cycles which covers the edge set) is moot. Note that in order to unambiguously name the entire vertex and edge set only 21 of the coplanar 32 faces of buckminsterfullerene would have to be used.

Another idea of importance arises when revisiting the various possible Moebianes. At this time, however, focus is shifted to a planar variety of the Moebius strip. The $C_{48}H_{24}$ compound illustrated in Fig. 33 is locally nearly coplanar throughout the entire molecule; however, because of the connectivity, it requires a three dimensional embedding space. Such a compound, which would be named as either:

$$(C\beta)_{48}:^{(1\text{-}45,3\text{-}51,7\text{-}55,9\text{-}61,13\text{-}65,15\text{-}71,19\text{-}75,21\text{-}81,25\text{-}85,31\text{-}87,35\text{-}91,41\text{-}93)}\beta;_{(5,11,17,23,27,29,33,37,39,43,47,49,53,57,59,63,67,69,73,77,79,83,89,95)}(1H) \quad (41)$$

or

$$[(C\beta)_2(\underline{C}\beta)]_4C\beta\underline{C}\beta[(\underline{C}\beta C\beta)_2\underline{C}\beta]_5\underline{C}\beta C\beta(\underline{C}\beta C\beta\underline{C}\beta)_2\underline{C}\beta:_{(1\text{-}45,3\text{-}51,7\text{-}55,9\text{-}61,13\text{-}65,15\text{-}71,19\text{-}75,21\text{-}81\text{-}25\text{-}85,31\text{-}87,35\text{-}91,41\text{-}93)}(\beta) \quad (42)$$

is classified in the Taylor [56] - Goodson [57] nomenclature scheme as fisular, in contradiction to reticular for its untwisted isomer (Fig. 34). This latter compound is named as either:

$$(C\beta)_{46}:^{(1\text{-}55)}[\beta(\underline{C}\beta)_2];^{(1\text{-}91,5\text{-}87,11\text{-}85,15\text{-}81,21\text{-}79,25\text{-}75,31\text{-}73,35\text{-}69,41\text{-}67,45\text{-}63,51\text{-}61,55\text{-}57,57\text{-}91)}(\beta);_{(3,7,9,13,17,19,23,27,29,33,37,39,43,47,49,53,59,65,71,77,83,89)}(1H) \quad (43)$$

or using the abbreviated form:

$$[(C\beta\underline{C}\beta)_2C\beta]_5(C\beta\underline{C}\beta C\beta)_7:^{(1\text{-}55)}[\beta(\underline{C}\beta)_2];^{(1\text{-}91,5\text{-}87,11\text{-}85,15\text{-}81,21\text{-}79,25\text{-}75,31\text{-}73,35\text{-}69,41\text{-}67,45\text{-}63,51\text{-}61,55\text{-}57,57\text{-}91)}(\beta) \quad (44)$$

Also, it is referred to, in the newer terminology, as corona-condensed [58] vs. the simpler peri-condensation that is usually used when categorizing polycyclic aromatic hydrocarbons. Again, because of the focus solely on the bridges and a lack of interest in rings, the question is moot as to whether Figure 34 is a 12 ring multiply-connected region or a 13 ring simply-connected region — with the 13-th ring being formed by the 18 "interior" carbon atoms.

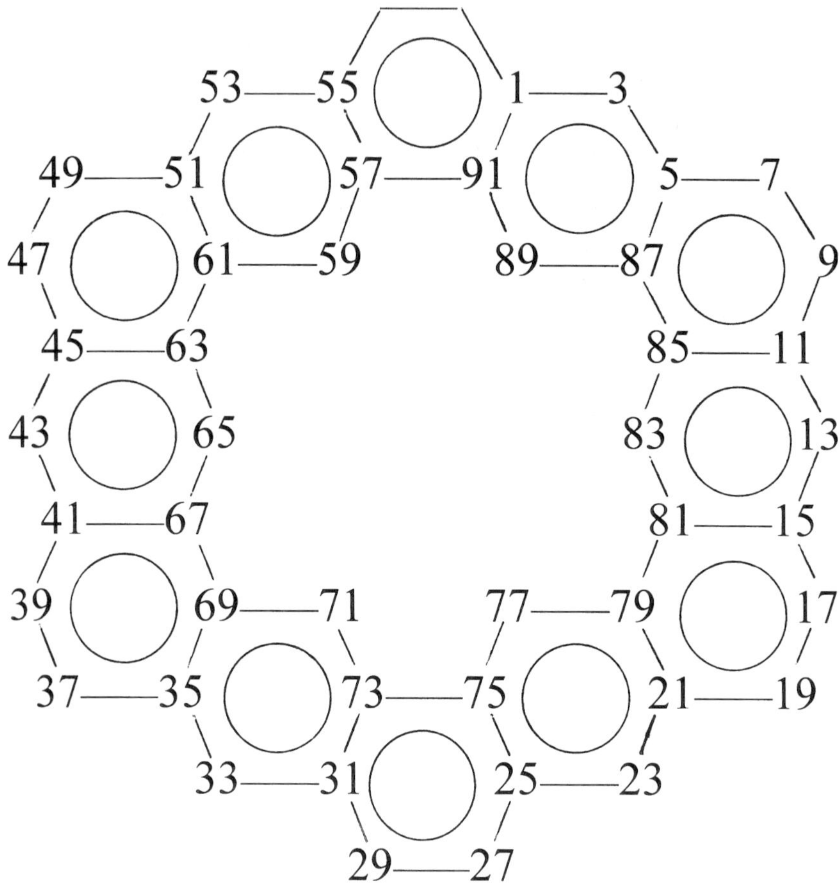

Fig. 34: Untwisted (corona-condensed) Isomer of Planar Moebiane

In addition to this discussion of "planar Moebiane" and the one earlier in this chapter of "linear Moebiane" and its cylindrical counterpart, attention is now directed to some other selected molecules of mathematical interest. Although the existence of molecules, formed by the edge fusion of benzene modules (generally referred to as "benzenoids"), in which it is not possible to assign a coherent system of conjugated single and double bonds that span the molecule was illustrated in Chapter 2, these have always had an odd number of "triple points" [59] ; e.g., see phenalene (#11 in Table 1 of

Chapter 2). For those benzenoids encountered so far that have an even number of triple points it has always been possible to find such a spanning set. For example, reiterating from the discussion in item 12 of Chapter 2, of seven mathematically possible four ring benzenoids, five are cata-condensed, one is peri-condensed and the seventh has a single triple point and thus can not be spanned by a conjugated system of single and double bonds. Similarly, as illustrated in [60] of the 22 possible five ring benzenoid systems, 12 are cata-condensed, 3 are peri-condensed (all with 2 triple points) and 7 can not be spanned by any combination of conjugated bonds. Of these 7, 6 have a single triple point and the seventh has 3 triple points.

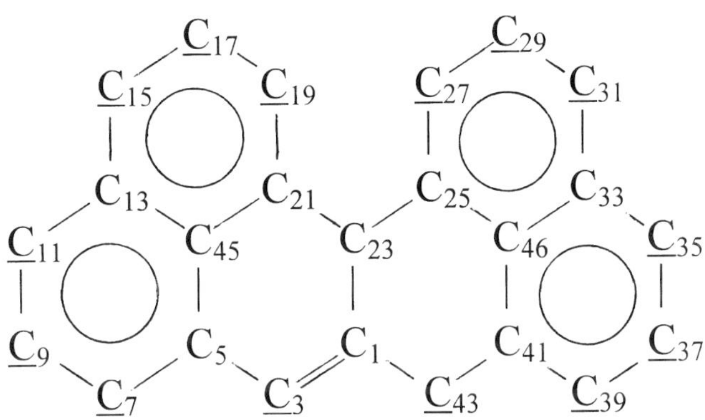

Fig. 35: Twisted zethrene with central single bond

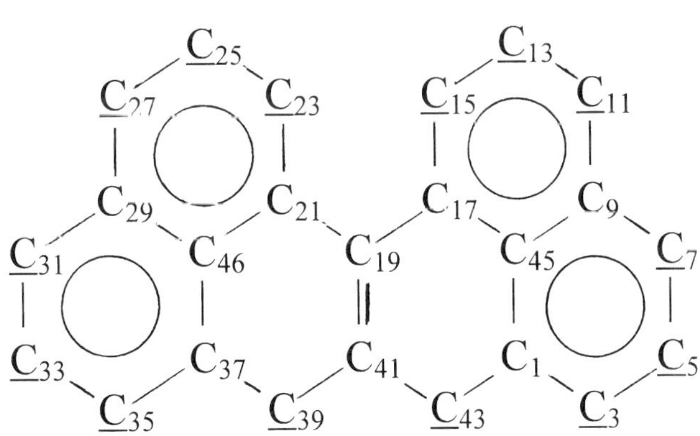

Fig. 36: Twisted zethrene with central double bond

Continuing to the 82 combinations of six benzenoid rings, one now encounters two examples of aggregations having an even number of triple points, but which still can not be spanned using a conjugate system. The first of these, referred to as "twisted zethrene", may be thought of as a molecule of zethrene (see Item 16 of Chapter 2) twisted along the four ring "axis" so that both off-line rings are on the same side of this axis (Figures 35 and 36). Now, consider both the zethrene and twisted zethrene molecules as the fusion of three naphthalene modules: two end naphthalene modules (one ring of each of these modules is the end of the four ring acene and the other is the off-horizontal ring) while the third one is the center two benzene rings. Observe that the four bonds emanating from the end naphthalene modules are constrained to be single bonds. The remaining three bonds yield a zigzag sequence of double bond - single bond - double bond for zethrene; i.e., a viable conjugation; and a Y shaped aggregation of three bonds for the twisted zethrene. For this second molecule note that regardless which of the three bonds is made into a double bond there will be left a sequence of four single bonds; thereby breaking the conjugation. Figure 35 has extended the conjugation along the perimeter by setting a double bond there. This leaves a four single bond sequence (C_{21}-C_{23}-C_1-C_{43}-C_{41}) passing through the center bond of the center naphthalene module. Such an interpretation of twisted zethrene would have systemic name:

$$\mathrm{C2C1[C\beta(\underline{C}\beta)_3]_2(C1)_3(\underline{C}\beta)_3C\beta(\underline{C}\beta)_3(C1)_2{:}^{(5\text{-}13)}(\beta C^{(=45)}\beta);^{(25\text{-}31)}(\beta C^{(=46)}\beta);}\ {}^{(21\text{-}45,33\text{-}46)}(\beta);^{(1\text{-}23)}(1) \qquad (45)$$

Alternately, as illustrated in Figure 36, one could set a double bond between the two central rings thereby leaving a sequence of four single bonds on the perimeter (C_{37}-C_{39}-C_{41}-C_{43}-C_1). This interpretation produces as the systemic name:

$$\mathrm{\{[C\beta(\underline{C}\beta)_3]_2(C1)_2\}_2(C1)_2{:}\ ^{(1\text{-}9)}(\beta C^{(=45)}\beta);^{(21\text{-}29)}(\beta C^{(=46)}\beta);^{(1\text{-}23)}(2);^{(17\text{-}45,37\text{-}46)}(\beta)} \qquad (46)$$

In both instances, the interruption of conjugation is a major feature of the chemistry of these molecules and thus should be included in the nomenclature.

An important attribute of benzenoids is next noted by differentiating between carbon atoms having one direction of electron spin (referred to as "marked") and those with opposed spin ("unmarked"). One measure of the

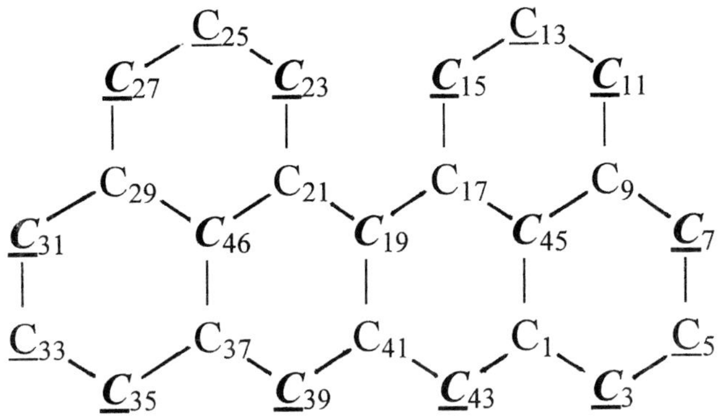

Fig. 37: Twisted zethrene with marked carbon atoms

stability of the molecule depends on whether the difference between these two numbers is an even or odd integer — a property referred to as "even-alternant" or "odd-alternant" [61]*. Before encountering twisted zethrene all of the viable benzenoid molecules known had an equal number of marked and unmarked carbon atoms. Similarly, all of the molecules that could not be spanned by a conjugated system of single and double bonds were odd-alternant benzenoids. These required an extra hydrogen to achieve a balanced neutral atom Twisted zethrene is the first example presented of an even-, but NOT zero-, alternant benzenoid. It also can not be spanned by a conjugated system of single and double bonds; consequently, different bond assignments resulted in the different canonical names given above. In Figure 37, the marked carbon atoms are indicated using italized bold font.†

Other even-, but not zero-, alternate molecules will also have this same limitation. For example, there exists exactly one other combination of six fused benzene rings that also can not be spanned by a conjugated system of single and double bonds. This molecule is known by the common name "triangulene". In Figure 38, if one marks the "top" carbon atom, a set of 10 marked carbon atoms and 12 not marked atoms are created. Again there is

---

* A typographical error is noted in [61]. In their figure XXIV on their page 111, the bottom two carbon atoms are incorrectly marked (yielding 15, rather than the correct 13 marked atoms); similarly in XXIII, the bottom three atoms are reversed — the bottom atom should not have been marked, but its two neighbors should have been.

† Note that marked atoms are completely independent of whether there is a hydrogen atom attached to that particular carbon atom (indicated with the same underscored convention that has been a standard abbreviation in this nomenclature system).

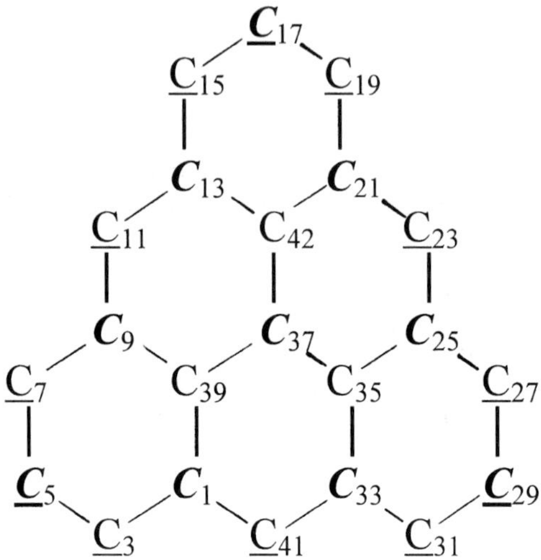

Fig. 38: Triangulene

not a spanning set of conjugated single and double bonds, and every assignment of double and single bonds will leave two atoms with only three single bonds. The best that can be done nomenclature-wise is to have 20 beta and 5 single carbon to carbon bonds, along with 12 single carbon hydrogen bonds. The canonical name for this molecule is thus either:

$$[C\beta(\underline{C}\beta)_3C\beta\underline{C}\beta]_2C\beta(\underline{C}\beta)_3(C\beta)_4\text{:}\,^{(1\text{-}33)}(\beta\underline{C}^{(=41)}\beta);^{(13\text{-}21)}(1C^{(=42)}1);^{(9\text{-}39,25\text{-}35,37\text{-}42)}(1) \qquad (47)$$

using the underscore convention, or

$$(C\beta)_{20}\text{:}^{(1\text{-}33)}(\beta C^{(=41)}\beta);^{(13\text{-}21)}(1C^{(=42)}1);^{(9\text{-}39,25\text{-}35,37\text{-}42)}(1);_{(3,5,7,11,15,17,19,23,27,29,31,41)}(1H) \qquad (48)$$

without it.

REFERENCES

[1] S.B. Elk, J.Chem.Inf.Comput.Sci., 37 (1997) 162.

[2] N. Lozac'h, A.L. Goodson and W.H. Powell, Angew.Chem.Int.Ed.Eng., 18 (1979) 887.

[3] S.B. Elk, J.Chem.Inf.Comput.Sci., 37 (1997) 835.
[4] S.B. Elk, J.Chem.Inf.Comput.Sci., 25 (1985) 17.
[5] S.B. Elk, J.Chem.Inf.Comput.Sci., 24 (1984) 203.
[6] S.B. Elk, THEOCHEM, 489 (1999) 189.
[7] S.B. Elk, Graph Theory Notes, 36 (1999) 14.
[8] International Union of Pure and Applied Chemistry, Nomenclature of Organic Chemistry: Section A, Pergamon Press: Oxford, U.K., 1979, 23-24
[9] A.T. Balaban and F. Harary, Tetrahedron, 24 (1968) 2505.
[10] S.B. Elk, J.Chem.Inf.Comput.Sci., 27 (1987) 70.
[11] S.B. Elk, MATCH, 8, (1980) 121.
[12] S.B. Elk, MATCH, 13, (1982) 263.
[13] S.B. Elk, MATCH, 17, (1985) 255.
[14] S.B. Elk, J.Chem.Inf.Comput.Sci., 26 (1986) 126.
[15] S.B. Elk, Polycyclic Aromatic Hydrocarbons, 1 (1990) 109.
[16] a. A.T. Balaban and F. Harary, Tetrahedron, 24 (1968) 2505.
b. A.T. Balaban, Tetrrahedron 25 (1969) 2949.
[17] D. Bonchev and A.T. Balaban, J.Chem.Inf.Comput.Sci., 21 (1981) 223.
[18] a. J. Brunvoll, B.N. Cyvin and S.J. Cyvin, J.Chem.Inf.Comput.Sci., 27 (1987) 14.
b. J. Brunvoll, B.N. Cyvin and S.J. Cyvin, J.Chem.Inf.Comput.Sci., 27 (1987) 171.
[19] J.R. Dias, Tetrahedron 44 (1993) 9207.
[20] W.C. Herndon and A.J. Bruce, "Studies in Physical and Theoretical Chemistry. 51. Graph Theory and Topology in Chemistry" (1987) 491.
[21] W.R. Muller, K. Szymanski, J.V. Knop, S.Nikolic and N. Trinajstic, J.Comp.Chem. 11 (1990) 223.
[22] D.W. Matula, S.I.A.M. Rev., 10 (1968), 273.
[23] S.B. Elk, Graph Theory Notes of N.Y., XVIII (1989) 40.
[24] S.B. Elk, J.Math.Chem., 4 (1990) 55.
[25] I. Gutman, A. Ivic and S.B. Elk, J.Serb.Chem.Soc., 58 (1993) 193.
[26] S.B. Elk & I. Gutman, J.Chem.Inf.Comput.Sci., 34 (1994) 331.
[27] S.B. Elk, J.Chem.Inf.Comput.Sci., 35 (1995) 233.
[28] Ibid #8, p. 7-8
[29] S. Fujita, , J.Chem.Inf.Comput.Sci., 28 (1988) 1.
[30] F.L. Taylor, Ind.Eng.Chem., 40 (1948) 734.
[31] G.J. Leigh, G. J., editor, "Nomenclature of Inorganic Chemistry Recommendations 1990", Blackwell Scientific Publications, London, 146.
[32] Ibid, 147.
[33] H.S.M. Coxeter, Regular Polytopes, 2-nd Ed., Macmillan Co., New York, 1963, 120.
[34] A. Greenberg and J.F. Liebman, "Strained Organic Molecules", Academic Press, New York, 1978, p. 64.
[35] S.B. Elk, J.Chem.Inf. Comput.Sci., 25 (1985) 11.
[36] M.Y. Redko, M. Vlassa. J.E. Jackson, W. Misiolek, R.H. Huang and J.K. Dye, J.Am.Chem.Soc., 124 (2002) 5928.
[37] Ibid #2

[38] D.W. Walba, R.M., Richards and R.C. Haltiwanger, J.Am.Chem.Soc., 105 (1982) 3219.
[39] S.B. Elk, J.Chem.Inf.Comput.Sci., 30(1990) 69.
[40] K.B. Wiberg and F.H. Walker, J.Am.Chem.Soc., 104 (1982) 5239.
[41] Ibid #34, 344-369.
[42] www.library.ucsb.edu/classes/chem 184/184lecj.html
[43] H. Prinzbach, A. Weller, P.,Landenberger, F. Wahl, J. Woerth, L.T. Scott, M. Gelmont, D. Olevano, B. Issendorff, Nature 407 (2000) 60.
[44] S.B. Elk, J.Chem.Inf.Comput.Sci., 35 (1995) 152.
[45] Ibid #1.
[46] M. Saunders, Science, 253 (1991) 330.
[47] F.A. Cotton and B.A. Frenz, Tetrahedron, 30 (1974) 1587.
[48] S. Wei, B.C. Guo, H.T. Deng, K. Kerns, J. Purnell, S.A. Buzza, A.W. Castleman, J.Am.Chem.Soc., 116 (1994) 4475.
[49] B. Olenyuk, M. Levin, J.A. Whiteford, J.A. Shield, P.J.Stang, J.Am.Chem.Soc., 121 (1999)10434.
[50] S.B. Elk, THEOCHEM, 453 (1998) 29.
[51] A.L. Goodson, J.Chem.Inf.Comput.Sci., 20 (1980) 172.
[52] Ibid #6.
[53] S.B. Elk, Graph Theory Notes 20 (1991) 42.
[54] E.J. Corey. G.A. Peterson, J.Am.Chem.Soc., 94 (1972) 460
[55] Ibid #25.
[56] Ibid #30.
[57] Ibid #51.
[58] A.T. Balaban and F. Harary, Tetrahedron, 24 (1968) 2505.
[59] Ibid #11, p. 126.
[60] Ibid, pp. 133-142.
[61] Ibid #35, p. 107.

Chapter 4

Oxidation numbers

CHAPTER ABSTRACT:
This chapter was originally included for the sole reason that much of inorganic chemistry nomenclature evolved using the concept of oxidation number. Because of this historical usage, IUPAC chose this idea to be the cornerstone of it standardization of the nomenclature of that particular subdivision of chemistry. This is notwithstanding that the premises which underlie this idea have long outlived their usefulness. Upon reflecting on this concept and what it inflicts on chemistry and its nomenclature, one discovers problems that would not have arisen had a better understanding of the foundations been available at that time. Unfortunately, once a nomenclature has been accepted, the pragmatic problems with making needed corrections often seem so drastic to the users of the present system that many prefer to live with an old inefficient system, rather than to make the necessary changes to incorporate newer, more accurate ideas.

As indicated in Chapter 1, much of inorganic chemistry nomenclature is predicated on a concept that is admittedly empirical. It should be noted that one of the leading inorganic chemistry textbooks [1] advises:

> "Classically, formal oxidation numbers ... though useful in balancing redox equations ... have no physical significance."

Moreover, one can go a step further and assert that oxidation numbers also have little to no value in balancing equations. To the contrary, the usage of this antiquated concept in high school and college chemistry textbooks as "the" method to use when balancing equations is capricious and, for complicated reactants and products, grossly inferior to solving a simple matrix problem [2]. Nevertheless, IUPAC [3] advises:

> "The concept of oxidation number is interwoven into the fabric of inorganic chemistry in many ways, including nomenclature."

Although this 1970's statement might be viewed as being superceded (or at least amended) by the 1990's Recommendations [4]:

> "The oxidation number of a central atom in a coordination entity is defined as the charge it would bear if all of the ligands were removed along with the electron pairs that were shared with the central atom. It is represented by a roman numeral."*

its influence in the domain of "inorganic chemistry" is pervasive. Consequently, it is desirable to examine the history of that process called "oxidation-reduction".

First of all, note that the term "oxidation" is based on a historical premise that is not relevant from a more modern perspective; namely, the combining of another element with oxygen to form a simple binary compounds; i.e., an "oxide"; similarly, the removal of oxygen atoms from an oxide molecule leaving the "reduced" element was the concept intended for the term "reduction". Although this idea works fairly well for many of the more simple interactions of oxygen with both metal and non-metal elements, a better, more comprehensive, definition that includes similar reactions with other elements, such as fluorine and chlorine, evolved that was based on the transfer of electrons from one atom (or ion) to another.

Now, however, as very many additional, unanticipated, combinations of various elements with oxygen were discovered, the dissatisfaction with the entire concept became more egregious. For example, as well as more familiar alkali oxides Li_2O, Na_2O_2, and KO_2 (which are referred to as oxide, peroxide and superoxide) that are taught in freshman chemistry courses [5] and for which a simple extension from other known compounds is evident, the existence of molecules having formulas Li_3O, Li_4O, Na_3O, Na_4O, etc. are also known [6]. Moreover, this complexity becomes compounded when examining the heavier alkali oxides of rubidium and cesium [7-8], especially the geometry of these compounds. Rubidium, for example, forms the single oxide Rb_6O by enclosing one oxygen atom in an octahedral cage, and the double oxide Rb_9O_2 by fusing two such cages so that they share a common face. Cesium, being of larger size, similarly forms oxygen cages; however, with fewer alkali atoms in the cage; namely, Cs_4O and Cs_7O_2. The proposed systemic name for the simplest of these compounds, Cs_4O, is either:

---

* This is notwithstanding the fact that Roman numerals are defined ONLY for positive integers. Not only is there no Roman numeral for zero, or for any negative number, there are, also, no non-integer Roman numerals. In other words, the basis of much that follows introduces internal contradictions.

$(Cs1)_3Cs\alpha O\alpha{:}^{(1\text{-}7)}1;^{(3\text{-}9,5\text{-}9)}\alpha$ (1)

or else

$(Cs1)_3Cs0O0{:}^{(1\text{-}7)}1;^{(3\text{-}9,5\text{-}9)}0$ (2)

depending on whether the presence of a higher than zero-bond order bond between the cesium and oxygen atoms is measurable or not -- an idea analogous to the lack of a bond in [1,1,1]-propellane, which was described in Chapter 3, as well as to endothelial compounds which will be nomenclated in Chapter 7.

At this point, it is important to reiterate that, despite the advances in knowledge of the relevant chemistry, large parts of the terminology have never been revised, or, when they have, the old has often been retained with the new. For example, Cahn and Dermer [9] organized the traditional ways in which names that can be assigned when given elements combine in different proportions into four categories: An example of these four names for $Fe_2(SO_4)_3$ is:

(a) the oldest nomenclature method uses -ic for the higher oxidation number and –ous for the lower, as well as the Latin, rather than the English, name: ferric sulfate
(b) **Stock system**: An oxidation number (Roman numerals in parentheses) is placed immediately after the name of the element: iron(III)sulfate
(c) **Ewens-Bassett system**: The charge on an ion is indicated by an Arabic numeral, followed by the sign of the charge: iron(3+)sulfate.
(d) **Stoichiometric:** Numeric prefixes show composition: diiron trisulfate

Moreover, the original concept of an integer number that quantified oxidation has remained the cornerstone of "inorganic chemistry" nomenclature – at least as it is practiced by IUPAC. [10] This is notwithstanding the fact that conceptual problems are attenuated when the familiar oxidation numbers that "work" for one compound (or ion) are used to determine the oxidation number of other compounds.

At this point, several examples of how the concept of oxidation number has impacted chemical nomenclature are presented.

(1) As described in Chapter 2, the thiosulfate ion contains two sulfur atoms which are constitutionally different. Consequently, one must choose between assigning +2 to an "average" sulfur atom (as is traditionally done when balancing equations involving thiosulfate), or else to give two different oxidation numbers to the same element in a

polyatomic compound; namely, let the central sulfur atom be assigned an oxidation number of +6, while the ligand sulfur is -2.

(2) The assignment of oxidation numbers to ions comprised of multiple atoms of a single type, such as the azide ion N_3^- or the tri-halide ions, such as I_3^-, as well as the superoxide ion O_2^-, might logically be interpreted to imply that the common atom has a fractional oxidation number (-1/3, -1/3 and -1/2 respectively). Such an assignment is tenuous at best and useless at worst. In practice, however, following upon the consensus that has arisen for many familiar atoms in which the oxidation number computed for each of the individual atoms is, in fact, an integer, I.U.P.A.C. [11] defines *oxidation number* as:

> "the charge which would be present on an atom of the element if the electrons in each bond were assigned to the more electronegative atom."

With such a definition in mind, one envisions that an electron will be transferred as a unit and thus reaches the conclusion that the resultant charge must be an integer. Alternately, Shriver, Atkins and Langford [12] seem to have no problem with *fractional oxidation numbers* and define the term as:

> "the effective ionic charge obtained by exaggerating electron drift".

Here the use of the word "effective" emphasizes the empirical nature of this concept. Next, one observes that, in selected cases, a "viable" description of the ion may be formulated by applying traditional oxidation values to the individual atoms of an ion. For example, one *could* describe an azide ion as having two "end" nitrogen atoms with a charge of -3 and a center nitrogen atom with the permissible oxidation value of +5. This, however, is the exception, rather than the rule. No similar justifiable assignments of oxidation number to the individual atoms in either the tri-iodide or the superoxide ions is evident: For the tri-iodide ion, **perhaps** by selecting an arcane combination* of +7 and -1 could a charge of -1 for the total ion be achieved. However, even by such a meaningless "playing with numbers", no similar "justification" of the observed value for the superoxide ion — which one anticipates must be some additive combination using only even integer values — has been achieved. Moreover, such a problem is exacerbated when trying to assign oxidation numbers to the individual

---

* For example: -1 = (7 + | -1 | + 7) mod 8, where the modulo was chosen to satisfy the octet rule

atoms in the many polyiodide ions that exist, such as I_5^-, I_7^-,I_8^{2-},I_9^-, etc. [13].

(3) Further inconsistency endemic to this approach is seen when examining one of the familiar oxides of iron, Fe_3O_4. For this compound, the "apparent" choices are:

(a) assign the non-integer oxidation number of +2 2/3 to the Fe in order to balance the generally accepted value of -2 for the O;

(b) assign the unusual value of -3 to the O and then +4 to the Fe thereby yielding integer values for both oxidation numbers;

(c) view this "compound" as a "mixture" of two familiar varieties of iron oxide, namely $FeO{\bullet}Fe_2O_3$.

Note that option (c) is what IUPAC seems to have in mind when they refer to this compound as "iron(II)diiron(III) oxide [14]. A more accurate picture, however, is one of a "mixed oxide with the inverse spinel structure having Fe^{II} in octahedral interstices and Fe^{III} ions half in tetrahedral and half in octahedral interstices of a cubic close-packed array of oxide ions." [15] In other words, applying a knowledge of the actual chemical composition is far more accurate than what one could deduce using the antiquated criterion of oxidation number. So much so that attempting to incorporate all of these relevant details into the nomenclature would be daunting, if not self-defeating. Instead, a more reasonable approach is to recognize that the concept of oxidation number is of historical value **only** and to use only the empirical formula (with maybe the words "mixed oxide") or, alternately, to append an paragraph description — when extreme detail is desired.

(4) As well as the two familiar oxides of carbon (CO and CO_2) with their assumed oxidation number of -2 for the oxygen and +2 and +4 respectively for the carbon, there are also two other stable oxides: C_3O_2 and $C_{12}O_9$. Each of these is incompatible with the simplistic idea upon which oxidation number is based; namely, assigning oxidation numbers of +4/3 and +3/2 respectively to the carbon atom distorts an understanding of the actual chemistry. Instead, note that C_3O_2 is a linear molecule with structure: O = C = C = C = O, which is readily named in the proposed nomenclature as:

$$O2(C2)_3O \qquad (3)$$

while C_{12} O_9 is a tetracyclic compound traditionally depicted as shown in Figure 1, for which the name:

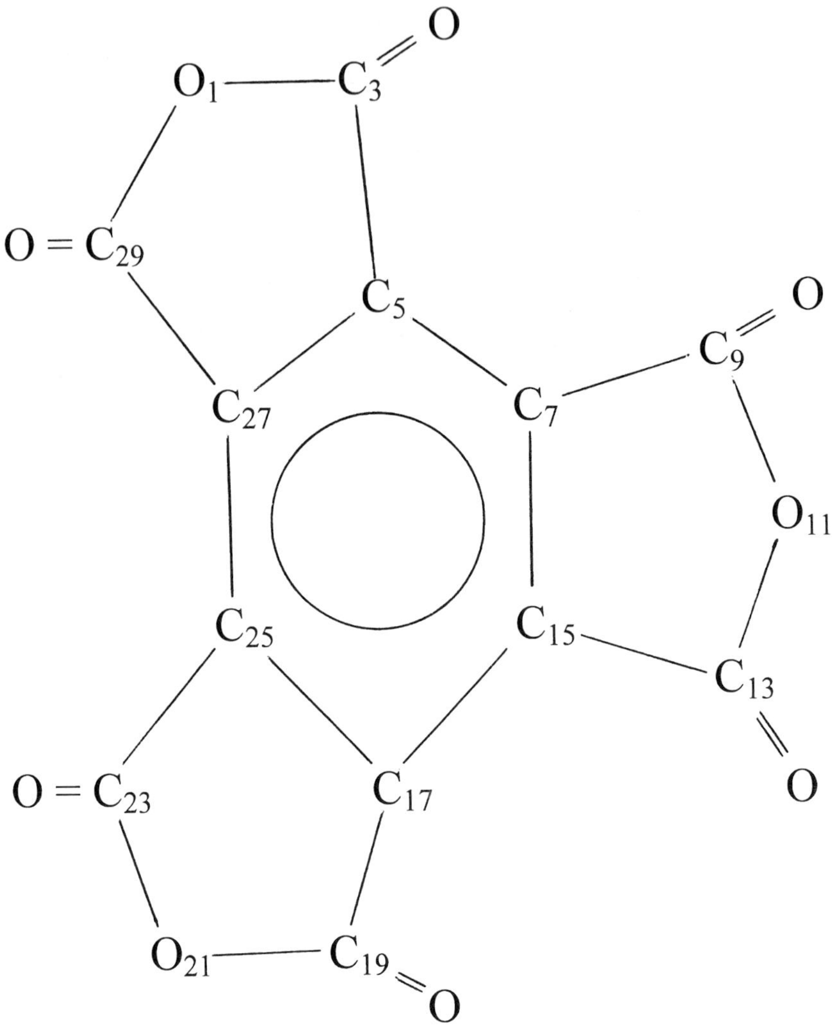

Fig. 1: Traditional ***Incorrect*** Picture of $C_{12}O_9$

$$[O1(C1)_4]_3{:}^{(5\text{-}27,7\text{-}15,17\text{-}25)}(\beta);^{(3,9,13,19,23,29)}(2O) \qquad (4)$$

would be appropriate.

Closer examination of the bonding pattern in this latter compound, however, reveals that, because of the doubly bonded oxygen atoms, the conjugation extends beyond the central (hexagonal) ring into the

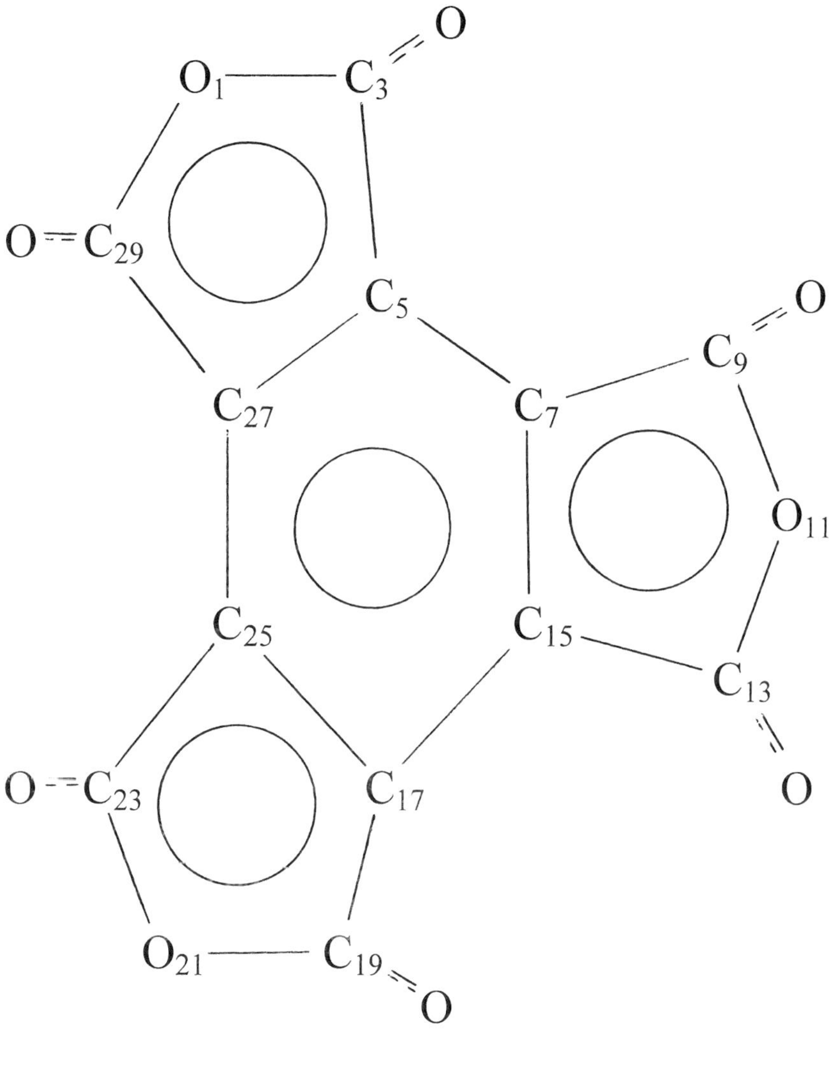

Fig. 2: Correct Picture of $C_{12}O_9$

three outer (pentagonal) rings. Moreover, because of the unpaired free electrons in the additional oxygen atoms in each of these outer rings the conjugation extends over the entire molecule; consequently, the use of Robinson ring symbols for all four rings is more appropriate than the indicated single bonds in the pentagonal rings (Figure 2). Now, however, instead of the Robinson symbol indicating a bond and a half on each side of the carbon atoms of the pentagonal rings, which combined with the indicated double bonded oxygen atoms would imply five bonds at these carbon atoms, one envisions the in-ring oxygen atoms as ℵ bonded to two carbons and the oxygen atoms of the carbonyl groups containing somewhat less than fully double-bonded character, which are better designated as β-bonded. Consequently, the name assigned to this compound becomes:

$$[O\aleph(C\beta)_3C\aleph]_3{:}^{(5\text{-}27,7\text{-}15,17\text{-}25)}(\beta);^{(3,9,13,19,23,29)}(\beta O) \qquad (5)$$

Furthermore, observe that not only are there these four stable oxides of carbon, there also exist various unstable carbon oxides [16], such as CO_3, C_2O, C_2O_3, C_3O, C_4O and C_6O. In particular, note that the geometry of the mono-oxides, such as C_4O (:C=C=C=C=O::) are nearly linear triplets [17] which would be named as:

$$O(2C)_n. \qquad (6)$$

(5) Not only is there interest in the oxides of carbon, also carbon has the capacity to form binary combinations with many other elements. In the historical development of chemistry, carbon, being in old column IV (new column 14) of the periodic table, was usually assigned the oxidation numbers of +4 and -4. This worked "satisfactorily" when combined with various familiar "more electronegative" elements such as chlorine in carbon tetrachloride CCl_4 and oxygen in carbon dioxide (C = +4) and "more electropositive" elements such as in silicon carbide SiC and in aluminum carbide Al_4C_3 (C = -4); however, its deficiency when applied to carbon monoxide (C here has an oxidation number of +2) was overlooked. Nowadays, a very large set of "unusual" oxidation numbers for binary carbon compounds are known [18]. In this set, only one sub-subset (the "methides") of the subset of "saline carbides" has the anticipated oxidation number of -4. Another sub-subset called the "acetylides" (or "dicarbides") contains the C_2^{2-} ion -- for example CaC_2. The systemic name for this set of ions is:

$Ca^{2+} (C3C)^{2-}$ (7)

The anion is formed from two carbon atoms being triply-bonded together, along with two electrons. In the simplistic manner that characterizes oxidation number, this implies each carbon atom has a coordination number of -1. However, even in this highly limited sub-subset, there is lack of uniformity. The carbon-carbon bond in the lanthanide carbides is substantially longer than a triple bond and would more realistically be represented by either a single or an aleph bond. Consequently, a useful nomenclature **must** reflect the observed bonding, rather than some "logical" extrapolation based on history, instead of observation. Additionally, yet another sub-subset is of interest: Graph intercalation compounds with formulas KC_8, KC_{16}, etc. are known. [19] Once again, extrapolation to a small fractional oxidation number for carbon or to large integer oxidation numbers for potassium is worse than useless.

(6) In the same manner as the existence of other, less familiar oxides of carbon (see 4 above) abound so similarly other oxides of sulfur (see 1 above) are worth noting [20]. As well as one of the common allotropes of sulfur, whose form is a puckered eight member ring —

$(S1)_8$ (8)

in the proposed nomenclature, one of the sulfur atoms in the ring may contain a doubly-bonded oxygen atom; namely:

$(S1)_8{:}^{(1)}2O$ (9)

Additionally, as well as the two familiar hydrates of these sulfur oxides, sulfuric and sulfurous acids, many other sulfur acids are known. Extreme examples of these include the sulfanesulfonic acids HS_nSO_3H and sulfanedisulfonic acids $HO_3SS_nSO_3H$ [21].* From these the oxidation number for an "average" sulfur atom can assume values from +5 down to as near zero as desired.

(7) Similar types of problems, with new twists, exist for many of the boron fluorides: First of all, boron trifluoride (BF_3), which is amenable to a simple oxidation number description, is reduced to

* Although the acids are mostly unstable, stable salts of some of these acids are known.

boron monofluoride when heated with crystalline boron at 2000°C. Boron monofluoride, like its isostere carbon monoxide, is meta-stable at room temperature. Although, superficially, it appears that this compound should be named B1F, the bond between the boron and fluorine is not a traditional single bond; consequently, a more appropriate name would be: BℵF; thereby indicating the nebulousness of the bond order. Furthermore, observe that this molecule, upon cooling reacts immediately with a molecule of BF_3 to add to this bond forming first $F_2B\text{-}BF_2$. Consequently, the systemic name for this molecule is:

$$F1(B1)_2F:^{(3,5)}(1F) \qquad (10)$$

Similarly, upon inserting another BF module between the two boron atoms yielding $F_2B\text{-}BF\text{-}BF_2$ is reflected in the proposed nomenclature as:

$$F1(B1)_3F:^{(3,5,7)}(1F) \qquad (11)$$

Note that, in contradistinction to the bonds in boron monofluoride, all of the bonds in (10) and (11) are single bonds.

(8) Returning the focus back to the discussion at the end of item (3) above, there exist many other known aggregations of atoms, which have a simple empirical form, but for which that form does not adequately describe the relevant geometry. Probably the most flagrant of these aggregations are those associated with the boron atom. Wells [22] advises: "... it is not possible to account for the formulae* of the borides in terms of ordinary conceptions of valency, ... as may be seen from the following examples:

CaB_6	AlB_2	B_4C	Cr_3B_4	Fe_2B	UB_4
SrB_6				FeB	UB_{12}
BaB_6 "					

Meanwhile note a specific example of this is found in magnesium boride, which following in the column of the periodic table one should expect to have an empirical formula of MgB_6. To the contrary, the most common empirical formula of magnesium boride is MgB_2 while the geometry is one of alternating layers of magnesium and boron

* See footnote on next page for the semanitcal difference in the two different plural forms of the word "formula".

atoms, arranged in six atom ring cycles [23]. This is an example of a grouping of atoms in what is commonly called an "intercalated" compound. Such compounds have a regular arrangement of distinct modules, with each module constrained to a single plane and these planes alternately stacked one on top of the other. Moreover, idiosyncrasies associated with the boron hydrides form another major area of study. The nomenclature consequences of this will be described in the next chapter.

(9) Another example of an intercalated compound was recently formulated by firing molten lithium nitride into an iron block at between 850 to 1050°C for 12 hours. For this compound, the use of non-integer values for the subscripts is appropriate. This process produces a rechargeable lithium ion battery with 10% of the lithium atoms being replaced by iron atoms. Such an aggregation of atoms should now be represented by the formula $Li_{2.7}Fe_{0.3}N$. A noteworthy comment about this practice is that this is in contradistinction to the historical use of **only** integers to express multiplicity in individual chemical formulas.* Such a distinction is important mathematically inasmuch as a decimal form indicates a range of values; namely, 2.7 **should** be interpreted with an error of $\pm$ 0.05; i.e., between 2.65 and 2.75. Additionally, one can NOT multiply by 10 to clear the decimal; i.e., $Li_{27}Fe_3N_{10}$ has a far greater precision as to the ratio of atoms, as well as describing a very different chemistry. Just as 1 mole of Li_3N contains Avagodro's number, N, of molecules or 4N atoms, similarly 1 mole of $Li_{27}Fe_3N_{10}$ weighs ten times as much.

As a further general comment it should be noted that, many of the researchers in this area, such as Shriver, accept the importance of oxidation number despite all of its inconsistencies; however, even Shriver admits [24]:

> "Oxidation numbers are of less importance in the organometallic chemistry of the d-block metals".

Nevertheless they further rationalize their unwavering support for this concept by listing some of its pragmatic utility, such as:

> "...help to synthesize reactions such as oxidative addition..."

and to: "...bring out analogies between the chemical properties of organometallic complexes and Werner complexes."

---

* A semantical difference between the two accepted plurals of the term "formula" is herein noted. When the focus is on a group the plural ends with "-ae", whereas when more than one of the individual members is being described, the desired plural is "-as".

We, on the other hand, rather than following the Ptolemaic principle of adding, first, eccentrics, then epicycles, then etc. to force a pre-conceived, but inadequate, system to describe some phenomenon*, believe that it is prudent to view any usage of oxidation numbers as grossly distorting the chemistry and to view the premise that underlies IUPAC inorganic chemistry nomenclature as fatally flawed. To the contrary, in order for any proposed change in nomenclature to be of worthy of the inconveniences that will be caused by its introduction, it must be adapted to the constitution of the individual molecule, ion, or polymer being named.

Consistent with the chemistry described in item (9) above, attention is directed to the formulae given for various of the cuprate superconductors [25]. Here, again, as is implied by the use of the variable x, the accuracy indicated in item (9) is one of a range of values. In particular, three of the most common classes of these moieties are expressed in decimal form as:

$$La_{2-x}M_xCuO_4 \quad (11)$$

$$YBa_2Cu_3O_{6+x} \quad (12)$$

and

$$Nd_{2-x}Ce_xCuO_4 \quad (13)$$

Adrian and Cowan [26] advise:

"the lanthanum cuprates are superconductive when x =0.06 to 0.3"

and that at the maximum temperature (36K) at which superconduction occurs, the appropriate formula is:

$$La_{1.85}Sr_{0.15}CuO_4 \quad (14)$$

REFERENCES:

[1] F.A. Cotton, G. Wilkinson, C.A. Murillo and M. Bochmann, "Advanced Inorganic Chemistry", 6-th Ed., Wiley, New York, 1999, p.310.

---

* See footnote on page 123 (Chapter 3).

[2] V.L. Fabishk, Chemistry, 40 (1967) 18.

[3] International Union of Pure and Applied Chemistry, Nomenclature of Inorganic Chemistry, 2-nd Ed., DefinitiveRules, 1970, Butterworth, London, U.K., p.5.

[4] International Union of Pure and Applied Chemistry, Nomenclature of Inorganic Chemistry Recommendations 1990, Blackwell Scientific Publications, Oxford, U.K.,p.148.

[5] R. Chang, Chemistry, 5-th Ed., McGraw-Hill, N.Y., 1994, p.316.

[6] E-U. Wurthwein, P.v.R. Schleyer, J.A. Pople, J.Am.Chem.Soc., 106 (1984) 6973.

[7] D.F. Shriver, P. Atkins and C.H. Langford, Inorganic Chemistry, 2-nd Ed., W. H. Freeman, New York, 1994, p.323.

[8] A. Simon, Angew.Chem.,Int.Edn.Engl. 27 (1988)159.

[9] R.S. Cahn and O.C. Dermer, Introduction to Chemical Nomenclature, 5-th Ed., Butterworths, London, U.K., 1979, 14-16

[10] Ibid #3.

[11] Ibid

[12] Ibid #7, p.54.

[13] R. Steudel, Chemistry of the Non-Metals, deGruyter Textbook (Engl. Ed.,) by F.C. Nachol and J.J. Zuckerman), New York, 1977, p.167.

[14] G.J. Leight, (Ed.): Nomenclature of Inorganic Chemistry Recommendations 1990, Blackwell Scientific Publications, London, p.66.

[15] Ibid #1, p.777.

[16] Ibid, p.224.

[17] R.J. Van Zee, G.R. Smith, and W. Weltner, Jr., J.Am.Chem.Soc., 110 (1988) 609.

[18] Ibid #7, p.489-492.

[19] Ibid #17.

[20] Ibid, p.210-229.

[21] Ibid. p. 227.

[22] A.F. Wells, Structural Inorganic Chemistry, 3-rd Ed., 1962, Oxford University Press, 820.

[23] J. Akimitsu, Nature, 410 (2001) 63.

[24] Ibid #7, 665.

[25] F.J. Adrian and D.O. Cowan, Chem& Eng.News, 12/21/92, p.24.

[26] Ibid.

Chapter 5

The boranes and related aluminum compounds

CHAPTER ABSTRACT:
A close scrutiny of the seminal report on the boranes reveals that the equivalence traditionally presented between the solid geometrical and the topological pictures of these compounds is incomplete. Because of this mathematical disparity, an alternate geometrical description of the bonding in the boron hydrides is introduced, as well as a new interpretation for the boron bridge. This results in the need for a concomitant development of nomenclature to more accurately describe this type of bonding. Meanwhile, awaiting reaction of the chemistry community to such structures, canonical names in both the IUPAC and the proposed system are supplied for this set of compounds. A corresponding description and nomenclature for related aluminum compounds is presented.

In Chapter 2, the use of α bonds for one special bonding that occurs in one particular boron compound, which was called "diborane", was introduced. An examination of other boron compounds and the nomenclature associated with them is now undertaken. First of all, it should be noted that despite that boron has three electrons in its outer shell and that a trigonal planar bonding pattern is common in molecules such as BF_3, etc., the simplest hydride of boron that is normally encountered is **not** BH_3. To the contrary, under normal conditions of temperature and pressure, not only is the smallest boron hydride a dimer, but also two distinct diboron hydride molecules are encountered.

As well as the analog to the alkanes in which one of the hydrogen atoms is removed from two methane molecules and the two methyl groups created combine to form ethane, two BH_2 modules may be viewed as fused together with a covalent bond between the boron atoms; namely B_2H_4, traditionally referred to as diboron(4). The systemic name* for this compound is:

$H1(\underline{\underline{B}}1)_2H$ (1)

Additionally, there exists the above mentioned second diboron hydride, which does not subscribe to the "traditional" picture of two atoms sharing a pair of electrons to form a covalent bond. In other words, as well as the familiar two-center two-electron bond (abbreviated as [2c-2e]), which is the main type of covalent bond, the Council of the American Chemical Society [1] also introduced two new types of "electron-deficient" three center - two electron (abbreviated [3c-2e]) bonds:

(1) The first, referred to as a "hydrogen bridge" (written as B/H\B) is equivalent to the bond depicted in Figure 1 of Chapter 2. In the nomenclature system being developed, the hydrogen bridge of interest for the boranes is viewed as a single hydrogen atom alpha bonded to two boron atoms. It may be represented in a linear fashion as:

$B\,\alpha\,H\,\alpha\,B$ (2)

This representation is compatible with a total coordination of 1 (1/2 for each α) for each hydrogen atom. Furthermore, for diborane(6), which had been nomenclated in Chapter 2 as: $(\underline{\underline{B}}\alpha H\alpha)_2$, attention is directed to the observation that each boron atom is singly-bonded to two hydrogen atoms and alpha-bonded to two other hydrogen atoms; thereby yielding the traditional valence of 3 for each boron atom.

(2) The second type, referred to as a "boron bridge", (written as 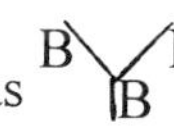

) contains three boron atoms, each connected to the other two, but with a more complex coordination than would exist in a simple cycle. Now, instead of the usual description of this bridge, as given in [1], one may regard this bond as the two dimensional analog of the hydrogen bridge: Instead of a three center bond involving three boron atoms, consider a central boron atom with

---

* In conformity with the convention that terminal hydrogen atoms are significant and that the longest chain in this molecule is four atoms long, the simpler name $\underline{\underline{B}}1\underline{\underline{B}}$ is eschewed. In a manner similar to the methyl group not being represented by a triple underscore, rather being nomenclated as $1\underline{\underline{C}}1H$, no terminal double underscored boron atoms will be encountered in the proposed nomenclature. Note, however, that there is not a similar limitation for cyclic boron compounds.

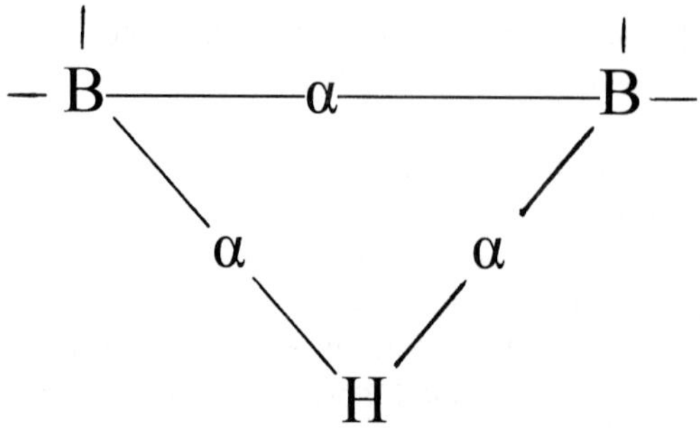

Fig. 1a: Module used in forming a boron bridge

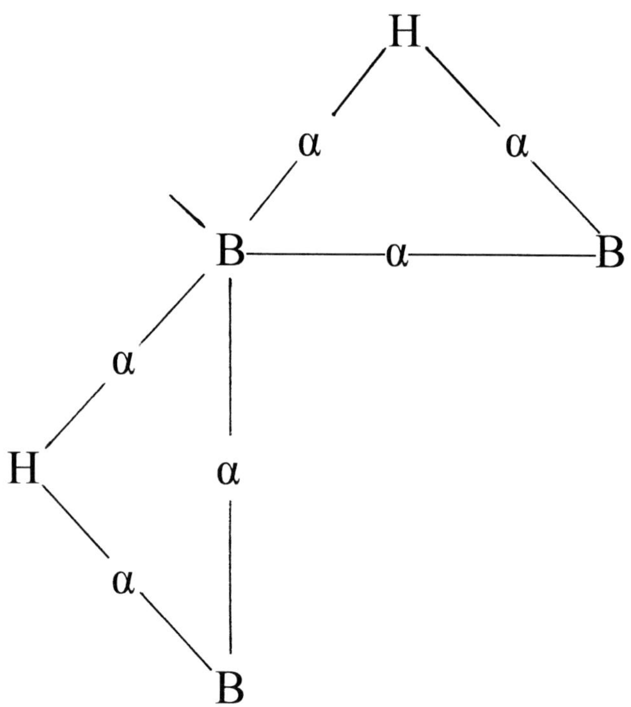

Fig. 1b: Combination of 2 modules to form a boron bridge

four alpha bonds (to two boron and two hydrogen atoms) and either a single covalent bond to a fifth atom (Figure 1) or else two more alpha bonds to a fifth and sixth atom [2].* By such a perspective there is compatibility of hydrogen-bridged boron atoms (in which there is not an additional alpha bond directly between two boron atoms) and boron-bridged boron atoms (in which there is); thereby producing either two or three 3-member rings at each "central" boron atom. Consequently, in the nomenclature being developed, each boron bridge may be viewed as part of the largest cycle using the doubly-alpha bonded hydrogen atoms augmented with an alpha-bridge between the two boron atoms. This will apply to the various boranes, as well as in the common nomenclature that will ensue for aluminum (same column in the periodic table as boron) compounds that have (3c-2e) bonds.

At this point it is important to observe that the seminal report [1] contained an appendix depicting a graph theoretical structure on the left and the corresponding solid geometry structure on the right for 17 of the simpler borane molecules and/or ions and seven molecules containing other types of ligand, such as carbon. However, after careful analysis of [1], the lack of complete equivalence between the "geometrical" and "topological" pictures included for some of the boranes was noted. For example, there is the possibility of visualizing a very different geometric picture than the one presented. Not only will this necessitate the formulation of two distinct nomenclatures — one for the traditional (ACS sponsored) picture and a different one for the potential structures that are being proposed; also, there is a **not inconsequential** difference in the model being nomenclated using the two ACS-sponsored (geometrical vs. topological) pictures. In particular, in order to be able to nomenclate the topological picture, the specification of a special symbol for the boron bridge, that is in addition to the symbols used so far, will be required. **However**, because this interpretation of non-isomorphicity between the two parts of the ACS-sponsored picture has never been acknowledged, the question as to whether there even exists the capacity to formulate such a symbol for the ACS topology, let alone (assuming the existence) whether such a formulation is desired, has never been broached by the ACS in their nomenclature. Instead, only common names were presented in [1].

Although the formulation of a systematic nomenclature in line with existent IUPAC nomenclature was included in the 1990 Recommendations [3], serious problems still abound. This is recognized by their statement: [4]

* Prior to the publication of [2], preliminary reports on this material were presented at Graph Theory Day 39, New York Academy of Sciences, May 13, 2000, and at Second Indo-U.S. Workshop on Mathematical Chemistry, Duluth, Minnesota, May 30, 2000.

> "The boron compounds ... include structures which can not be readily dealt with by any of the classical concepts and procedures of organic or inorganic chemistry founded on assumptions concerning the localization of bonding electrons."

Nevertheless, it is precisely because the same procedures as the 1968 report [1] are still employed that this chapter (in the 1990 book of recommendations) is merely a "band-aid". Although the chapter starts off by acknowledging the features contributing to nomenclature complexity [5]:

(a) Connectivity
(b) Triangular association of boron atoms
(c) Hydrogen bridges
(d) Three-center bonds involving boron atoms
(e) Linkage of polyhedral moieties
and (f) skeletal replacement

its two methods of nomenclating add little to an understanding of the inherent geometry of the molecules:

(i) the stoichiometric approach serves merely to addend the number of hydrogen atoms to that of the number of boron atoms; i.e., it is purely an empirical formula approach without any consideration of structure.
(ii) the structural-descriptor based nomenclature fails to have a geometrical basis, relying instead on what is admittedly only a semi-systematic set of characteristic structural prefixes which are formula-based (n = closo; n-2 = nido; n-4 = arachno, etc.).

Meanwhile, despite that the accuracy of the proposed picture must await acceptance or rejection by the chemistry community, this approach to nomenclature incorporates all six of the features listed above. Furthermore, it is important to remember, as stated in Chapter 1, that any nomenclature has the capacity ONLY to nomenclate a model, rather than any actual chemical moiety. Consequently, it is important to pick the best model that can be formulated.

Turning the focus to the various boranes described in [1], the next highest homologue, B_4H_{10} (which, using the conventions established in [1], is depicted as having only hydrogen bridges) is examined. Figure 2 is a reproduction of Figure 2 in [1]. ***Were this figure to be accurate***, the name of this compound in the proposed system would be:

$$(\underline{B}\alpha H\alpha\underline{\underline{B}}\alpha H\alpha)_2{:}^{(1\text{-}9)}(1) \qquad (3)$$

At this point, excluding the possibility of being extremely liberal in interpreting the geometrical half of the model portrayed in Figure 2 (which

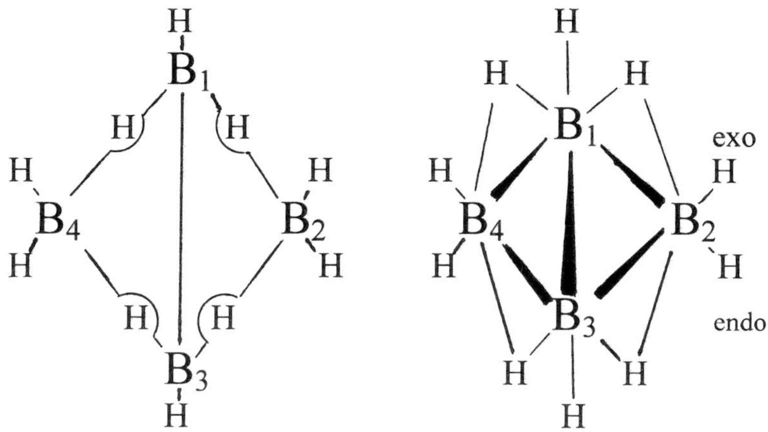

(a) Topological picture (b) Geometrical picture

Fig. 2: ACS council's representation of B_4H_{10}

is then not consistently maintained in the rest of this illustration), there is a difference in the bonding of the topological and geometrical pictures of this molecule. The geometric picture has covalent, as well as hydrogen, bonds between the four peripheral edges versus there being only a hydrogen bond in the topological picture. It is the topological picture that was nomenclated by (3). Such an interpretation is compatible with boron having a valence of 3. This is in contradistinction to the geometrical picture implying an

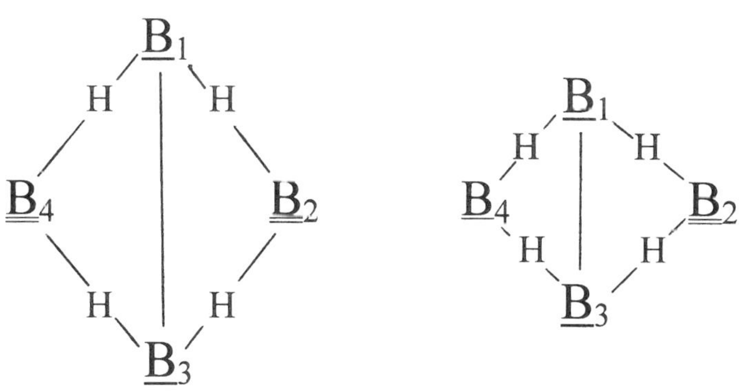

Fig. 3: Simplification of Fig. 2 using the underscore convention illustrates the equivalence of the topological and geometric pictures of B_4H_{10}

"impossible" valence of 5. The alternate interpretation of these "peripheral bonds" as phantom lines to aid in deciphering the drawing creates the problem how one is to distinguish these "phantom bonds" from the desired bond between B_1 and B_3.

Using the underline convention introduced in Chapter 1 and grouping those hydrogen atoms which function as a unit with specific boron atoms produces a clarified picture of the relevant geometry and gives reason to question whether this representation of B_4H_{10} is flawed. As illustrated in Figure 3, now the two halves of Figure 2 are seen to be identical, and the importance of the endo and exo designation on the B_2 atom (which is suppressed by the underline convention of Figure 3) is explained. Meanwhile one observes that the configuration depicted is one with a single bond between two boron atoms (B_1 and B_3 in their numbering scheme) and no bond between their B_2 and B_4.

Instead of this essentially planar focus, consider the molecule in three dimensions with its geometry being that of a regular tetrahedron with a boron-hydrogen module at each of the vertices (the hydrogen atom is directed away from the centroid) of the tetrahedron and a hydrogen atom centered on each of the six edges of the tetrahedron (Figure 4). In this configuration the longest Eulerian path follows along the edges of a skew quadrilateral.* Were this path to be augmented with two αHα bridges to

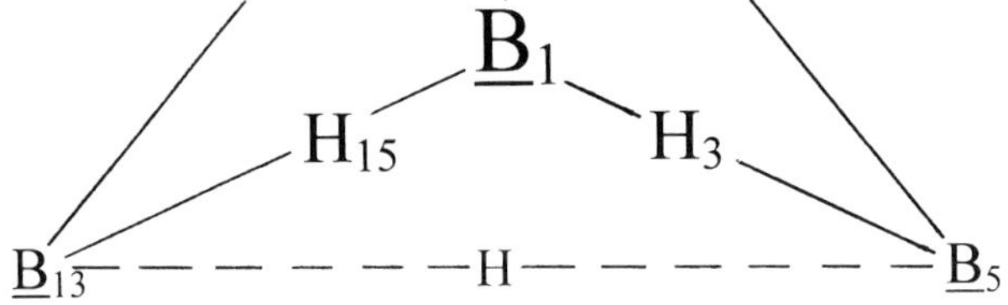

Fig. 4: Systemic representation of B_4H_{10} - Uncorrected

* Note that all three possible skew quadrilaterals (ABCD = ADCB, ABDC = ACDB, ACBD = ADBC) are equivalent and thus any one of them can be selected for the canonical nomenclature.

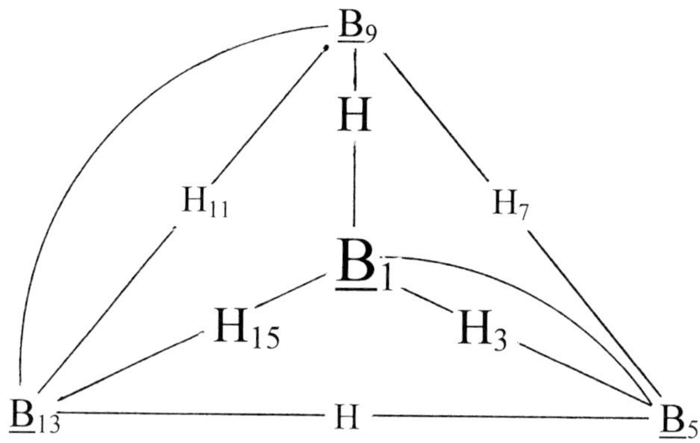

Fig. 5: Systemic representation of B_4H_{10} - with fluxional correction

complete covering the edge set of the tetrahedron, the systemic name for this highly symmetric compound would become:

$$(\underline{B}\alpha H\alpha)_4{:}^{(1\text{-}9,5\text{-}13)}(\alpha H\alpha) \qquad (4)$$

However, Figure 4 does not take into consideration questions of valence. Consequently, the name just assigned (4) contains the same shortcoming. In other words, once again, as desired, the proposed name is completely dependent on the geometric description of the molecule. In the case of this particular molecule, the problem is minor. As suggested in [2], it is readily corrected by introducing a fluxional model containing two additional non-contiguous alpha bonds (Figure 5):

$$(\underline{B}\alpha H\alpha)_4{:}^{(1\text{-}9,5\text{-}13)}(\alpha H\alpha);^{(1\text{-}5,9\text{-}13)}(\alpha) \qquad (5)$$

The systematic development of the nomenclature for the next smallest borane: B_5H_9, continues with an examination of the representation given as Figure 3 in [1] (reproduced here as Figure 6), along with the corresponding modular simplification in Figure 7). The topological picture on the left side of Figure 6 (also Figure 7) has B_2 and B_5 connected both through this new "boron bridge" with B_1, as well as with a direct hydrogen bridge. A similar scenario exists for B_3 and B_4 forming some type of bridge with B_1. What one can conclude from what appears to be a deliberate difference in the choice of symbols for connecting to B_1 in Figure 3 of [1] is, at best, highly ambiguous.

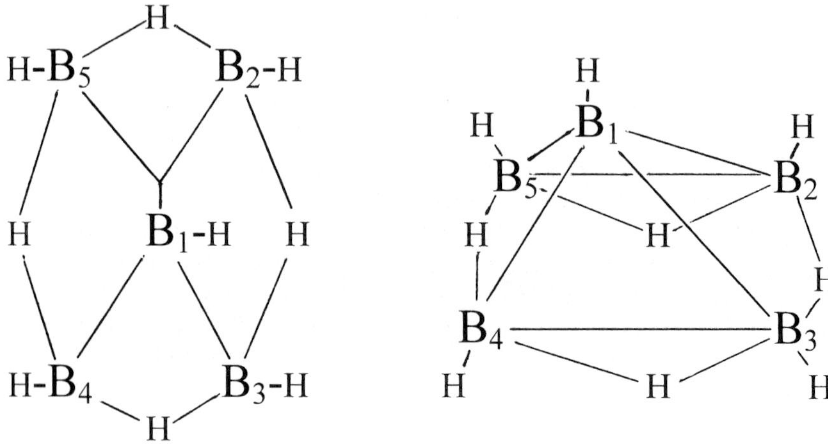

(a) Topological Picture (b) Geometrical Picture

Fig. 6: ACS Council's Representation of B_5H_9

On the other hand the traditional "geometrical" representation (the right side of Figure 6) has the five boron atoms in the configuration of a square pyramid with all five of the boron atoms having an axially directed hydrogen

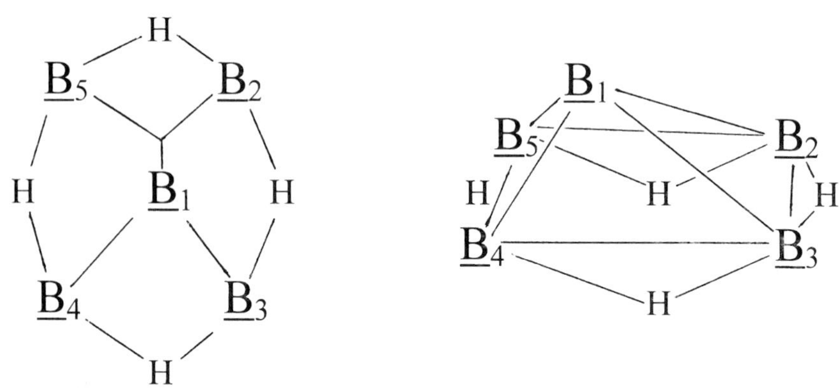

(a) Topological Picture (b) Geometrical Picture

Fig. 7: Using the underscore convention, the topological picture of B_5H_9 in Figure 6 is seen to be just the planar projection of the geometrical picture

atom and each of the four base boron atoms having a hydrogen bridge to its two base boron atom neighbors, but not to either the apex or to the diagonally opposite boron atoms. The isomorphism between the topological and the geometrical picture (IF IT EXISTS AT ALL) is now purely a function of this as to yet be described "boron bridge". This is in addition to the problem that such a representation suffers a comparable bond order "fallacy" as did the earlier picture for B_4H_{10} (Figure 4); namely: although B_1 has the requisite bond order of 3 (4 times 0.5 for the 4 alpha bonds to each of the other boron atoms plus 1.0 for a single bond to a hydrogen atom), B_2, B_3, B_4 and B_5 do NOT. Each of these base boron atoms have α bonds (with traditional bond order = 0.5) to three other boron atoms and to two hydrogen atoms, plus a single bond to a hydrogen atom; thereby indicating a total bond order of 3.5. In other words, if one is to follow tradition, the present system also fails to satisfy the Lewis structure.

At this point, however, Gillespie [6] advises:

> "bonding in a five-coordinated molecule may be described in terms of a set of sp^3d hybrid orbitals.

He further notes:

> "there is no reasonable way of deciding whether the $d_z{}^2$ (directed toward the corners of a trigonal bipyramid) or $d_x{}^2\text{-}_y{}^2$ (directed toward the corners of a square pyramid) orbital should be chosen."

This comment serves as justification for selecting some "optimal" geometry* as the basis for nomenclating several of the boranes. In particular, rather than using the square pyramid model of Figures 6 or 7, which in the proposed nomenclature would have the name:

$$\underline{B}1\underline{B}\,(\alpha H\alpha\underline{B})_3 1{:}^{(3\text{-}15)}(\alpha H\alpha);^{(1\text{-}7,1\text{-}11,3\text{-}7,3\text{-}15,7\text{-}11,11\text{-}15)}(1) \qquad (6)$$

use as the geometrical model for B_5H_9 a trigonal bipyramid, which has five vertices and nine edges. In this model, one has uncoupled the boron-hydrogen module, which is prevalent in all of [1] and has equated all of the hydrogen atoms by setting a boron atom at each vertex of such a figure and a hydrogen atom in the middle of each edge (Figure 8) thereby maximizing the heuristic concept of symmetry. This heuristic is further enhanced by pseudo-rotation so that a particular boron atom may be axial part of the time and equatorial others [7]. Now, even without a correction to accommodate

---

* Unfortunately, the word "optimal" is heuristic and is related to maximizing "simplicity" (again a heuristic concept) and symmetry.

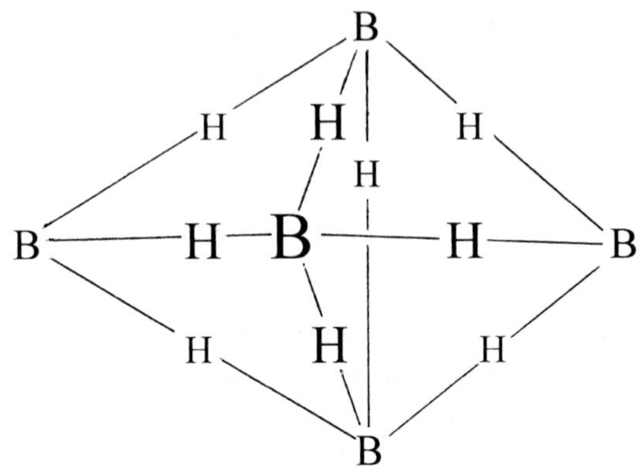

Fig. 8: Systemic representation of B_5H_9 – Uncorrected

the valence problem (exactly as described above for B_4H_{10}), the proposed name for this compound:

$$(B\alpha H\alpha)_5{:}^{(1\text{-}13,5\text{-}13,5\text{-}17,9\text{-}17)}(\alpha H\alpha) \tag{7}$$

gives a better representation of the desired five-fold symmetry than does (6) (Figure 5).

Further improvement is effected by incorporating into the name six additional alpha bonds (from the two axial boron atoms to each of the three

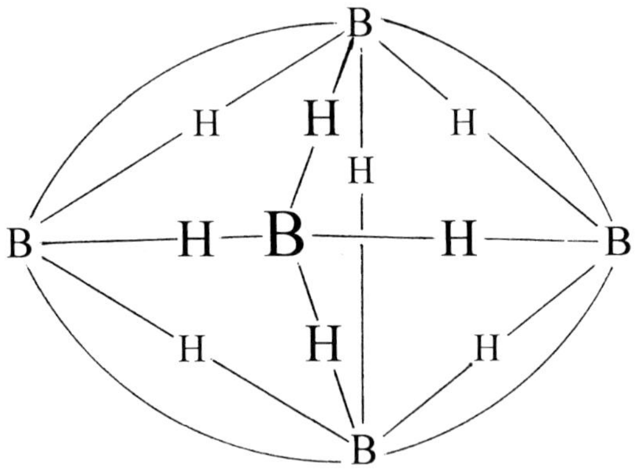

Fig. 9: Systemic representation of B_5H_9 - Corrected

equatorial ones). As well as this being precisely the above described picture of the boron bridge, this model produces the desired valence of 3 at each boron atom. Thus the picture for this molecule is as shown in Figure 9 and its name is:

$$(B\alpha H\alpha)_5{:}^{(1\text{-}13,5\text{-}13,5\text{-}17,9\text{-}17)}(\alpha H\alpha);^{(1\text{-}5,1\text{-}13,1\text{-}17,5\text{-}9,5\text{-}13,5\text{-}17)}(\alpha) \qquad (8)$$

Meanwhile, returning to the topological representation that is usually presented for B_5H_9 (part a of Figure 6), note that "traditional" systems of nomenclature do not even attempt to address such questions as the bonding. In fact, were one to want to do this, they would be confronted with the need for a new symbol (such as the central dot) to represent the boron bridge, with all of the graph theoretical complications that would follow such a shrinking of a three node subgraph to a point.

The fourth simplest borane in the ACS scheme, B_5H_{11}, has both of the same complications as did the earlier ones. The ACS geometrical picture of this molecule has one boron atom in an apical position with four additional boron atoms forming a bent linear path in a "open" pyramidal arrangement rather than as a cycle analogous to B_5H_9 (see Figure 6 and 7). Figure 4 of [1], is reproduced here as Figure 10. This open arrangement manifests itself in the topological picture having two boron bridges and three hydrogen bridges. However, the vagaries of the boron bridges in the topological picture become more pronounced using the modularized picture containing $\underline{B}$ and $\underline{\underline{B}}$ in Figure 11. One very salient feature of the topological picture is the lack of a traditional (single covalent) bond between B_3 and B_4 that **IS** present in the geometrical picture. Can one attribute this to being simply a typographical error that has gone unnoticed? In any event, the complexity of the traditional model for the boron bridge will be obviated by the proposed nomenclature.

One particular model, which MAY be appropriate for B_5H_{11}, is similar to the above picture for B_5H_9 in that it supplements the geometric arrangement of that molecule by modularizing the two axial boron atoms with hydrogen atoms. The other nine hydrogen atoms are again localized along the edges of a trigonal bipyramid. The additional two hydrogen atoms are attached by traditional single bonds to the apical boron atoms (Figure 12). For such a structure, the uncorrected name is:

$$(B\alpha H\alpha)_5{:}^{(1\text{-}9,,5\text{-}13,5\text{-}17,9\text{-}17)}(\alpha H\alpha);^{(1,13)}(1H) \qquad (9)$$

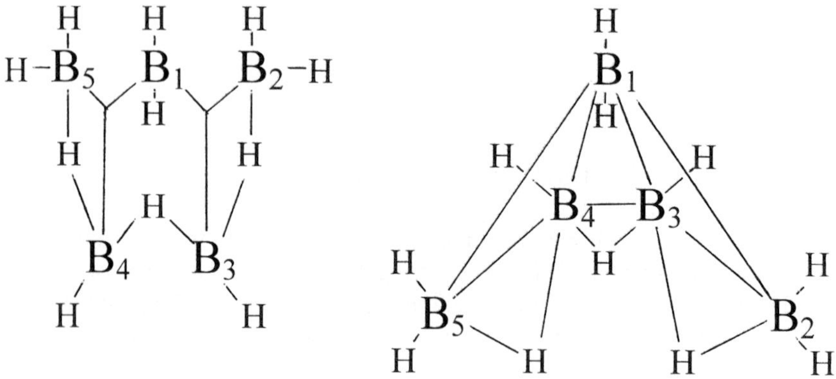

(a) Topological Picture (b) Geometrical Picture

Fig. 10: ACS Council's Representation of B_5H_{11}

while with additional alpha bonds for the boron bridges (Fig. 13), becomes:

$$(B\alpha H\alpha)_5{:}^{(1\text{-}9,,5\text{-}13,5\text{-}17,9\text{-}17)}(\alpha H\alpha);^{(1\text{-}5,1\text{-}9,5\text{-}9,13\text{-}17)}(\alpha);^{(1,13)}(1H) \qquad (10)$$

An analogous description of each of the higher boranes can be produced by selecting ***not the traditional pyramids favored by ACS***, but rather those solid geometry models which heuristically have a concept of "regularity" and, when that is not attainable, of "semi-regularity". For example, exactly as the tetra-boron molecule minimized Coulomb repulsion

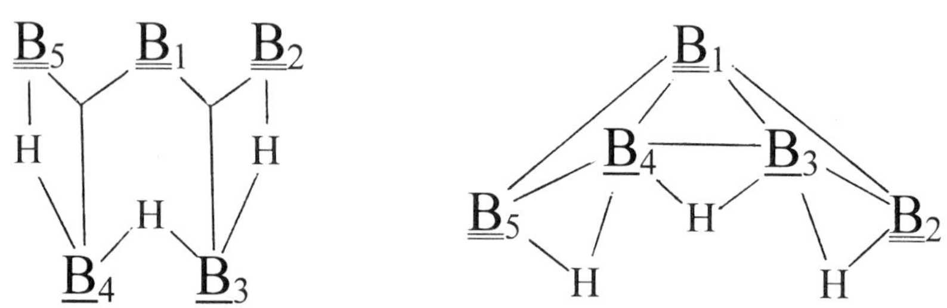

(a) Topological Picture (b) Geometrical Picture

Fig. 11: Using the underscore convention, the topological and geometrical pictures of B_5H_{11} in Figure 10 are compared

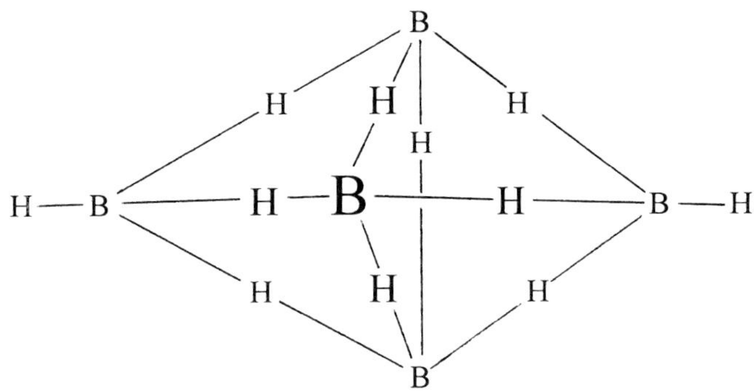

Fig. 12: Systemic representation of B_5H_{11} - Uncorrected

by positioning the boron atoms at the vertices of a regular tetrahedron, a hexa-boron atom would locate these atoms at the vertices of a regular octahedron and an octa-boron atom at the vertices of a cube. Once the boron atoms are "fixed", the hydrogen atoms are now located at "equilibration points" that optimize symmetry.

Upon examining the geometrical structure selected for B_6H_{10} (see Figure 5 in [1]), reproduced here as the right hand half of Figure 14 (as well as the modularized version shown in Figure 15), the predilection for beginning with a planar image and superficially extending it into the third dimension, rather than actually thinking in the third dimension, is evident. One important coplanar model of six of one atom and ten of the other has as its graph theory picture a wheel [8]. (See Figure 16 below.) Here the six

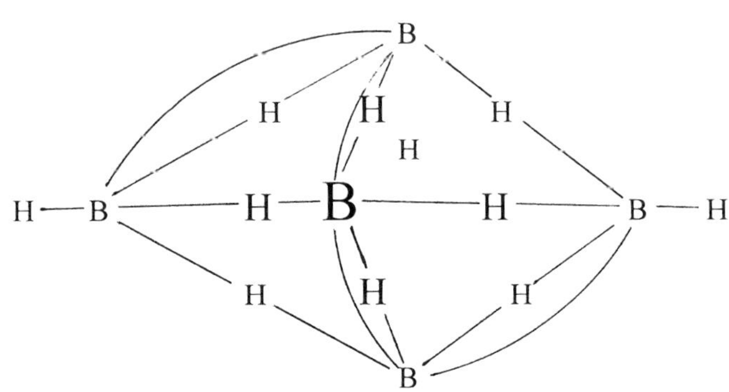

Fig. 13: Systemic representation of B_5H_{11} - Corrected

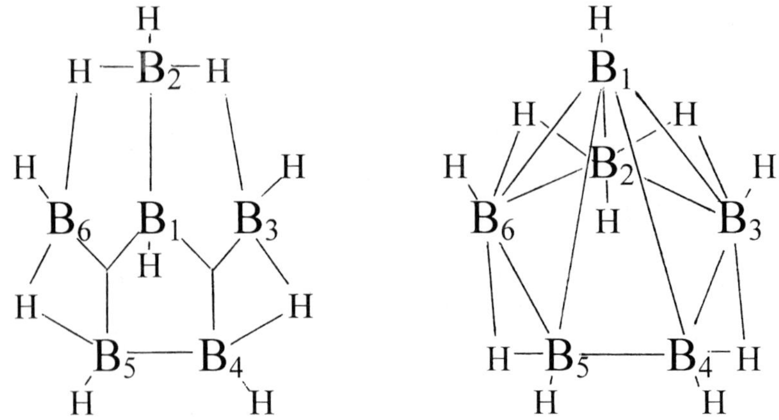

(a) Topological Picture (b) Geometrical Picture

Fig. 14: ACS Council's Representation of B_6H_{10}

nodes are the boron atoms and the ten edges are hydrogen atoms. A pseudo three-dimensional extension, which uses the identical graph, is produced by raising the center boron atom out of the plane thereby producing a pentagon-based pyramid. Such a configuration would be named in the proposed system as:

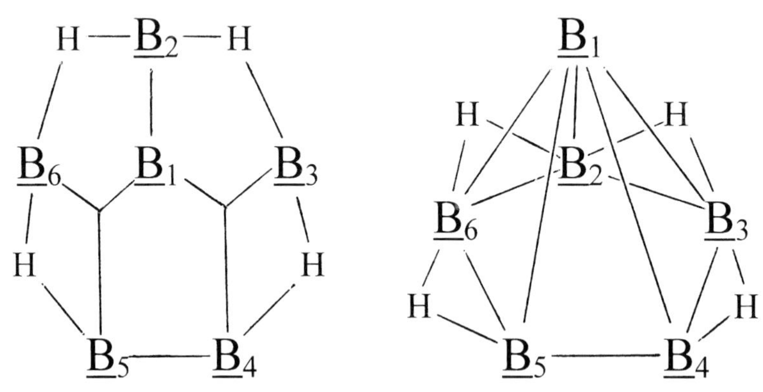

(a) Topological Picture (b) Geometrical Picture

Fig. 15: Using the underscore convention, the topological and geometrical pictures of B_6H_{10} in Figure 14 are compared

$\underline{B}1(\underline{B}\alpha H\alpha)_2\underline{B}1\underline{B}\alpha H\alpha\underline{B}\alpha H{:}^{(1\text{-}7,1\text{-}11,1\text{-}13,1\text{-}17,3\text{-}7,3\text{-}17,7\text{-}11,13\text{-}17)}(1);^{(3\text{-}19)}(\alpha)$ (11)

Note in this name six of the hydrogen atoms are coupled with boron atoms whereas the remaining four hydrogen atoms are relegated to being hydrogen bonded between adjacent pairs of boron atoms in the pentagonal base of the pyramid. However, there is no rationale for the bond between B_4 and B_5 in this figure being different from the other four pentagonal bonds. Were it not for the historical bias of every boron atom having a covalently bonded hydrogen whenever there are more hydrogen than boron atoms (i.e., acting in concert as a B-H module), a more symmetric picture would had the apical boron atom without a covalently bonded hydrogen atom and all ten of the hydrogen atoms located along the edges of this configuration. Such a symmetric picture would accommodate both a coplanar and a pyramidal structure and would be named in the proposed system as:

$B\alpha H\alpha(\underline{B}\alpha H\alpha)_5{:}^{(1\text{-}3,1\text{-}4,1\text{-}5,2\text{-}6)}(\alpha H\alpha)$ (12)

In place of the pyramidal picture recommended by ACS [1] (Figures 14 or 15) or the mathematically more symmetric, but essentially still coplanar, one

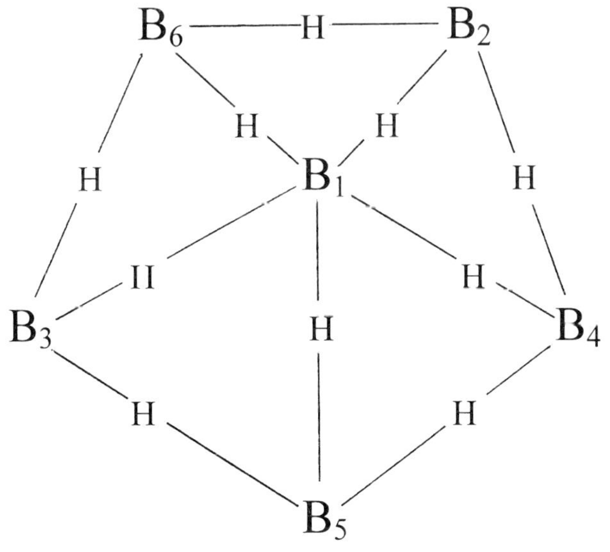

Fig. 16: Potential coplanar or pyramidal graph for B_6H_{10}

portrayed in Figure 16, attention is now focused on the three dimensional, geometrically simpler octahedron.

Using this geometric solid as the basis for assigning a canonical name, one starts by positioning the six boron atoms at the vertices of a regular octahedron. By this technique the Coulomb interaction between these atoms is minimized. Unfortunately, at this point one notes that finding "optimal locations" for the ten hydrogen atoms is neither easy nor unambiguous. There is no positioning of these ten hydrogen atoms that produces a static, stable equilibrium in three dimensional space. Instead, one has the option of continuing the IUPAC tradition of modularizing each of the boron atoms with a hydrogen atom and then trying to locate the remaining four hydrogen atoms along selected edges, or locating eight of the hydrogen atoms along the edges of the octahedron and trying to find desirable locations for the remaining two hydrogen atoms, some combination of these two, or possible even some of the other techniques that have been known to exist, such as either ionization such as PF_5 does in order to form the desired regular octahedron and cubic ions (PF_6^- and PF_4^+) or else dimerization (or polymerization) to produce a fused set that has the requisite three dimensional geometry [10].

Until experimental techniques become sufficiently refined as to be able to unambiguously assert which of these descriptions is correct, the best that can be done nomenclature-wise is to list all and let future studies determine which one prevails. In other words, the nomenclature techniques proposed herein are sufficiently flexible that whatever final structure is

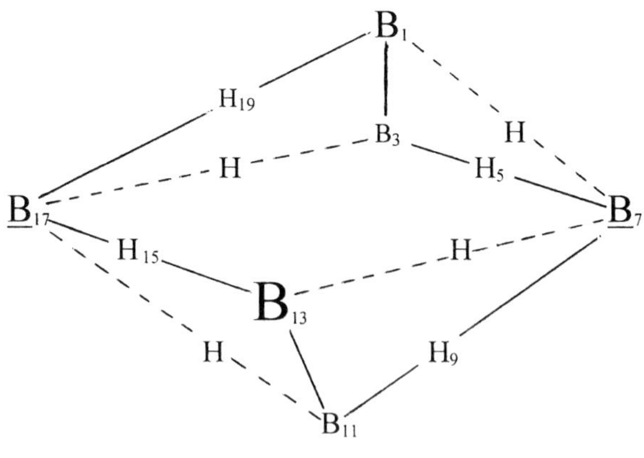

Fig. 17: Systemic representation of B_6H_{10}

decided upon, that structure can be conveniently nomenclated. The structure that this author leans toward is that of the boron atoms being at the vertices of a regular octahedron, two of the hydrogen atoms being modularized with a pair of non-adjacent boron atoms, which are located axially, the remaining eight hydrogen atoms form hydrogen bridges between each pair of a B̲ and a B atom, and the four unmodularized boron atoms are singly bonded as two pairs. This arrangement then undergoes pseudo-rotation between the three figures that can be formed by selecting a different pair of the boron atoms as being the "axial" pair. The proposed structure and canonical numbering is shown in Figure 17. This produces as the name for this molecule:

$$(\mathrm{B1B\alpha H\alpha \underline{B}\alpha H\alpha})_2{:}^{(1\text{-}7,3\text{-}13,7\text{-}13,11\text{-}17)}(\alpha\mathrm{H}\alpha) \qquad (13)$$

Whether this proposed octahedral structure for B_6H_{10} is, in actuality, the structure of this molecule awaits further laboratory research. Meanwhile note that the proposed model for the next larger borane, B_8H_{12}, again resumes geometrical simplicity with the eight boron atoms at the vertices of a cube and the twelve hydrogen atoms centered on the edges (Fig. 18). The proposed name for such a structure is thus:

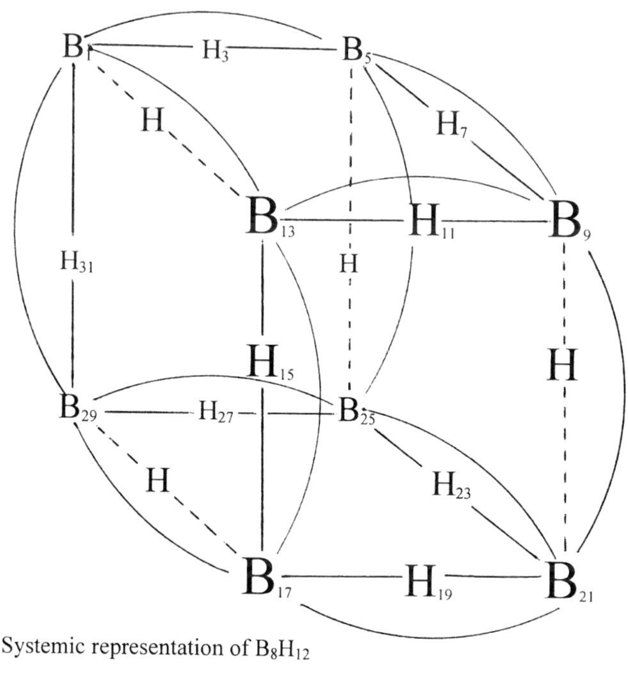

Fig. 18: Systemic representation of B_8H_{12}

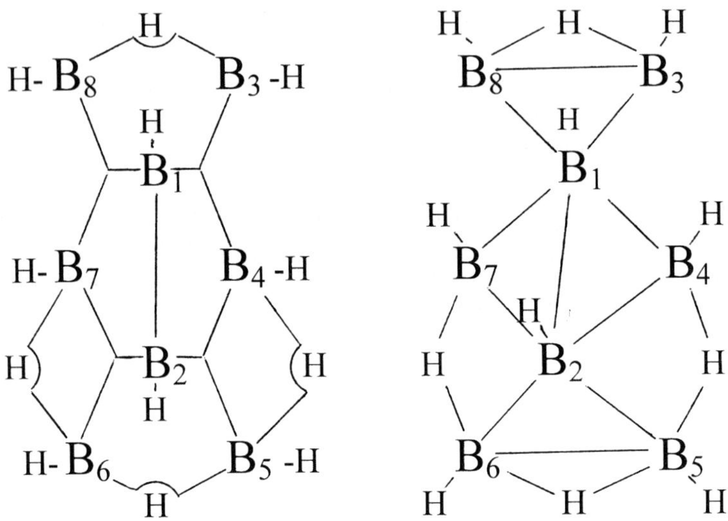

Fig. 19: ACS Council's Representation of B_8H_{12}

$$(B\alpha H\alpha)_8{:}^{(1\text{-}13,5\text{-}25,9\text{-}21,17\text{-}29)}(\alpha H\alpha);^{(1\text{-}5,1\text{-}13,1\text{-}29,5\text{-}9,5\text{-}25,9\text{-}13,9\text{-}21,13\text{-}17,17\text{-}21,17\text{-}29,21\text{-}25,25\text{-}29)}(\alpha) \quad (14)$$

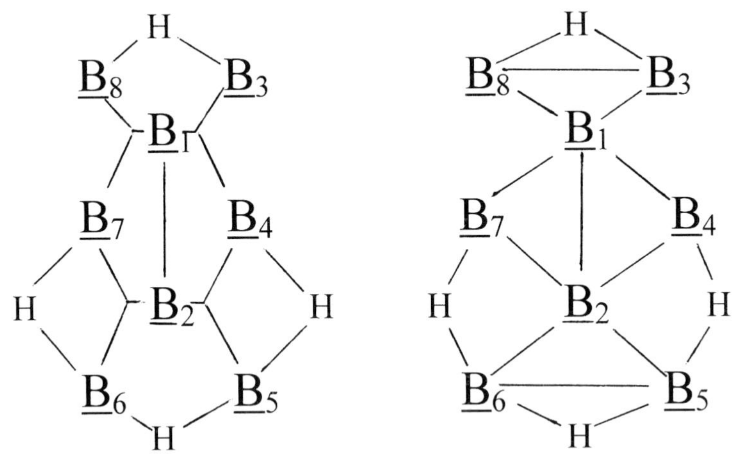

Fig. 20: Using the underscore convention, the topological and geometrical pictures of B_8H_{12} in Figure 19 are compared

A note of special interest for this proposed structure is its difference from cubane (Figure 16 in Chapter 2); namely, none of the boron atoms are modularized with a hydrogen atom; i.e., there are no B groups in this name. This is in contradistinction to the "geometrical" structure portrayed by ACS in Fig. 19, which would be named as:

$$(\underline{B}1)_6(\underline{B}\alpha H\alpha)_2:^{(1-5,3-9,3-13,3-15,9-13)}(1);^{(1-5,7-11)}(\alpha H\alpha) \quad (15)$$

Replacing the B-H combinations with B in Figure 19, presented as Figure 20, clarifies the pentagonal pyramidal basis that underlies the ACS concept of the geometry of this molecule, as well as highlighting the inadequacy of the boron bridge symbol traditionally used. In particular, whether there is supposed to also be a covalent bond between B_3 and B_8 as indicated in the right hand (geometric) picture or only a hydrogen bond as in the left hand (topological) picture.

A further comment on boron nomenclature is that for molecules in which the number of boron atoms is not 4, 6, 8, 12 or 20, and thus can not be located at the vertices of a regular polyhedron (also 30, which can be located at the center of the edges of a regular polyhedron), one looks to solid figures which deviate as little as possible from one of the five regular polyhedra. For example, as shown above for five boron atoms, the optimal choice was either the square pyramid or the trigonal bipyramid. Now, unlike the traditional perspective, it is our belief that the selection of geometry is strongly influenced by the number of hydrogen atoms, especially when there are two or more geometrical structures of "nearly equal" approximation to regularity -- as was the case for the B_5 compounds. If a stable pentaborane with an even number of hydrogen atoms is discovered, we anticipate that the square pyramidal structure would prevail over the trigonal bipyramid. This speculation is predicated on the fact that although the hydrogen atom is smaller than the boron atom, it is not insignificant, in much the same manner as cyclobutane is NOT coplanar [11] — despite that the hydrogen repressed model would suggest such coplanarity.

Continuing in this manner, this same technique can be applied to each of the other boron hydrides (B_mH_n); i.e, to select the most expeditious geometry to use when formulating the relevant nomenclature model. From the perspective of solid geometry, this means select that geometric solid having **m** vertices and either **n** edges or faces which has some heuristic concept of "maximum symmetry". However, it is important to remember that although mathematically many combinations of m and n may be

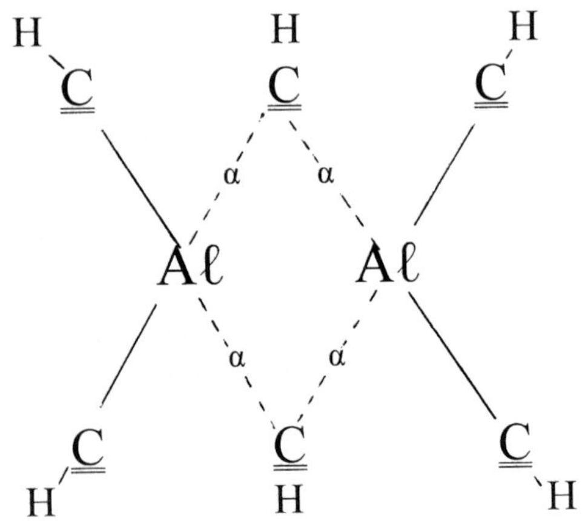

Fig. 21: Trimethylaluminum

possible, not all of them will, necessarily, correspond to boranes or other related compounds such as carboranes, etc. In other words, it is anticipated that there exists some hierarchy of "target" polyhedra which fits this geometric "ideal". The first of these is, most likely, the five convex regular (Platonic) solids. The next level is the semi-regular polyhedra in which the number of vertices or edges deviate as little as possible from one of the Platonic solids. When one is unable to find such a polyhedron, other strategies to equalize Coulomb attraction between like ligands, such as polymerization and/or ionization, etc. are resorted to.

Although one could continue to demonstrate the virtues of the proposed nomenclature for the remaining boranes in [2], the models that have been postulated for each such structure seem to be getting further removed from what is truly known about the relevant geometry and topology of the respective compounds. Consequently, although the proposed system of nomenclature addressed the problem, which the presently accepted system did only belatedly and with an extremely convoluted set of terms, locant numbers, etc., it's further development must await until more laboratory results are accumulated. In the present state of knowledge, it is more speculative than is desirable for any canonical system of "scientific" nomenclature.

As well as the various [3c-2e] covalent compounds involving boron, similar principles are next extended to compounds of the next element in Group 13 of the periodic table, aluminum, and to extended alpha bonds that

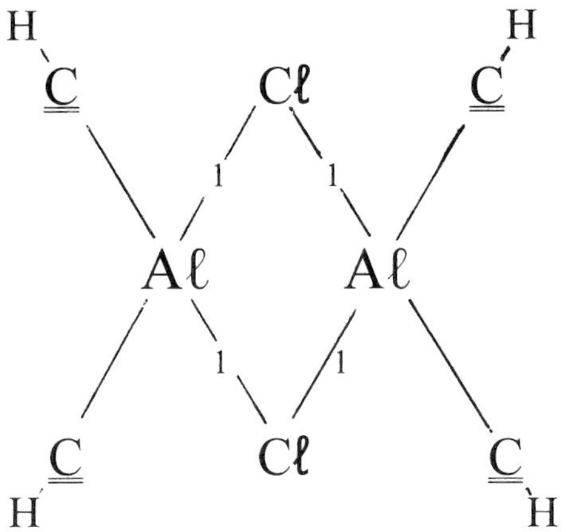

Fig. 22: Dimethylaluminum chloride

arise in such compounds [9-10]. For example, trimethylaluminum (Figure 21) occurs primarily as a dimer with the formulation of alpha Aℓ-C bonds. The resulting compound formed is a four member ring. On the other hand, dimethylaluminumchloride (Figure 22) dimerizes with the formation of ordinary 2c-2e bonds. These differences are indicated in the nomenclature by the presence of "α" in the former vs. "1" in the latter compound:

$$(A\ell\alpha\underline{\underline{C}}\alpha)_2{:}^{(1,1,5,5)}(1\underline{\underline{C}}1H);^{(3,7)}(1H) \quad \text{vs.} \quad (A\ell 1\underline{\underline{C}}\ell 1)_2{:}^{(1,1,5,5)}(1\underline{\underline{C}}1H) \qquad (16)$$

Moreover, there is an apparent lack of a simple predictable relationship that characterizes the geometry of many of the group 13 compounds, especially aluminum. This is illustrated with tri(2,4,6-trimethylphenyl)aluminum, common name trimesitylaluminum (Figure 23) which occurs as a monomer vs. triphenylaluminum (Figure 24) which forms a four member alpha bonded ring with the two aluminum atoms and carbon atoms from two of the benzene rings and occurs as a dimer vs. diisobutylaluminum hydride (Figure 25) which occurs as a trimer with hydrogen and aluminum atoms forming a six member alpha bonded ring. The choice as to whether to use single vs. alpha bonds between elements is determined by the valence of three for each of the aluminum atoms. The respective names assigned to these compounds are:

For tri(2,4,6-trimethylphenyl)aluminum (Figure 23):

$$A\ell 1(C\beta)_5C{:}^{(3\text{-}13)}(\beta);^{(5,9,13)}(1\underline{\underline{C}}1H);^{(7,11)}(1H);^{(1,1)}(1(C\beta)_5C{:}^{(3\text{-}13)}(\beta){:}^{(5,9,13)}(1\underline{\underline{C}}1H);^{(7,11)}(1H) \qquad (17)$$

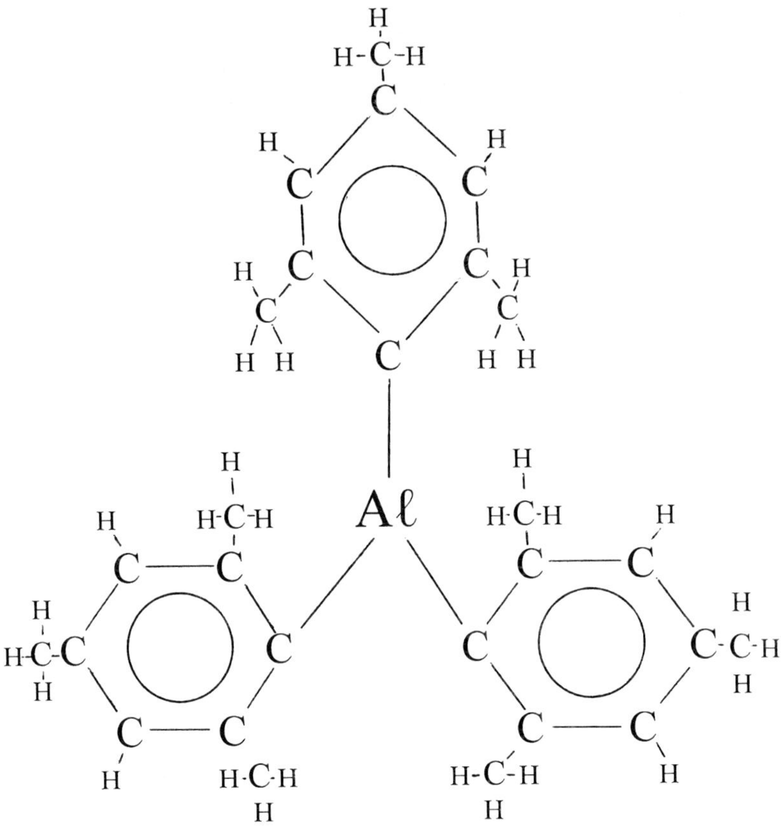

Fig. 23: Trimesitylaluminum

For triphenylaluminum (Figure 24):

$$(A\ell\alpha C\alpha)_2{:}^{(1,1,5,5)}(1Ph);^{(3.7)}(\beta(\underline{C}\beta)_5) \quad (18)$$

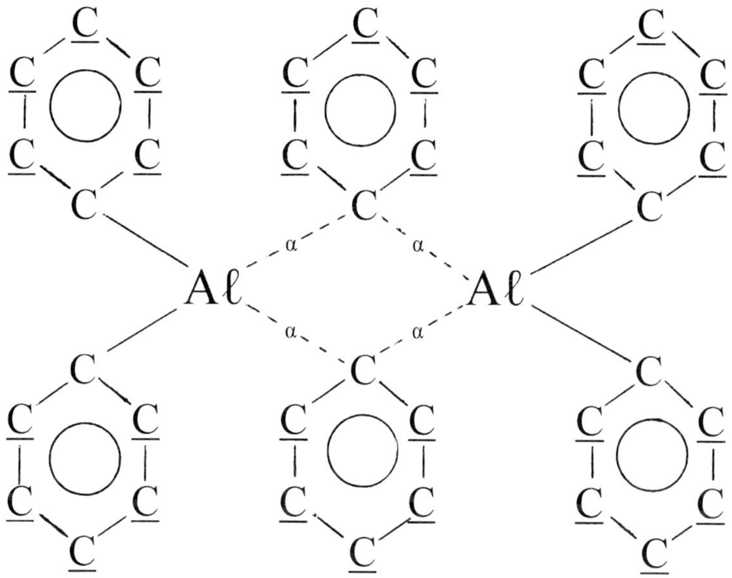

Fig. 24: Triphenylaluminum dimer

and for diisobutylaluminum hydride (Figure 25):

$$(A\ell\alpha H\alpha)_3{:}^{(1,1,5,5,9,9)}(1\underline{\underline{C}}1\underline{\underline{C}}1\underline{\underline{C}}1H){:}^{(5)}(1\underline{\underline{C}}1H) \qquad (19)$$

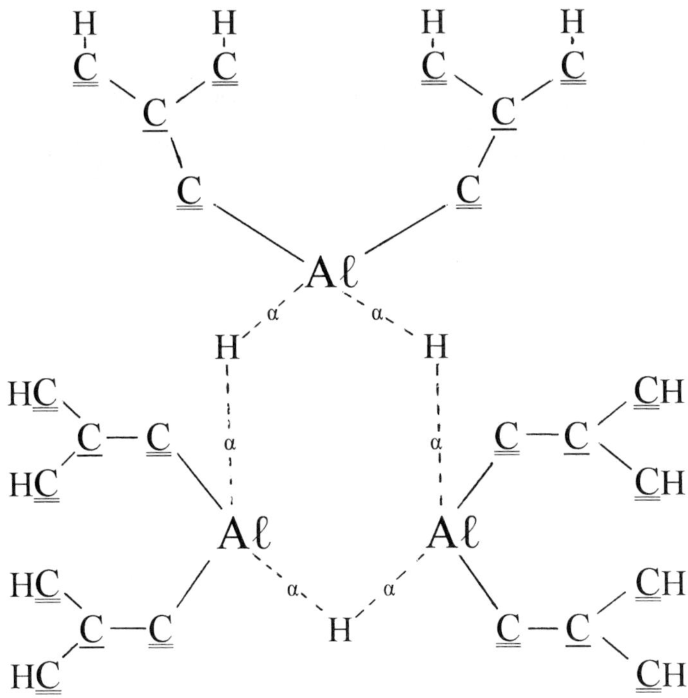

Fig. 25: Diisobutylaluminum hydride trimer

The assignment of "better" names for these compounds using spherical names will be undertaken in Section 6.

REFERENCES

[1] Council of the American Chemical Society, Inorg.Chem., "The Nomenclature of Boron Compounds", 7 (1968) 1945.

[2] S.B. Elk, THEOCHEM,548 (2001)143.

[3] International Union of Pure and Applied Chemistry, Nomenclature of Inorganic Chemistry Recommendations 1990. Blackwell Scientific Publications. Oxford, U. K., 207-237.

[4] Ibid, 207.
[5] Ibid, 208.
[6] R.J. Gillespie, J.Chem.Soc., 1963, 4672.
[7] S.B. Elk, J.Chem.Inf.Comput.Sci., 35 (1995) 858.
[8] F. Harary, Graph Theory, Addison-Wesley, Reading, Mass., 1969, 46.
[9] Ibid, 17.
[10] Ibid #7.
[11] F.A. Cotton and B.A. Frenz, Tetrahedron, 30 (1974) 1587.
[12] F.A. Cotton, G. Wilkinson, C.A. Murillo and M. Bochmann, Advanced Inorganic Chemistry, 6-th Ed.; John Wiley, New York, 1999, 194.
[13] Encyclopedia of Inorganic Chemistry, Editor, R. B. King, 1994, John Wiley, Chichester, U.K., Vol. 1, 119.

Chapter 6

Spiro and related compounds

CHAPTER ABSTRACT:
In order to extend canonical nomenclature to compounds in which the union of two or more cycles is either a single atom, a single edge, or when a succession of edges connect otherwise unconnected cycles, an expansion of the concept of "bridging" is undertaken. Four important different protocols suggested by such an extension include:

(1) A system of nomenclature modeled on spherical coordinates. This extension is useful for molecules in which two or more congruent rings emanate from a single atom;
(2) A modification of (1) called "sandwich" nomenclature for that class of organometallic compounds called "metallocenes";
(3) Consideration of a cyclic compound in terms of its largest spanning tree, which is then supplemented by the set of edges that had been removed to form that tree;
(4) Use of redundant paths to cover all of the atoms and bonds of the molecule.

These latter two techniques are important both for spiro molecules in which the rings containing the common atom are of different constitution and for ring assembly compounds. Additionally, note that in many instances the other distinct types of nomenclature that had been introduced earlier [Cartesian (in Chapter 1) and cylindrical (in Chapter 3)] may be interconverted in much the same way as the different types of coordinate systems in geometry may be interconverted. The choice of an optimal name for a given molecule is based on the heuristic principles of shortness and simplicity.

Up until this point, all of the cyclic compounds that have been examined have the property that removal of a single atom may disrupt the cycle, but will still leave a connected graph. In graph theoretical terms this is expressed in terms of "blocks" [1], where a "block" is defined as a subgraph in which there is one or more contiguous paths from every node to every other node of the subgraph. In this chapter, attention is focused on formulating canonical names for cyclic compounds containing atoms whose deletion would increase the number of blocks.* Extending the terminology traditionally used in "organic" chemistry to all chemical moieties, when removal of a single atom (and all bonds emanating from it) increases the number of blocks, these atoms shall be referred to as "spiro atoms" [2]. Note that "spiro organic compounds" are limited to compounds comprised of two rings joined at a single carbon atom; however, "spiro inorganic compounds" do not have the restriction of four bonds to a single atom. In the inorganic case, traditionally the focus has been on that single atom which is "coordinating" other single atoms or chains of atoms, rather than on the cycle that has been formed. In the proposed nomenclature, as a means of standardization, all such moieties (both organic and inorganic) will be viewed in terms of their largest graph theoretical cycle. By this perspective, the same criteria will apply to both "simply-spiro" and "multiply-spiro" compounds, and the canonical name will regard all additional (besides the longest graph theoretical cycle) chains as bridges which start and end at the spiro atom. A singly- and a multiply-spiro compound are illustrated in Figures 1 and 2. These may be named respectively as:

$$C1(\underline{\underline{C}}\ 1)_5:^{(1\text{-}1)}[(1\underline{\underline{C}})_4 1] \qquad (1)$$

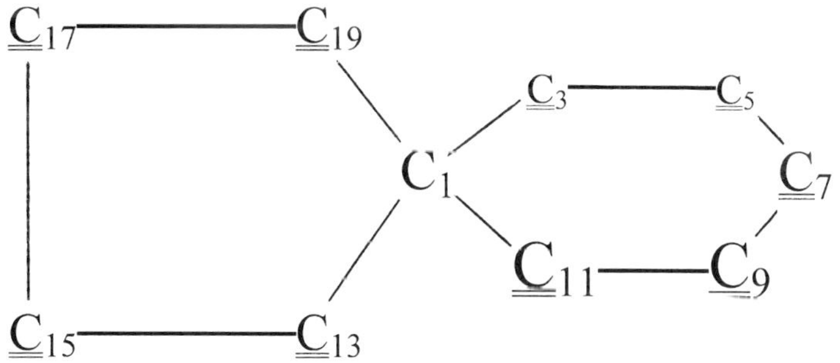

Fig. 1: A singly-spiro compound

* Note that for acyclic compounds larger than a single atom, deletion of an interior atom will always increase the number of blocks.

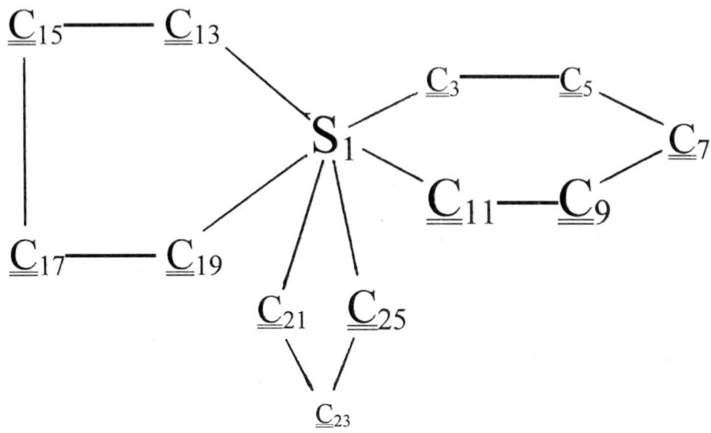

Fig. 2: A multiply-spiro compound

and

$$S1(\underline{\underline{C}}\ 1)_5{:}^{(1\text{-}1)}[(1\underline{\underline{C}})_4 1];^{(1\text{-}1)}[(1\underline{\underline{C}})_3 1] \qquad (2)$$

Moreover, when the cycles are of equal size, in a manner similar to the cylindrical system that was introduced in Chapter 3, there exists a somewhat different simpler nomenclature, which will be referred to as "spherical":*

---

* Whereas the cylindrical system had been envisioned in Chapter 3 as having two atoms viewed as the bases of a "cylinder" with chains connecting these two atoms lying on curved "lateral faces" of a "cylindrical surface", the spherical system is envisioned as having connecting chains closing back on a single atom. The nomenclature thus associated with these two forms differs only in that the focus is on these different initial and final atoms for the cylindrical system versus being on a single atom (which is both the initial and the final atom of the chain) for a spherical system. Moreover, although some may consider this choice of names as capricious, inasmuch as "toroidal" and "ellipsoidal" are geometrically closer to the three dimensional shape swept out by the connecting chains emanating from the defining one and two points respectively, these names were selected based on their analogy to the three dimensional coordinate systems with these names. Of the other orthogonal coordinate system in three dimensions [3] (bipolar, confocal, confocal paraboloidal, conical, ellipsoidal, elliptic cylindrical, parabolic, parabolic cylindrical, prolate spheroidal, oblate spheroidal, and toroidal), as well as some selected non-orthogonal ones, there is a similar capacity to set up special nomenclatures systems that will simplify naming selected molecules; however, because of the general lack of familiarity with these other systems, attention has been limited to only these three (Cartesian, cylindrical and spherical) names.

$C[1(\underline{\underline{C}}\ 1)_5]_2$ (3)

and

$S[1(\underline{\underline{C}}\ 1)_5]_3$ (4)

Having introduced these three methods of assigning canonical names to chemical moieties, an examination is now undertaken of "nearly similar"[*] names that are likely to could cause confusion ***IF*** meticulous attention is not paid to details of punctuation.[†] Although this scenario may seem to be analogous to the discussion on pages 41 through 43 in Chapter 1 — wherein the presence or absence of a single blank space in IUPAC nomenclature made drastic differences in the compounds being named, these "near similarities" are nowhere nearly as flagrant as inclusion or omission of a blank space; however, they do bear paying close attention to.

$S1[O1(\underline{\underline{C}}1)_2]_3S$ (5a)

$S[1O1(\underline{\underline{C}}1)_2S]_3$ (5b)

$S[1O1(\underline{\underline{C}}1)_2]_3S$ (6)

$S[1O1(\underline{\underline{C}}1)_2]_3$ (7)

$S[1O1(\underline{\underline{C}}1)_2S{:}]_3$ (8)

First of all, observe that the basis of all naming so far has been that of an atom, followed by a bond, followed by an atom, etc. Moreover each sequence has begun and ended with an atom. This is true even though some conventional coding for repeated units has been introduced. For example, (5a) and (5b) refer to "linear" sequences with a repeated module. The first of these started with **O** — an atom, and ended with **1** — a bond (5a); while the other started with **1** — a bond and ended with **S** —an atom. These are illustrated in Figure 3. Note that the positioning of the brackets determined

---

[*] See footnote on page 20 of Chapter 1.

[†] Whether the specific combination of atoms with the bonds given are actually found in the laboratory is irrelevant to the formulation of a nomenclature. What is important is the ability to assign a unique canonical name to every possible combination of atoms and bonds not specifically excluded by theory.

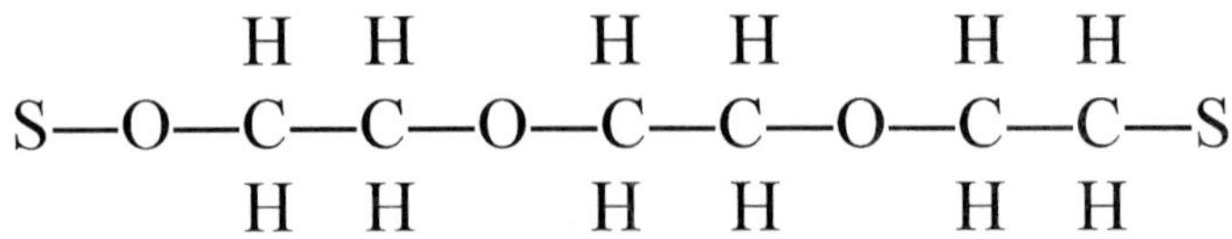

$S1[O1(\underline{\underline{C}}1)_2]_3S$

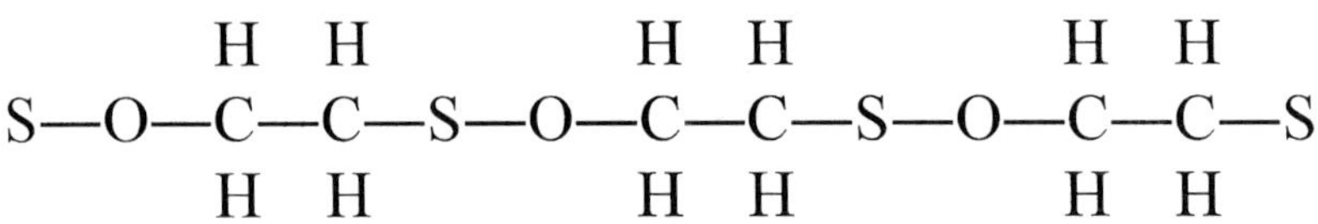

$S[1O1(\underline{\underline{C}}1)_2S]_3$

Fig. 3: Comparison of "similar" names -- two theoretical "linear" molecules

the starting and ending point of the repeating module.* Secondly, there exists a conventional meaning assigned to sequences that start and end with a bond. One such sequence was introduced in Section 3 as the naming

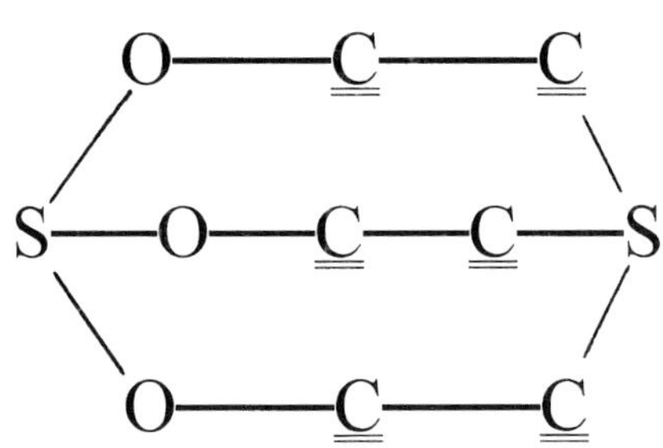

$S[1O1(\underline{\underline{C}}1)_2]_3S$

Fig. 4: Comparison of "Similar" Names - A Theoretical "Cylindrical" Molecule

* Additionally, it should be noted that formula 6-5b, while linguistically correct, is NOT the preferred name. When there is more than one way to position parentheses indicating repetition, the protocol selected for equal length chains is: name the atom first and then the bond: $[S1O1(\underline{\underline{C}}1)_2]_3S$.

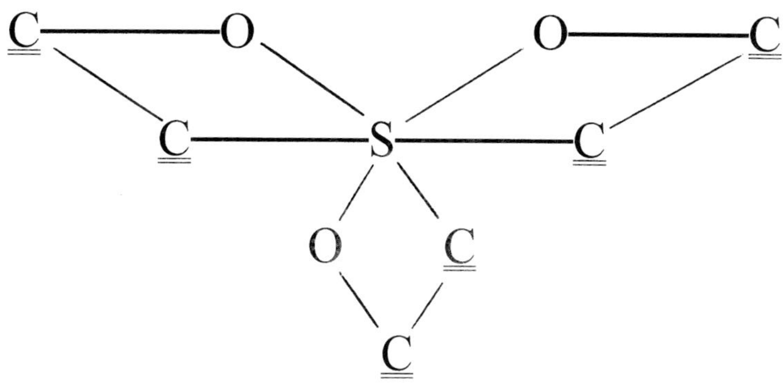

$S[1O1(\underline{C}1)_2]_3$

Fig. 5: Comparison of "Similar" Names – A Theoretical "Spherical" Molecule

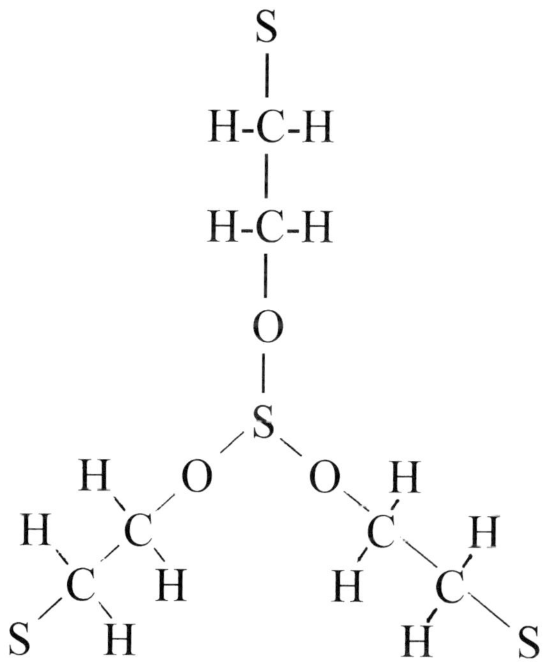

$S[1O1(\underline{C}1)_2S:]_3$

Fig. 6: Comparison of "Similar" Names - A Theoretical "Dendritic" Molecule

scheme for cylindrical nomenclature. This will be the case ONLY when the repeated sequence is contained between two atom designators. See (6) and Figure 4. When the repeated sequence follows an atom designator but is the terminus of the name, as in (7) and Figure 5, this shall be interpreted as a spherical name. Additionally, by the addition of marks of punctuation other combinations may be described. One such punctuation mark of utility is the colon to indicate the end of a specific sequence, but not the end of the code. (8), which differs from (5b) only by such a colon, refers to a special type of spherical symmetry, known as "dendritic". In its simplest (but not particularly useful)* form, selected whole molecules, such as methane, carbon tetrachloride and 2,2-Dimethylpropane, could be spherically named as:

$$C[1H:]_4 \qquad (9)$$

$$C[1C\ell:]_4 \qquad (10)$$

$$C[1\underline{\underline{C}}1H:]_4 \qquad (11)$$

etc. This idea, however, will be advantageous in the naming of terminal molecular segments, such as the tertiary butyl group†, which may now be abbreviated to:

$$1C[1\underline{\underline{C}}1H:]_3 \qquad (12)$$

---

* Pragmatically, this usage for most "organic" compounds is not encouraged. In its place, continued use of Cartesian names, especially when supplemented with the underlining convention is recommended. The three compounds which could be represented by equations 6-9 thru 6-11 are better named as: $H1\underline{\underline{C}}1H$ (9a); $C\ell1C1C\ell:^{(3,3)}(1C\ell)$ (10a) and $H1\underline{\underline{C}}1C1\underline{\underline{C}}1H:^{(5,5)}(1\underline{\underline{C}}1H)$ (11a). Even for the last of these, which does have significant simplification in spherical form, the spherical symmetry is not the chemically defining feature of this molecule and so usage of spherical nomenclature, while correct, may seem pedantic. Note this is in contradistinction to the desire to emphasize this spherical symmetry in selected "inorganic" molecules, such as iodine heptafluoride IF_7, for which, in the proposed system, the spherical name $I[1F:]_7$ is preferred over the Cartesian name $F1I1F:^{(3,3,3,3,3)}(1F)$. Observe that without the colon $I[1F]_7$ would designate a chain of 7 fluorine atoms singly bonded to each other and only one (at the front end) bonded to the iodine

† Note that because of the spherical symmetry of this aggregation of atoms, this abbreviation despite being longer than the abbreviation **tBu** introduced in Section 2, would probably be preferred by many mathematical and theoretical chemists in that a new symbol need not be memorized.

as well as for dendrites (a special type of polymer). Meanwhile formula (8) has been included here as an example of a first level dendrite (Figure 6). Higher level dendrites would be formed by having each of the terminal S atoms augmented with three similar $[1O1(\underline{\underline{C}}\ 1)_2S]$ chains. For example, a second level dendrite would be named as:

$$S\{1O1(\underline{\underline{C}}\ 1)_2S{:}[1O1(\underline{\underline{C}}\ 1)_2S{:}]_3\}_3 \quad (13)$$

Similarly, higher level dendrites would continue this "nesting" process using the grammatical symbol of ellipsis, such as:

$$S\{1O1(\underline{\underline{C}}\ 1)_2S{:}\ \ldots\ [1O1(\underline{\underline{C}}\ 1)_2S{:}]_3\}_3 \quad (14)$$

This will be developed in Chapter 8.

Some earlier scenarios in which the use of spherical names would shorten the nomenclature for symmetric dimers (in some cases only very slightly, while in others significantly) are herein noted:

(1) Instead of: $(O\beta C\beta O1H\alpha)_2{:}^{(3,11)}(1R)$ — see (9) in Chapter 2, use:

$$[O\beta C\beta O1H\alpha{:}^{(3)}(1R)\,]_2 \quad (15)$$

and instead of $(O2C1O1H\alpha)_2{:}^{(3,11)}(1R)$ — see (10) in Chapter 2, use:

$$[O2C1O1H\alpha{:}^{(3)}(1R)]_2 \quad (16)$$

In other words, there is no need to use the higher locant numbers (11 in this case) for the repeated parts of the name.

(2) The bis iron complex shown in Figures 21 and 22 of Chapter 2 can be assigned a spherical name by locant numbering the two halves with the same digit -- one without a prime and one with. Now, instead of the eight member cycle with the iron atoms exterior (Figure 2-21), by appropriately numbering the atoms in Figure 2-22 a ten node "circle" (Figure 7) (with two congruent five node parts) can be formed. By including a bridge between a primed and an unprimed node (3 and 9), the two parts of the name need not be repeated; namely:

$$[Fe\aleph(\underline{C}\beta)_3\underline{C}\aleph{:}^{(3\text{-}9')}\beta;^{(1\text{-}5,1\text{-}7)}\aleph;^{(1,1,1)}(1C3O)]_2 \quad (17)$$

emphasizes the symmetry, as well as creating a name that is shorter

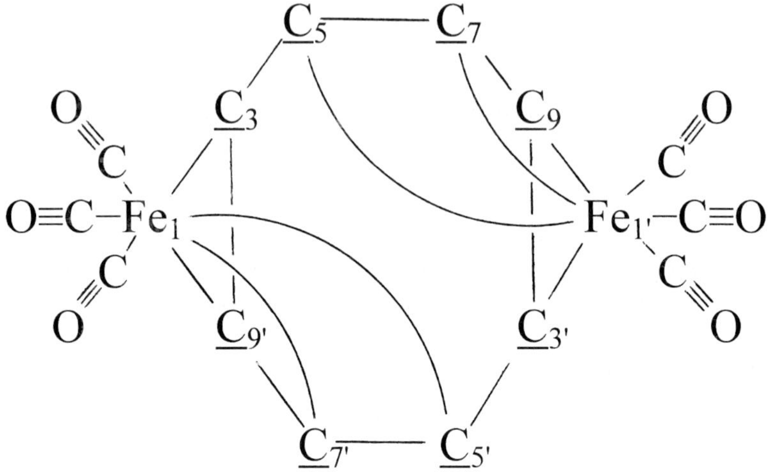

$[\text{Fe}\aleph(\underline{C}\beta)_3\underline{C}\aleph\text{:}^{(3\text{-}9')}\beta\text{;}^{(1\text{-}5',1\text{-}7')}\aleph\text{;}^{(1,1,1)}(1\text{C3O})]_2$

Fig. 7: Systemic spherical name for bis iron complex of Figures 2-21 and 2-22

than the Cartesian name assigned in Chapter 2 (Formula 30):
$[\text{Fe}\aleph(\underline{C}\beta)_3\underline{C}\aleph]_2\text{:}^{(3\text{-}19,9\text{-}13)}(\beta)\text{;}^{(1\text{-}15,1\text{-}17,5\text{-}11,7\text{-}11)}(\aleph)\text{;}^{(1,1,1,11,11,11)}(1\text{C3O})$

(3) Consider the three aluminum dimers named and illustrated in the previous chapter, as formula (14), Figure 15; formula (15), Figure 16 and formula (16), Figure 18. The respective spherical names are:

Trimethylaluminum: $[\text{A}\ell\alpha\underline{\underline{C}}\alpha\text{:}^{(1,1)}(1\underline{\underline{C}}1\text{H})\text{;}^{(3)}(1\text{H})]_2$ (18)

Dimethylaluminum chloride: $[\text{A}\ell 1\text{C}\ell 1\text{:}^{(1,1)}(1\underline{\underline{C}}1\text{H})]_2$ (19)

Triphenylaluminum: $[\text{A}\ell\alpha\text{C}\alpha\text{:}^{(1,1)}(1\text{Ph})\text{;}^{(3,7)}(\beta(\underline{C}\beta)_5]_2$ (20)

Additionally, the molecules portrayed in Chapter 5 as Figures 17 and 19 have 120°, rather than 180°, symmetry and, using spherical nomenclature, would be canonically named as:

Trimesitylaluminum: $\text{A}\ell[1(\text{C}\beta)_5\text{C}\text{:}^{(3\text{-}13)}(\beta)\text{;}^{(5,9,13)}(1\underline{\underline{C}}1\text{H})\text{;}^{(7,11)}(1\text{H})]_3$ (21)

and

Diisobutylaluminum hydride: $[(A\ell\alpha H\alpha:^{(1,1)}(1\underline{\underline{C}}1\underline{C}1\underline{\underline{C}}1H):^{(5)}(1\underline{\underline{C}}1H)]_3$ (22)

Similarly, although spherical nomenclature creates the heuristically most efficient name for very many cyclic organo-metallic compounds, for multiple spiro compound, especially those that lack an atom at the geometric center, there is another technique, called the "redundant path" method, which gives a "better" canonical name. This will be described below.

At this point in the development and interpretation of the proposed nomenclature code, an alternate view of a cycle is considered. Namely, select any atom of a ring and regard it as a "bridge" between its two neighbors. For example, although they are not in canonical form, cyclohexane and benzene could be named respectively as:

$$(\underline{\underline{C}}1)_5\underline{\underline{C}}:^{(1\text{-}11)}(1) \quad (23)$$

and

$$(\underline{C}\beta)_5\underline{C}:^{(1\text{-}11)}(\beta) \quad (24)$$

Such a technique is the one traditionally done in I.U.P.A.C. "inorganic" nomenclature in which a selected atom becomes the focus of attention. One may, similarly, apply this procedure to spiro carbon atoms in "organic" chemistry, as well as to a "coordinating" metal atom, call it M, in organo-metallic chemistry. In this latter case the technique could be applied to acyclic, as well as cyclic, molecules. For the acyclic case one merely names the longest chain irrespective where M is in the chain. For cyclic compounds M may be viewed as the center of a spherical system, with the remaining atoms forming congruent cycles containing or similarly connected to M.

An important new class of molecules in which a metal atom was bonded to entire carbon rings, rather than just to a single carbon atom of a ring was discovered in 1951 [4-5]. The first example of this class, dubbed "ferrocene", undergoes electrophilic aromatic substitutions and was shown by x-ray crystallography to have an iron (II) ion located between two cyclopentadienyl rings (Figure 8). Moreover, the measured carbon-carbon bond lengths of this compound are 140 nm and the carbon-iron bond lengths 204 nm. Because a bond length of 133 nm is the value normally associated with a carbon-carbon double bond, while 154 nm is the value for carbon-

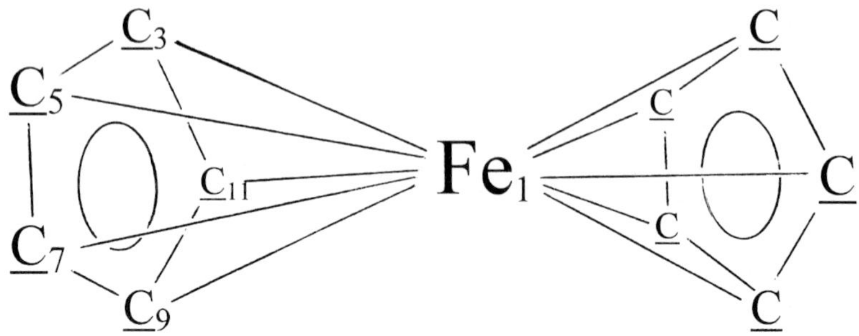

Fig. 8: Ferrocene: traditional connectors

carbon single bonds; the bond length between carbon atoms of 140 nm neatly dovetails with the above description of a beta bond. Additionally, the iron - carbon bond length of 204 nm, which is ascribed to the overlapping of 3d orbitals from the iron atom with p orbitals of the cyclopentadienyl rings forming a pi-delta, rather than a sigma-sigma, bond is substantially longer than a typical single bond. Consequently, being further motivated by the aromatic character of the entire aggregation, the relevant chemistry is best described by assigning aleph (ℵ) bonds as the connectors from the iron atom to the individual carbon atoms of each ring, rather than either using traditional single bonds or considering these bonds as alpha bonds.

Next note that although IUPAC names this compound: bis(η-cyclopentadienyl)iron(II); Chemical Abstract Services (CAS) has pointed out that such a name is, in fact, ambiguous. The ambiguity became evident when applying the same principles to the beryllium analog of ferrocene. Unlike the iron compound, two different beryllium compounds which IUPAC could name as: bis(η-cyclopentadienyl)beryllium are known. As well as the direct analog of Figure 8, in Figure 9 a compound in which only one of the Cp rings is aleph bonded to M while the other one is traditionally single-bonded from the beryllium atom to only one of the carbons of the other Cp ring is illustrated*. Consequently, CAS chose to modify the name to be assigned to any metallocene by addending a superscript to the

* Figure 9 has been designated as only a preliminary picture describing the structure of monohaptoberyllicene. In this picture it appears as though at locant 13 there is a valence of 5. An explanation and correction to this picture is supplied below.

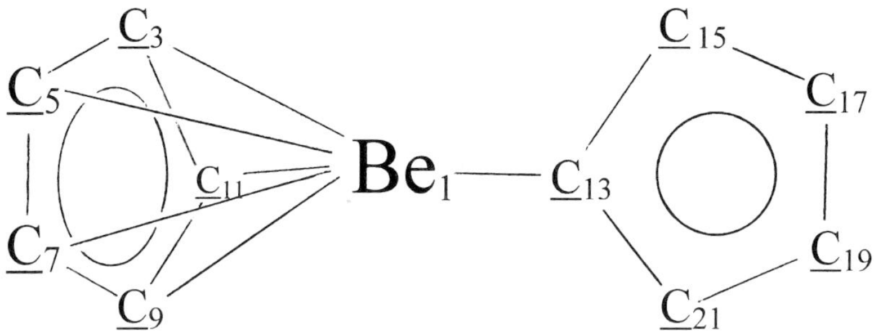

Fig. 9: A preliminary picture of monohaptoberyllicene

descriptor η; namely, bis(η^5-cyclopentadienyl)iron(II) for ferrocene; bis(η^5-cyclopentadienyl)beryllium for the direct analog, and η^5-cyclopentadienyl, η^1-cyclopentadienyl beryllium for what they call "monohapto-cyclopentadienylberyllium".

Cahn and Dermer [6] advises that the problem originally arose because in other instances, CAS had occasion to interpret the symbol η to denote "some or all unsaturated atoms in the chain or ring are bound to the central atom". This is in contradistinction to demanding that there are ten bonds from the iron atom, one to each of the five carbon atoms in the two rings. Because of this "some or all" in their definition, a modification was necessary.* The proposed nomenclature, on the other hand, includes precisely those bonds which are relevant and thus is unaffected by any such ambiguity:

$$Fe[\aleph(\underline{C}\beta)_4\underline{C}\aleph{:}^{(3\text{-}11)}\beta{;}^{(1\text{-}5,1\text{-}7,1\text{-}9)}\aleph]_2 \qquad (25)$$

At this point an implicit assumption made which underlies not only (25), but also all of the other names that have been allocated so far, is noted; namely, that all chemical moieties can, and should, be named using **only** the principles of graph theory. In other words, that there is a complete equivalence between a bond in chemistry and an edge in graph theory. Ignoring the philosophical challenges raised (which may be viewed in the same light as were the topological vs. geometrical description of the boranes in Chapter 5), this assumed isomorphism is questioned by the description allocated to selected molecules. Of these, ferrocene was the first to be

---

* This addending of additional descriptors in order to salvage an existing system was akin to the system of Ptolemaic astronomy described in the footnote on page 123 in Chapter 3.

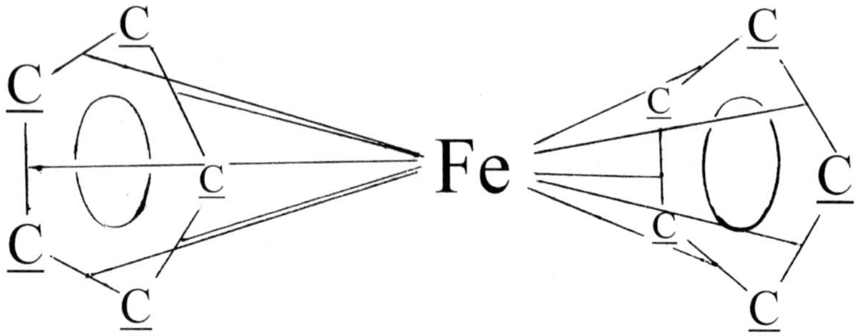

Fig. 10: Ferrocene: atom- bond connectors

discovered. Namely, are the bonds from the iron atom to the two cyclopentadienyl rings to be viewed as being both binary and localized as is traditionally listed in textbooks [7], or is this bonding spread out over either an entire edge or even over a solid segment so that this assumed isomorphism is no longer applicable. This is equivalent to asking the question: Might a chemically more relevant description be either:

(a) a set of ten bonds from the iron atom directly to the beta bonds of the rings (each of which sweeps out a planar sector), rather than being linear connections to the carbon atoms,

or

(b) a set of only two bonds from the iron directly to a point interior to each of the two rings (such as to the center of the rings)? This is in accordance with an interpretation [8] that views aromatic compounds as being mathematically represented as two dimensional, whereas there is a one-dimensionality associated with aliphatic compounds. By such a perspective, the bond from the iron to the rings may be regarded in the shape of a two-napped cone.

In other words, instead of the individual triangles (comprised of atom 1, bond 1, atom 2, bond 2, atom 3, bond 3) that are the logical consequence of thinking in terms of SSSR [9] such as $Fe_1\aleph C_3\beta C_5\aleph$, $Fe_1\aleph C_5\beta C_7\aleph$, etc. which has been the basis for the assigning of canonical names up to this point, one could consider that the iron atom was joined directly to the pentadienyl module -- rather than to the individual carbon atoms. Descriptor (a) for such a cycle would now abandon the alternation of atom and bond, along with the knowledge that in the primary cycle atoms had odd locant numbers and bonds even. In other words, one **could** picture ferrocene as shown in Figure 10; however, in order to be able to now assign a name to such an aggregation one would need to define a descriptor designating a bond from a given atom to another bond, say δ and also a bond from one bond to a second bond that includes an atom, say ε(oA) where o is the bond order and A is the symbol for the atom; in this case with beta bonds through a carbon atom ε(β$\underline{C}$β). Although such a technique, which places emphasis on individual bonds connected by an atom in some instances and on atoms connected by a bond in others, is viable; it is sufficiently tedious to make it not worthwhile. Choice (b), being predicated on a solid geometry vs. a plane geometry description, similarly, presents a challenge for devising a nomenclature. Instead, the description problems may be by-passed by affixing a subscript to the bond symbol (in contradistinction to its, until now exclusive, use on a chain of bonds and atoms). This now denotes that there is a bonding with bond order aleph from the preceding atom to the aggregate (which is enclosed within parentheses). This aggregate may now be viewed as either all of the atoms, or all of the bonds of the succeeding group of atoms and bonds. For example, the denotation that would be ascribed to the sequence:

$$Cr\aleph_6(PhH) \tag{26}$$

is that of an aleph bond between the preceding chromium atom and the benzene ring*. Additionally, such a symbol could be further modified to bond to a selected set of bonds of the ring, by augmenting this aleph symbol

* Attention is directed to the fact that the abbreviation Ph for phenyl involves six carbon and only five hydrogen atoms with the carbon atom not bonded to a hydrogen atom being the one that is bonded to an adjacent atom by a single bond. Consequently, in order to indicate this new type of bonding, one requires either the creation of a new symbol for a C_6H_6 module (as well as additional new symbols for other similar combinations such as the cyclopentadienyl grouping) or, far better, the augmenting of two atom symbols without a separating bond, namely PhH, to denote this particular atom/ bond combination. This composite symbol following the ℵ indicates the presence of aleph bonds from the previously named atom to all six of the carbon atoms of benzene. In other words, there has been a deliberate eschewing of Ph1H, as such a combination would imply that one particular hydrogen atom was a focus of bonding. **To the contrary** all of the bonds to the central metal atom are equivalent. Moreover, although the traditional total valence of the individual carbon atoms appears to be 5, rather than 4 (this is what it would be using a bond order of 1.5 for each of the two β bonds in the ring, a bond order of 1 for the carbon-hydrogen bond plus an additional bond order of 1 for the ℵ bond between C and metal), there is, **ASSUMING** that such a limitation is deemed to be important, enough leeway in the range of bond orders for each of the two beta and the one aleph bond to be less than their median value and thus the sum to still yield exactly four.

with a superscript. For example, if one wanted aleph bonding from a metal atom to alternating carbons (or bonds) of the ring, the name assigned would be:

$$\mathrm{Cr}\aleph_3^{(1,3,5)}(\mathrm{PhH}), \qquad (27)$$

Similarly, using the Cp symbol for the cyclopentadienyl module*, augmented with an H, (25) simplifies to the spherical form:

$$\mathrm{Fe}[\aleph_5(\mathrm{CpH})]_2 \qquad (28)$$

At this point, a new descriptor for nomenclature (comparable to the postulation of cylindrical and spherical names) is introduced; namely, consider a cascade of more than one aleph subscripted symbols, such as:

$$(\mathrm{CpH})\aleph_5\mathrm{Fe}\aleph_5(\mathrm{CpH}) \qquad (29)$$

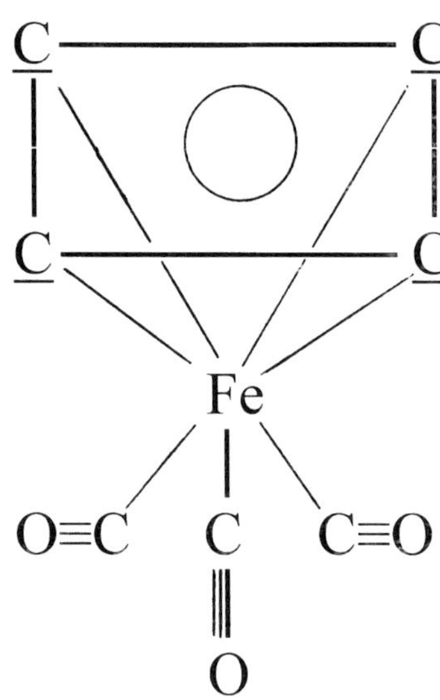

Fig. 11: Half sandwich $[(\underline{C}\beta)_4]\aleph\mathrm{Fe}\aleph[\mathrm{C3O}:]_3$

* Introduced in Chapter 2.

Because this newly specified nomenclature has the individual layers as self-contained modules, it is referred to as a "sandwich compound". Terminology useful for sandwich compounds includes "bread" for the outer layers (cyclopentadienyl rings in formula (29) and "meat" for the "sandwiched atom or group of atoms [the iron atom in (29)]. By the use of this sandwich nomenclature, the decision as to whether Figure 8 or 10 is the chemically correct one has been sidestepped. Furthermore, there is a straight-forward expansion of such a nomenclature to both "open" (also called "half") and also to "multiple" (also called "Dagwood") sandwiches:

(1) Figure 11, IUPAC name: tricarbonyl(η-1,3-cyclobutadiene)iron, is an example of a half-sandwich. The systemic Cartesian name that would have been assigned to this molecule:

$$Fe\aleph(\underline{C}\beta)_3\underline{C}\aleph{:}^{(3\text{-}9)}(\beta);^{(1,1,1)}(1C3O) \qquad (30)$$

may now be replaced by the more intuitive sandwich name:

$$[(\underline{C}\beta)_4]\aleph Fe\aleph[C3O{:}]_3 \qquad (31)$$

where the subscript inside the brackets indicates the size of the cycle and the subscript outside the independence of three C3O aggregates aleph connected to the iron atom, which continuing the food analogy may be referred as "garnishes". Also, in conformity with the convention established at the beginning of this chapter a colon has been included following the O in (31) to indicate the end of a sequence.*

(2) The mercury compound [10] (only part of which is shown in Figure 12) is an example of a Dagwood sandwich. In order to formulate the complete sandwich name for this compound, one notes that each layer of bread has formula $Hg_3C_{18}F_{12}$ and each layer of meat is benzene; consequently a name that superficially appears to be adequate for this compound is:

$$\{[Hg1(C\beta)_5C1{:}^{(5,7,9,11)}(1F)]_3\aleph[PhH]\aleph\}_n[Hg1(C\beta)_5C1{:}^{(5,7,9,11)}(1F)]_3 \qquad (32)$$

---

* This was done in order to eliminate the appearance that the O of one chain was bonded to the C of a second chain. Note such an omission of the colon would have indicated that, unlike the systemic meaning given to aggregations such as the PhH of formula (26) etc., two atoms were somehow fused together but were not separated by a bond.

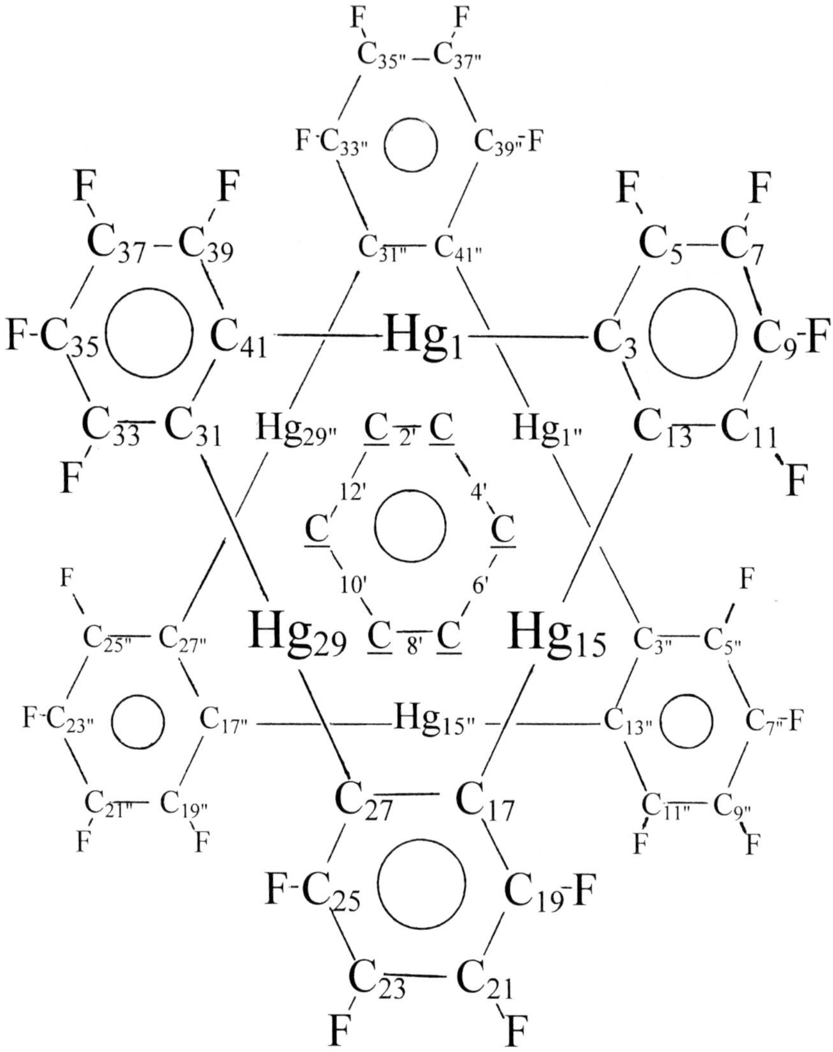

Fig. 12: Example of a Dagwood sandwich

where the subscript n indicates multiple possible layers of bread followed by meat followed by bread, etc. The final square bracketed term indicates that the molecule begins and ends with the same bread as the two "outer layers". Such a name, while necessary, is not sufficient. The problem with this name is that it lacks the assignment of locant superscripts to the aleph designator in order to delineate which of the mercury atoms is bonded to which edges of the internal benzene ring. Instead, the actually observed arrangement of atoms is supplied by the formula:

$$\{[\mathrm{Hg1(C\beta)_5C1}:^{(5,7,9,11)}(\mathrm{1F})]_3\aleph_3{}^{(1\text{-}2',15\text{-}6',29\text{-}10')}[\mathrm{PhH}]\aleph_3\}_n{}^{(4'\text{-}1'',8'\text{-}15'',12'\text{-}29'')}$$
$$[\mathrm{Hg1(C\beta)_5C1}:^{(5,7,9,11)}(\mathrm{1F})]_3 \qquad (33)$$

Here the assignment of odd integer locant numbers to atoms and even integers to bonds is important inasmuch as the individual mercury atoms bond to specific edges of the sandwiched benzene rings above and below it. This is indicated in the superscripts. Without such locant numbering, one does not know whether to assume that there is free rotation between the layers in much the same manner as shall be evident for the topological compounds described in Chapter 7.

An interesting adaptation of spherical nomenclature occurs when the "central" atom is smaller than one or more of the "ring" atoms. Because carbon is larger than beryllium, consistency requires that the canonical Cartesian name starts with a carbon. In the spherical name, however, this is not a requirement and there is no ambiguity introduced by naming bis(η^5-cyclopentadienyl) beryllium as:

$$\mathrm{Be}[\aleph_5(\mathrm{CpH})]_2. \qquad (34)$$

nor is there in the sandwich name:

$$(\mathrm{CpH})\aleph_5\mathrm{Be}\aleph_5(\mathrm{CpH}) \qquad (35)$$

The nuances in which details of nomenclature become important occur in those molecules in which not all of the atoms or bonds of a ring are connected to the "spiro" metal atom. For example, returning the focus to Figure 9, Margl, Schwarz and Bloechl [11] advise: "Experimental studies find that beryllocene prefers a so-called "slip sandwich" conformation of C_s symmetry in the crystal as well as the gas phase; i.e., Be has one η^1- and one η^5-coordinated ring."

The systemic Cartesian name assigned to this compound is:

$$(\underline{\mathrm{C}}\beta)_4\underline{\mathrm{C}}\aleph\mathrm{Be}\aleph(\underline{\mathrm{C}}\beta)_4\underline{\mathrm{C}}:^{(1\text{-}9,13\text{-}21)}(\beta);^{(1\text{-}11,3\text{-}11,5\text{-}11,7\text{-}11)}(\aleph) \qquad (36)$$

where the second aleph was chosen, in lieu of a traditional single bond, in order to indicate that aromaticity extend across the entire molecule. The connectivity of the two rings to the beryllium atom may now be viewed as an integrated entity in both the mono-hapto- and the penta-hapto molecules. This is in contradistinction to the disjoint pockets of aromaticity which will be showcased later in this chapter for molecules that are classified as ring assembly compounds, such as the biphenyl molecule. By this choice, the nomenclature is based on the observed chemistry; namely, the extended aromaticity exhibited by the mono-hapto connected ring is described using the CpH abbreviation (with its specification of beta bonds, rather than fixed single and double bonds, etc.)*

Similarly, the spherical name becomes:

$$Be[\aleph_5(CpH);\aleph_1(CpH)] \tag{37}$$

where the $\aleph_1$ symbol indicates aleph bonding to only one of the carbons in that particular cyclopentadienyl ring. Meanwhile, in this same study by Margl et al:

> "the fluxional dynamics of beryllocene at 400K over several periods totaling about 15 picoseconds"

revealed the existence of two isomerization mechanism:

> "In the gear mechanism, the bond between Be and η^1 ring migrates from one carbon atom to the next, while preserving the interactions with the η^5 ring. ... The transition state is $\eta^2;\eta^5$-coordinated."

The systemic name for this transition state thus becomes:

$$Be[\aleph_5(CpH);\alpha_2(CpH)] \tag{38}$$

where the alpha, rather than aleph, indicates a bond order substantially lower than a single bond and the subscript 2 denotes bonding to two (adjacent unless indicated to the contrary) atoms of the cyclopentadienyl ring.

For the second of the isomerization mechanisms described by Margl:

---

* This explains why Figure 9 was drawn using a Robinson circle in the mono-hapto-ring, instead of opting for fixed double bonds between C_{15}-C_{17} and C_{19}-C_{21} and single bonds everywhere else, including the connection to the beryllium atom. Using the traditional pattern for bonding this ring to the beryllium atom, while simultaneously maintaining conjugation in the ring, would require that there be a double bond between the beryllium atom and the connecting carbon (C_{13}) of the ring. Such a bonding pattern would cause the beryllium atom to have the incorrect valence of 3.

> "The molecular inversion mechanism interchanges the role of the η^1 and η^5-rings by a motion of the Be atom parallel to the ring planes from the centrally bonded position of one ring to that of the other ring. The transition state for this mechanism is an η^3,η^3 configuration of C_{2h} symmetry."

The spherical name that would be assigned to this transition state is:

$$Be[\aleph_3(CpH);\aleph_3(CpH)] \tag{39}$$

where the aleph, rather than alpha, bonding, is retained for each connection from the central beryllium atom to the two rings. Similarly, the sandwich name is:

$$(CpH)\aleph_3Be\aleph_3(CpH) \tag{40}$$

Next, returning the focus of this development of nomenclature to the above discussion of a "coordinating" atom, there is an "organo-metallic chemistry" correlation to monocyclic "spiro" atoms. A molecule may be considered as "semi-spiro" when the focus is directed to a "central" atom, whose removal would disconnect the graph at that point; however, this particular atom is not a member of the largest ring in the molecule. In Figure 23 of Chapter 2, for example, the molecule which IUPAC called: tetracarbonyl(η-1,5-cyclooctadiene) molybdenum and which CAS augmented with an oxidation number of zero: tetracarbonyl(η-1,5-cyclo-octadiene) molybdenum(0), had been named, using Cartesian nomenclature as formula (31):

$$[C2\underline{\underline{C}}1(\underline{\underline{C}}1)_2]_2{:}^{(1\text{-}11)}(\aleph Mo^{(=17)}\aleph);^{(17,17,17,17)}(1C3O).$$

Now, instead of the Cartesian protocol (of counting all atoms in a ring as equal when selecting the principal cycle) that had resulted in the

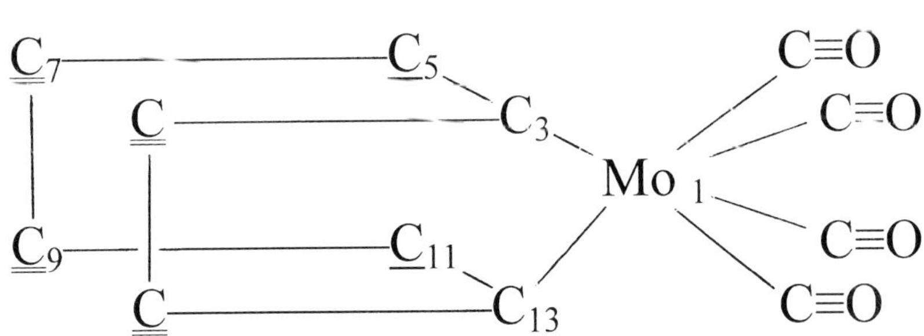

Fig. 13: Systemic spherical name for half-sandwich compound described in Chapter 2

molybdenum atom being relegated to a bridge, (despite that it is the coordinating atom in most, if not all, inorganic and organo-metallic nomenclature systems), using the ideas inherent in spherical nomenclature, one produces a spherical canonical name, that is closer to traditional inorganic nomenclature, by viewing this molecule as a seven member ring containing the molybdenum atom, augmented by a two atom bridge and four ligands (Fig. 13).

$$\mathrm{Mo1C2\underline{C}1(\underline{\underline{C}}1)_2\underline{C}2C1:^{(3\text{-}13)}[1(\underline{\underline{C}}1)_2];^{(1,1,1,1)}(1C3O)} \qquad (41)$$

An even more intuitive name is supplied using sandwich nomenclature (Figure 14). Here a half-sandwich with an eight atom bread, a molybdenum meat and four carbonyl garnishes is named:

$$\mathrm{[C2\underline{C}1(\underline{\underline{C}}1)_2\underline{C}1C1(\underline{\underline{C}}1)_2]\ \aleph^{(1\text{-}11)}Mo\ \aleph[C3O:]_4} \qquad (42)$$

Having laid the foundation of nomenclature in terms analogous to those used in formulating coordinate systems in geometry, attention is now directed to a synergy from the proposed systems of nomenclature (in particular, features from the cylindrical and spherical forms) that suggests a very different geometrical structure for a supramolecular cluster than the traditional one [12-13]. Consider the tri-ruthenium cluster traditionally illustrated as Figure 15. Note that, although the representation in which each benzene ring, in metallocene fashion, is bonded to a single ruthenium atom is adequate, the existence of major differences in the connectivity **not only** of the three ruthenium atoms **but also** of the three hydrogen and two oxygen atoms, is disconcerting. This, we believe, is in error; especially the

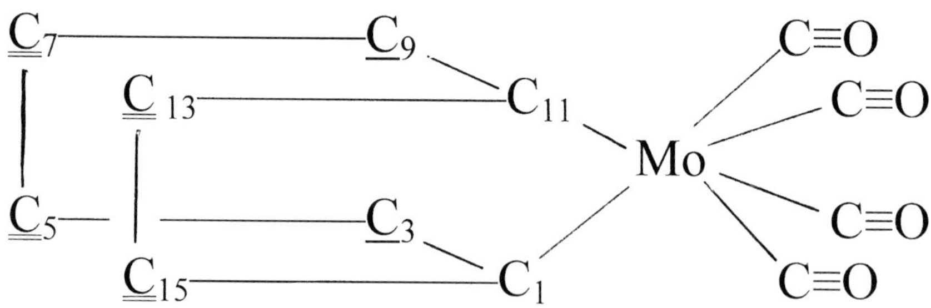

Fig. 14: Systemic sandwich name for half-sandwich compound described in Chapter 2

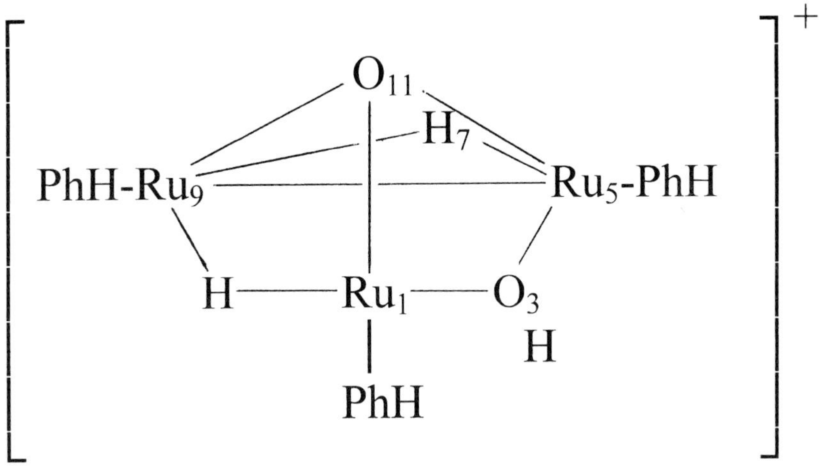

Fig. 15: Traditional representation of a ruthenium cluster ion

environments of the two oxygen atoms, each with a coordination of 3. Note that the oxygen atom to which locant number 3 has been assigned, would be viable by traditional methods **only** if it were viewed, like the boron atoms described in Chapter 5; namely, as being 3-center-2-electron (i.e., alpha) bonded to the two ruthenium atoms and single bonded to the hydrogen. However, even this is insufficient for the other oxygen atom (locant number 11). Nevertheless, ignoring whether or not Figure 15 illustrates the appropriate geometry*, such an aggregation of atoms and bonds would be nomenclated in the proposed (Cartesian) system as:

$$\mathrm{Ru\alpha\underline{O}\alpha Ru\alpha H\alpha Ru1O1}:^{(1\text{-}9)}(\alpha\mathrm{H}\alpha);^{(1\text{-}9,5\text{-}11)}(1);^{(1,5,9)}[\aleph_6(\mathrm{PhH})] \qquad (43)$$

Now, instead of the connectivity shown in Figure 15, there is the potential for far greater symmetry, and consequently simplicity, by focusing on the **two** oxygen atoms instead of the **three** ruthenium atoms; namely, let the oxygen atoms be the "apexes" of a propellane-like structure, as illustrated in

* The implications of this statement are that this nomenclature system, **as well as all other nomenclature systems**, assigns canonical names to a mathematical ideal, rather than to a chemical reality. Consequently, it is prudent to remember that any nomenclature is only as good as the science it describes.

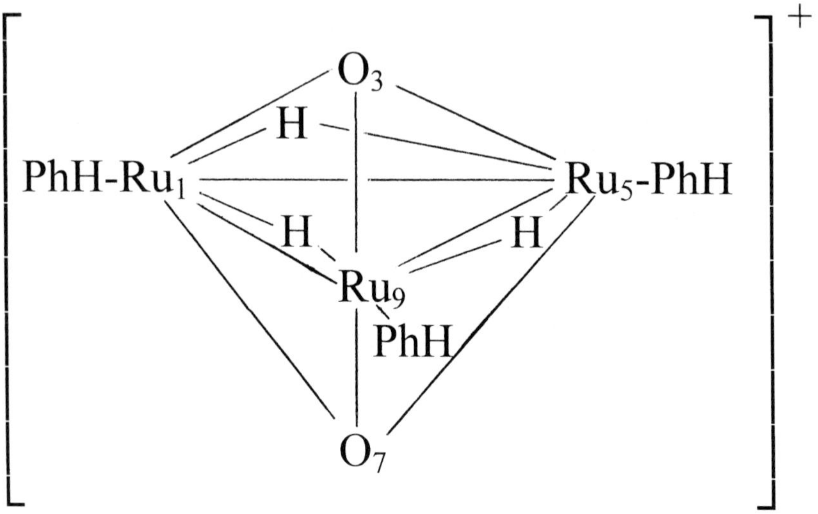

Fig. 16: Proposed representation of this ruthenium cluster ion

Figure 16. The respective Cartesian and cylindrical names for such an arrangement of atoms and bonds are:

$$(\mathrm{Ru}\aleph\mathrm{O}\aleph)_2\mathrm{Ru}\alpha\mathrm{H}\alpha{:}^{(1\text{-}5,1\text{-}7,1\text{-}9,5\text{-}9)}(1);^{(3\text{-}9)}(\aleph);^{(1\text{-}5,5\text{-}9)}(\alpha\mathrm{H}\alpha);^{(1,5,9)}\ [\aleph_6(\mathrm{PhH})] \qquad (44)$$

and

$$\mathrm{O}(\aleph\mathrm{Ru}\aleph)_3\mathrm{O}{:}^{(3\text{-}3',3\text{-}3'',3'\text{-}3'')}(1;\alpha\mathrm{H}\alpha);^{(3,3',3'')}\ [\aleph_6(\mathrm{PhH})] \qquad (45)$$

Note that, not only does the cylindrical name (45) have the desired heuristics of symmetry and simplicity, but also that the bonding selected, solves the problem of different environments for like atoms (ruthenium, hydrogen and oxygen). The further question: 'is a coordination of three for the second oxygen atom a problem?' is answered by assuming that between each pairs of ruthenium atoms there is a single bond and an alpha bond thru two adjacent hydrogen atoms; i.e., a bond identical to the boron bridge described in Figure 1b of Chapter 5. Moreover, each of the ruthenium atoms is bonded to the two oxygen atoms by an aleph bond. In other words, despite what may seem as the "unusual" scenario of a triply-coordinated oxygen atom, representation with an aleph bond is compatible with a bond order = 2/3.

As well as the cylindrical symmetry inherent in the above molecule, various clusters having spherical symmetry have also been formulated and can be nomenclated using spherical nomenclature. Figure 17 illustrates an organoiridium lithium ion recently in the news for the treatment of bipolar disorder [14]. The Cartesian name for this cluster is:

$$\{[\mathrm{Ir1O1(C\beta)_2(\underline{C}\beta)_3N1}]_3{:}^{(1\text{-}7,17\text{-}23,33\text{-}39)}(1\mathrm{O}1);\ ^{(5\text{-}15,21\text{-}31,37\text{-}47)}(\beta);\ ^{(3\text{-}19)}(1\mathrm{Li}^{(=49)}1);^{(35\text{-}49)}(1);^{(1,17,33)}(\aleph_5\mathrm{CpH})\}^{(+)} \tag{46}$$

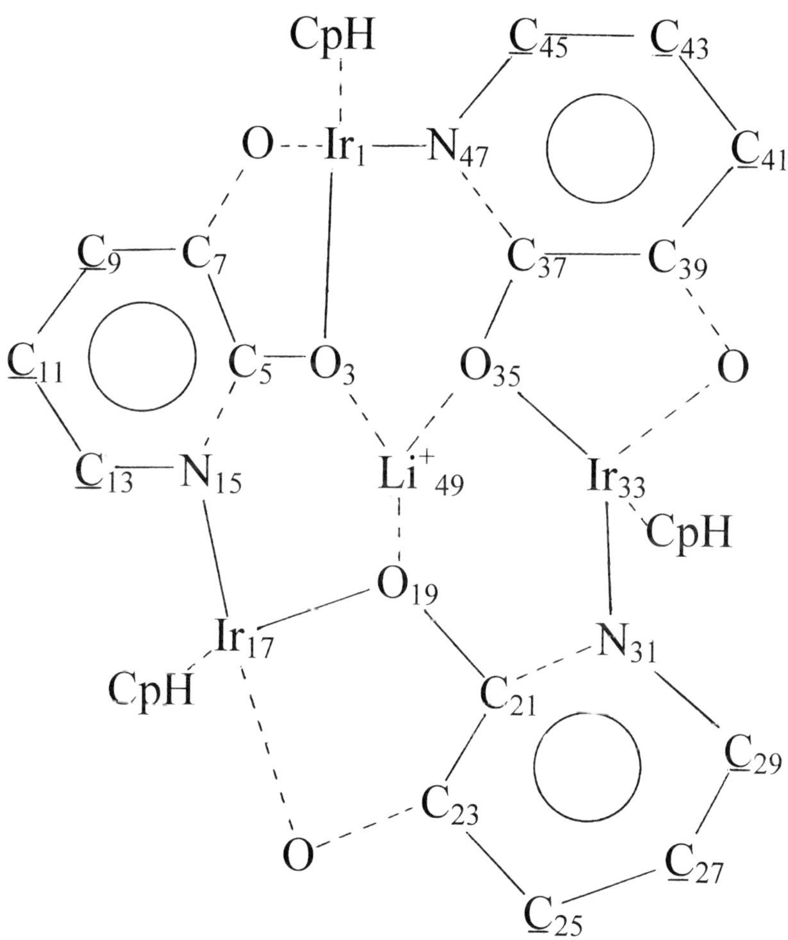

Fig. 17: A 12 ring organoiridium lithium ion

a canonical name that would be greatly simplified using spherical nomenclature (Figure 18); namely:

$$Li\{\alpha[O1Ir1O1(C\beta)(\underline{C}\beta)_3N\beta C1:^{(9\text{-}19)}(\beta);^{(5\text{-}17')}(1);\,^{(5)}\aleph_5(CpH)]\}_3\,^{(+)} \qquad (47)$$

Observe that in (47), the sub-cycle being nomenclated started at locant #3 and progressed through locant #20. Additionally, the connection between the sub-cycles is by means of an unprimed - primed bridge (5-17' for this cluster).

Before turning to another molecule of interest, a further word on the two different sets of oxygen atoms is in order at this time. Focusing attention on Figure 18, the oxygen atom with locant #7 has a coordination of two with a single bond to a carbon atom of the pyridine ring and another single bond to the iridium atom, while the oxygen atom with locant #3 has a coordination of three with similar bonds to the iridium atom and to a different carbon atom of this same pyridine ring, as well as a third bond to the coordinating lithium atom. Once again the desire for consistency in the assignment of nomenclature has raised some ***not inconsequential*** questions: What bond orders should be used in the nomenclature?; Should different valences be assigned to these oxygen atoms?; Is there some type of fluxional relationship that is needed in order to be able to accurately

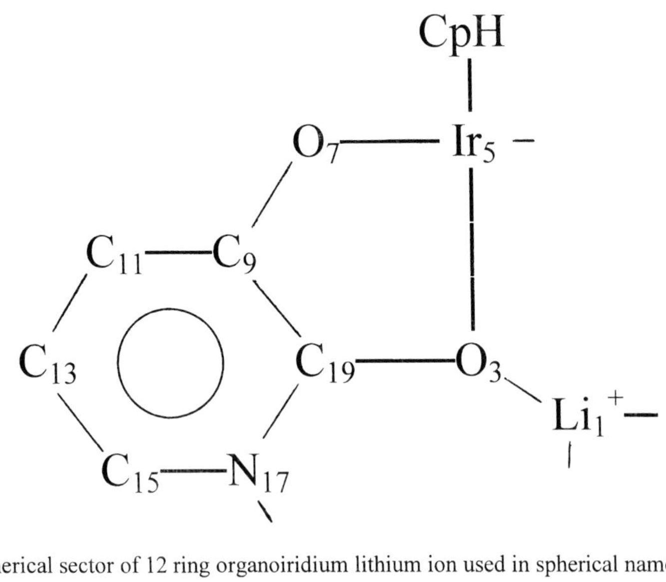

Fig. 18: Spherical sector of 12 ring organoiridium lithium ion used in spherical name

$$\begin{array}{l}
C_{19''} - \underline{\underline{C}}_{1} - O_{3} - \underline{\underline{C}}_{5} - \underline{\underline{C}}_{7} - O_{9} - \underline{\underline{C}}_{11} - \underline{\underline{C}}_{13} - O_{15} - \underline{\underline{C}}_{17} - C_{19} \\
\| \qquad\qquad\qquad\qquad\qquad\qquad\qquad\qquad\qquad\qquad\qquad\qquad\qquad \| \\
C_{21''} - \underline{\underline{C}}_{39} - O_{37} - \underline{\underline{C}}_{35} - \underline{\underline{C}}_{33} - O_{31} - \underline{\underline{C}}_{29} - \underline{\underline{C}}_{27} - O_{25} - \underline{\underline{C}}_{23} - C_{21}
\end{array}$$

Fig. 19: The module in the cylindrical counterpart to moebiane (see figure 5 in chapter 3) that is used in forming the spherical name

nomenclate this compound?; etc. Furthermore, had the focus been on Figure 17, one might speculate whether the three oxygen atoms of the bridges (which did not have locant numbers assigned to them) could function in the same manner as did the locant numbered oxygen atoms; namely, that they bond to another lithium ion; thereby polymerizing. Might such a perspective be useful in gaining a better understanding of the pharmacology of bipolar disorder?

Another example wherein spherical nomenclature introduces simplicity is the untwisted cylinder that was isomorphic with Moebiane (see Figure 6 in Chapter 3). By viewing this molecule as a triple repeat of the 20 atom sequence illustrated in Figure 19, the spherical canonical name for this molecule becomes:

$$[(\underline{\underline{C}}1O1\underline{\underline{C}}1)_3C2C1(\underline{\underline{C}}1O1\underline{\underline{C}}1)_3:^{(19\text{-}1',21\text{-}39')}(1)]_3 \qquad (48)$$

which is much simpler than the Cartesian name given as (11) in Chapter 3:

$$\{[O1(\underline{\underline{C}}1)_2]_2O1(\underline{\underline{C}}1C1\underline{\underline{C}}1[O1(\underline{\underline{C}}1)_2]_2O1\underline{\underline{C}}1C2C1\underline{\underline{C}}1\}_2: {}^{(37\text{-}81,39\text{-}79)}[1\underline{\underline{C}}1(O1(\underline{\underline{C}}1)_2)_2]O1\underline{\underline{C}}1;^{(17\text{-}59,37\text{-}39,79\text{-}81)}(2) \qquad (3\text{-}11)$$

Some further comments about such nomenclature include:

(1) As well as the spherical name being much shorter, by virtue of the priming convention introduced above, no locant numbers above 40 are required.

(2) Had the two "ladder side-rails" been connected by diradical oxygen atoms instead of double bonds, the resulting graph would have been a fusion of three 24crown8 ether modules, with the order of the fusion determining whether the molecule was cylindrical or unorientable (Moebiane).

(3) Although the third sequence of atoms in Moebiane (Figure 5 in Chapter 3) is the same as the first two, the set of three connecting double bonds are not parallel; therefore, there does not exist a simple primed-unprimed sequential connection of linear strands, which would lend itself to a simplified nomenclature. Instead, all attempts to set up an algorithm in which the desired strands are joined are strictly ad hoc, with different matings for triple repeat vs. quadruple repeat, etc. Consequently, no advantage in any spherical nomenclature devised to date has been discerned and only the Cartesian nomenclature described in Chapter 3 seems practical for such molecules.

Further examples wherein the synergy between the spherical name and the structure of selected molecules simplifies the description of a molecule include the calixarenes*. Before examining what is generally recognized as the smallest member of this set, calix[4]arene, one of the mathematical virtues, as well as physical liabilities, of any consistent nomenclature (namely, the object being nomenclated does not have to exist in the physical world) is reiterated. In order to be able to form a viable molecule, the metric distance between the two benzene rings in the two meta connected chains of calix[2]arene would have to be much longer than a single methylene unit; i.e., such a molecule is sterically impossible without either more methylene links in the chains (i.e., a more typical cyclophane) or replacement of the single bonds in the chain with acetylenic chain extendors (the code for such an extendor in the proposed nomenclature is $\mathbf{1(C3C1)_n}$ in place of a **1**). Nevertheless, despite this steric impossibility, there do exist consistent Cartesian (Figure 20) and spherical (Figure 21) names for such a construct:

$$\mathrm{O1(C\beta)_2(\underline{C}\beta)_3C1\underline{\underline{C}}1C\beta(\underline{C}\beta)_3C\beta C1O\alpha H\alpha{:}^{(5\text{-}25)}(1\underline{\underline{C}}1);^{(1\text{-}29)}(\alpha H\alpha);^{(3\text{-}13,17\text{-}27)}\beta} \quad (49)$$

and

$$\mathrm{[\underline{\underline{C}}1C\beta(\underline{C}\beta)_3C1{:}^{(3\text{-}11)}(\beta C^{(=13)}\beta);^{(13)}(1O\alpha H\alpha{:}^{(3)}\alpha]_2} \quad (50)$$

Note that in drawing the line of symmetry that will be used in formulating the repeating part of the spherical name (50), the oxygen atom is not a part

---

* The relation to the larger group of substituted cyclophanes (see Figure 15 in Chapter 2) is noted; namely, calix[2]arene is an abbreviated name for 2,2'-dihydroxy-[1,1]-meta, metacyclophane.

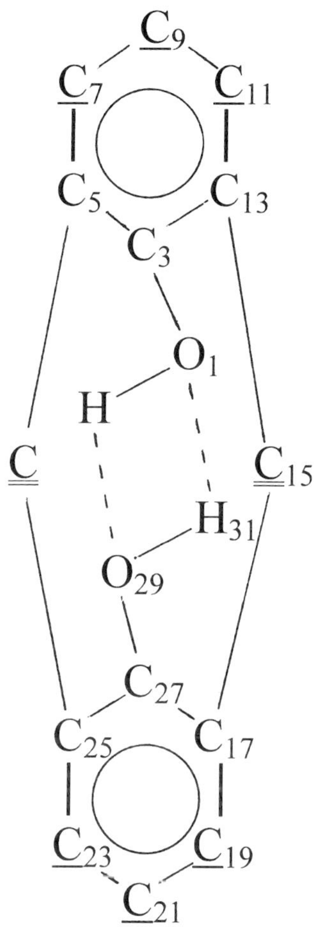

Fig. 20: Calix-2-arene with Cartesian locant designations

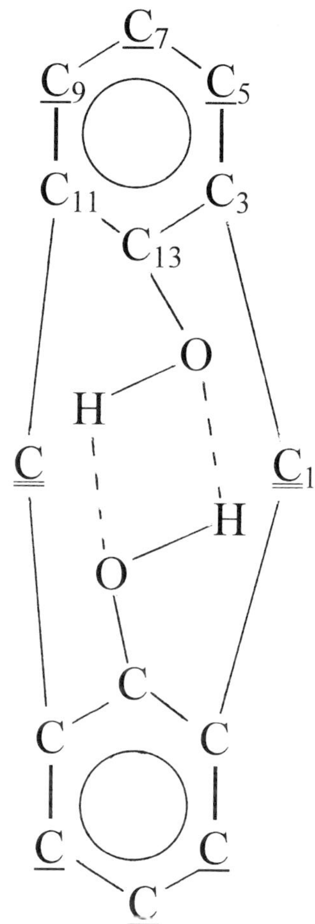

Fig. 21: Calix-2-arene with spherical locant designations

of any cycle and thus is relegated to a branch. This is in contradistinction to its prominent position in the Cartesian name.

Upon examining next that aggregation of atoms that would be known as calix[3]arene, nearly similar metric problems are in evidence. Namely, without the hydroxyl groups the molecule would approach coplanarity; however, the hydroxyl groups introduce Coulomb repulsion resulting in extreme lability. As above, despite the non-viability of such a molecule,

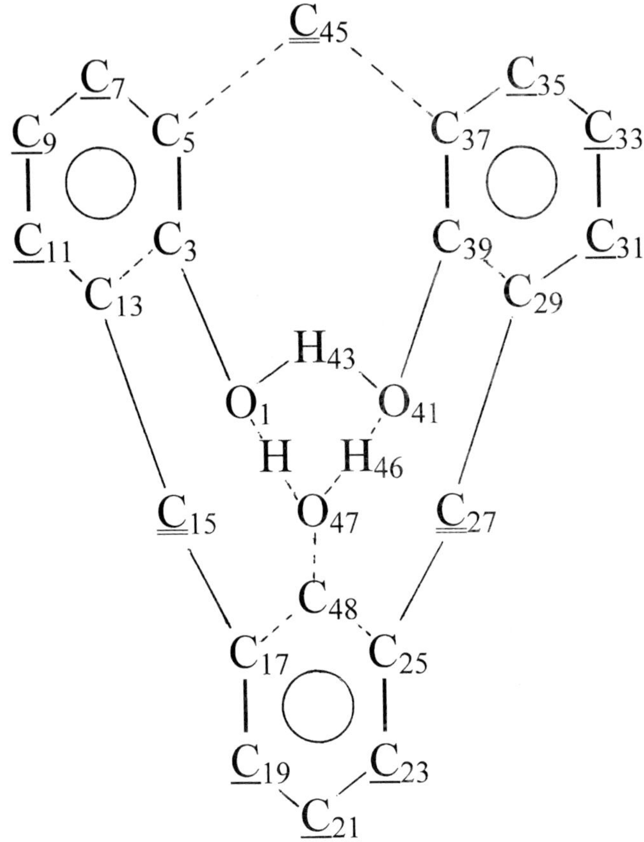

Fig. 22: Calix-3-arene with cartesian locant designations

both the Cartesian (Figure 22) and the spherical (Figure 23) names for this moiety are easily created; namely:

$$O1(C\beta)_2(\underline{C}\beta)_3C1\underline{\underline{C}}1C\beta(\underline{C}\beta)_3C\beta\underline{\underline{C}}1C\beta(\underline{C}\beta)_3C\beta C1O\alpha H\alpha{:}^{(5\text{-}37)}(1\underline{\underline{C}}^{(=45)}1);{}^{(1\text{-}17)}(\alpha H^{(=46)}\alpha O^{(=47)}1C^{(=48)}\beta);{}^{(41\text{-}47)}(\alpha H\alpha);{}^{(3\text{-}13,25\text{-}48,29\text{-}39)}(\beta) \qquad (51)$$

and

$$[H\alpha O1(C\beta)_2(\underline{C}\beta)_3C1\underline{\underline{C}}1{:}^{(7\text{-}17')}(1);{}^{(1\text{-}3')}(\alpha)]_3 \qquad (52)$$

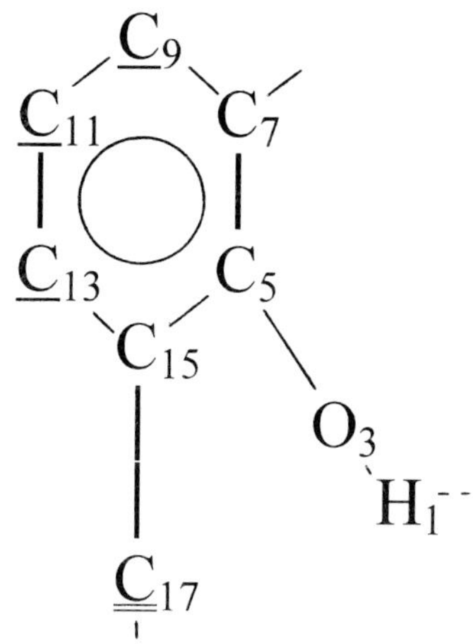

Fig. 23: Spherical sector of calix-3-arene used to form spherical name

respectively.

From a pragmatic perspective, it is customary to consider the smallest calixarene as calix[4]arene (Figure 24). This four benzene ring molecule has as its Cartesian name:

$$[\mathrm{O1(C\beta)_2(\underline{C}\beta)_3C1\underline{\underline{\underline{C}}}1C\beta(\underline{C}\beta)_3C\beta C1O\alpha H\alpha}]_2{:}^{(3\text{-}13,17\text{-}27,35\text{-}45,49\text{-}59)}(\beta);^{(1\text{-}29,33\text{-}61)}(\alpha\mathrm{H}\alpha);^{(25\text{-}37,57\text{-}5)}(1\underline{\underline{\underline{\mathrm{C}}}}1)] \qquad (53)$$

and the much shorter spherical name:

$$[\mathrm{H\alpha O1(C\beta)_2(\underline{C}\beta)_3C1\underline{\underline{C}}\alpha}{:}^{(7\text{-}17')}(1);^{(1\text{-}3')}(\alpha)]_4 \qquad (54)$$

Note that one did not have to draw the spherical sector for calix-4-arene as it is identical to the module illustrated in Figure 24. The only difference in the spherical name is the subscript denoting how many of these modules are combined to form the molecule. See spherical names (52) and (54). Upon comparing Cartesian names (51) and (53), on the other hand, one sees a familial, but not exact, duplicativity.

Further comments concerning symmetry and nomenclature include:

(1) Another molecule, described earlier, whose spherical name is more efficient than the Cartesian name prescribed was illustrated in Figures 21 and 22 in Chapter 2. IUPAC names this molecule as:

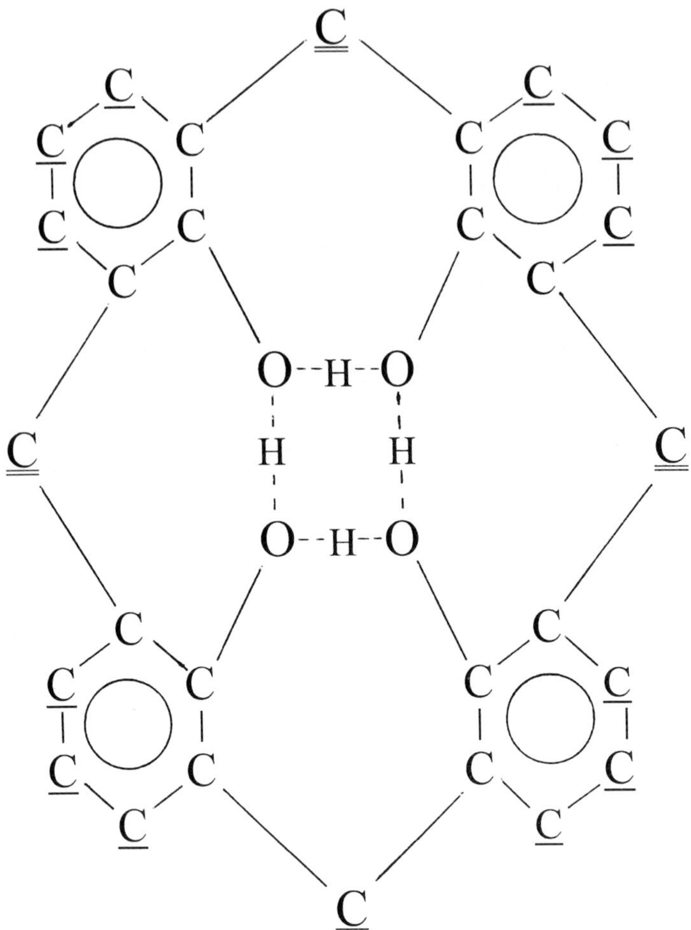

Fig. 24: Calix-4-arene with cartesian locant designations

[μ-(1,2,3,4-η-:5,6,7,8-η-cycloocta-1,3,5,7-tetraene)]bis (tricarbonyliron). The systemic Cartesian name (30) Chapter 2) was:

$[Fe\aleph(\underline{C}\beta)_3\underline{C}\aleph]_2{:}^{(3\text{-}19,9\text{-}13)}(\beta);^{(1\text{-}15,1\text{-}17,5\text{-}11,7\text{-}11)}(\aleph);\ ^{(1,1,1,11,11,11)}(1C3O)$

The systemic spherical name for this molecule is:

$$[\mathrm{Fe}\aleph(\underline{\mathrm{C}}\beta)_3\underline{\mathrm{C}}\aleph{:}^{(9\text{-}3')}(\beta);\,^{(1\text{-}5,1\text{-}7,1\text{-}9')}(\aleph);\,^{(1,1,1)}(1\mathrm{C}3\mathrm{O})]_2 \qquad (55)$$

(2) As indicated earlier, when symmetry is not evident, there is no advantage in either cylindrical or spherical nomenclature. An example of this is the ruthenium-carborane complex illustrated in Figure 25, which has as its Cartesian name:

$$\mathrm{Ru}1(\mathrm{C}1)_2\mathrm{B}\alpha\mathrm{H}\alpha\mathrm{Ru}\alpha\mathrm{H}\alpha\underline{\mathrm{B}}1\underline{\underline{\mathrm{B}}}1{:}^{(1\text{-}11,7\text{-}17)}(\alpha\mathrm{H}\alpha);^{(1\text{-}15,3\text{-}15)}(1);^{(3,5)}(1\underline{\underline{\mathrm{C}}}1\mathrm{H});\ ^{(1,11)}\aleph_5(\mathrm{CpH}) \qquad (56)$$

Note that the constitution indicated in Figure 25 has a hydrogen atom in the path between the two ruthenium atoms; consequently, there is an α bond between the two ruthenium atoms vs. a single bond between a ruthenium and a carbon atom. Examining the set of single bonds emanating from either of the ruthenium atoms, the lowest locant numbering has one of the carbon atoms as locant #3 in the canonical name. Whether this is the correct geometry or not is left for future study; however, inasmuch as no simplifying symmetry is evident, there is no reason to expect a different geometry would

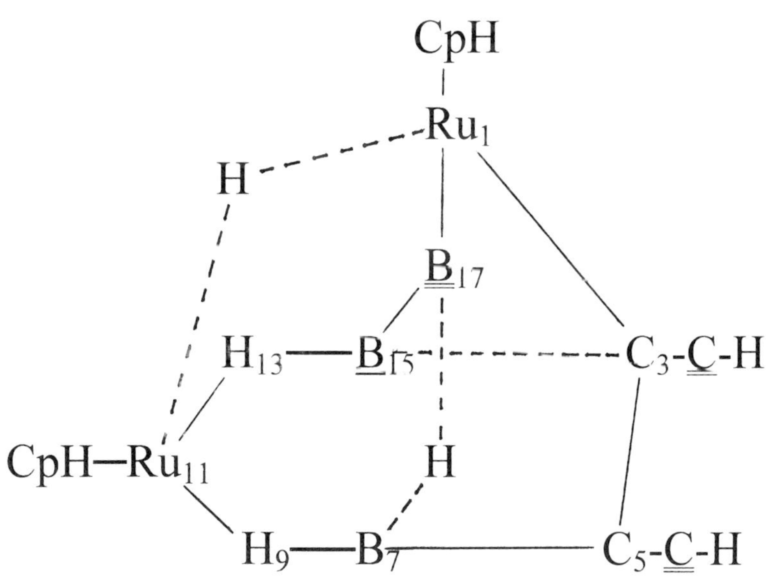

Fig. 25: A metallocarborane with no evident symmetry

minimize the Coulomb forces between atoms; therefore, the Cartesian name is recommended.

Returning the focus to spiro aggregations, the next class of molecules to be considered is those having more than one spiro atom. For these cases, expansion of the above methods often become confusing; consequently, an alternate method, called the "redundant path" method, is introduced. Note that, although this method **could** be used to canonically name all spiro compounds (single as well as multiple), pragmatically its main utility arises when other methods are cumbersome. A secondary value of this coding method is the immediate identification of a spiro atom by a single superscripted equal sign and of a bridge by a sequence of superscripted equal signs.

Just as an alternate view of rings, in terms of linear compounds with a "bridge" produced possible names for cyclic compounds [see formulas (23) and (24) for cyclohexane and benzene respectively], one can, similarly, name spiro compounds by cutting and re-pasting them at the spiro atom. Unlike bridging, however, for multiple spiro atoms, a single path would have to pass through some spiro atoms more than once. This is described in the nomenclature by a superscripted equality. In this process note that the particular atom repeated is not counted in determining the number of atoms in a chain. Moreover, in order to minimize locant numbering, the path chosen traces through the smallest ring first; rather than the longest one . Using this technique, the respective coding for the spiro compounds shown in Figures 1 and 2 become:

$$C1(\underline{\underline{C}}1)_5C^{(13=1)}1(\underline{\underline{C}}1)_5 \qquad (57)$$

and

$$S1(\underline{\underline{C}}1)_3S^{(9=1)}1(\underline{\underline{C}}1)_4S^{(19=1)}1(\underline{\underline{C}}1)_5 \qquad (58)$$

Comparing the three names developed so far for spiro compounds [(1) vs. (3) vs. (57)], and [(2) vs. (4) vs. (58)], one finds that the choice of which name is "better" is a strictly heuristic one. For example, choose the spherical name whenever the rings involved are congruent. At the other extreme, to use redundant paths for metallocenes would be, at best, pedantic, requiring the path to pass through the iron atom in ferrocene ten times. Additionally, as shall be demonstrated below, the redundant path technique is often the

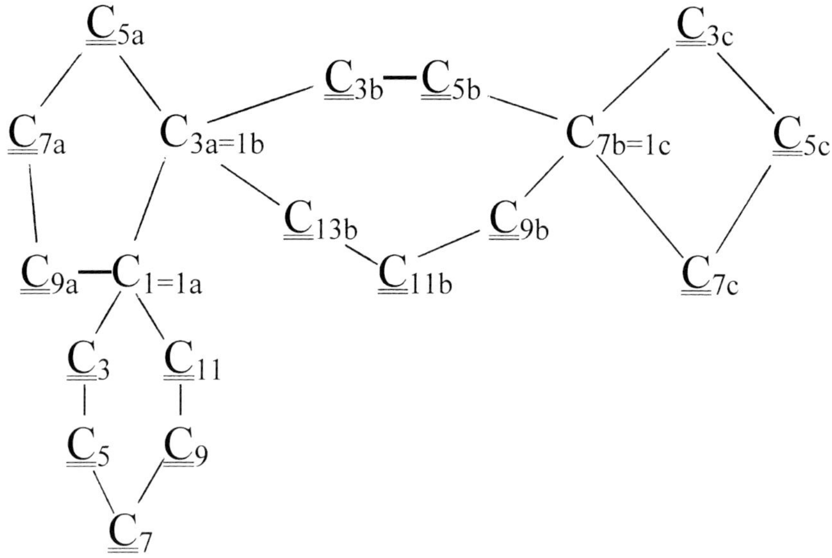

Fig. 26: A multiple singly spiro compound - bridged nomenclature

logical one to expand upon when nomenclating ring assemblies. This is done by including superscripted equalities for edges, as well as nodes.

Applying the above two protocols to the triply-spiro structure shown in Figures 26 and 27, locant numbers for the bridged method are minimized by starting at that spiro atom which connects the larger of the two terminal rings (6 atoms in this case) to the remainder of the molecule. In this ring the locant numbers are unprimed integers. The adjacent ring is now traversed along the shortest path to the next spiro atom. The locant numbers in this second ring are augmented with either a prime or the lower case letter **a**. A similar process is continued into the third ring where the locant numbers are followed with either a double prime or the lower case letter **b**; etc.* The resulting name for Fig. 26 is thus:

$$C1(\underline{C}1)_5{:}^{(1\text{-}1)}(1C1(\underline{C}1)_3{:}^{(3'\text{-}3')}\{1(\underline{C}1)_2C1(\underline{C}1)_3{:}^{(7''\text{-}7'')}[1(\underline{C}1)_3]\}) \qquad (59)$$

* The use of triple or higher primes is avoided and lower case letters are recommended, inasmuch as higher primes are both harder to read and also not part of the usual keyboard letters.

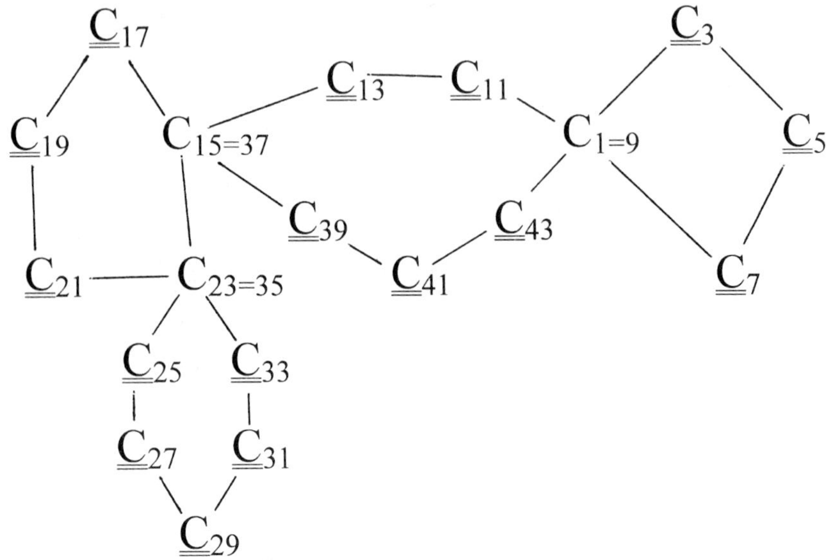

Fig. 27: A multiple singly spiro compound - redundant path nomenclature

if, as in this case, there is no desire to focus on atoms in a fourth, or higher, ring or else :

$$\mathrm{C1}(\underline{\underline{\mathrm{C}}}\mathrm{1})_5{:}^{(1\text{-}1)}(\mathrm{1C1}(\underline{\underline{\mathrm{C}}}\mathrm{1})_3{:}^{(3a\text{-}3a)}\{1(\underline{\underline{\mathrm{C}}}\mathrm{1})_2\mathrm{C1}(\underline{\underline{\mathrm{C}}}\mathrm{1})_3{:}^{(7b\text{-}7b)}[1(\underline{\underline{\mathrm{C}}}\mathrm{1})_3]\}) \qquad (60)$$

if there is. Alternately, the locant numbering and name using the redundant path method (Fig. 27) is determined starting from the spiro atom connecting the smaller terminal ring to the residue:

$$\mathrm{C1}(\underline{\underline{\mathrm{C}}}\mathrm{1})_3\mathrm{C}^{(9=1)}1(\underline{\underline{\mathrm{C}}}\mathrm{1})_2\mathrm{C1}(\underline{\underline{\mathrm{C}}}\mathrm{1})_3\mathrm{C1}(\underline{\underline{\mathrm{C}}}\mathrm{1})_5\mathrm{C}^{(35=23)}\mathrm{1C}^{(37=15)}1(\underline{\underline{\mathrm{C}}}\mathrm{1})_3 \qquad (61)$$

Meanwhile observe that if there is a bond, even an α bond, between two rings whose only other connection is through a spiro atom, this compound is now classified as either rectangular or fisular and **must** be named as such. For example, by the inclusion of a single bond between the two carbon atoms numbered 7 and 7c (or 7''') in Figure 26, there is a different locant numbering that is applicable for this 19 atom ring structure. In Figure 28 NONE of the previously spiro connections are regarded as important, no less significant. Instead the molecule would now contain three

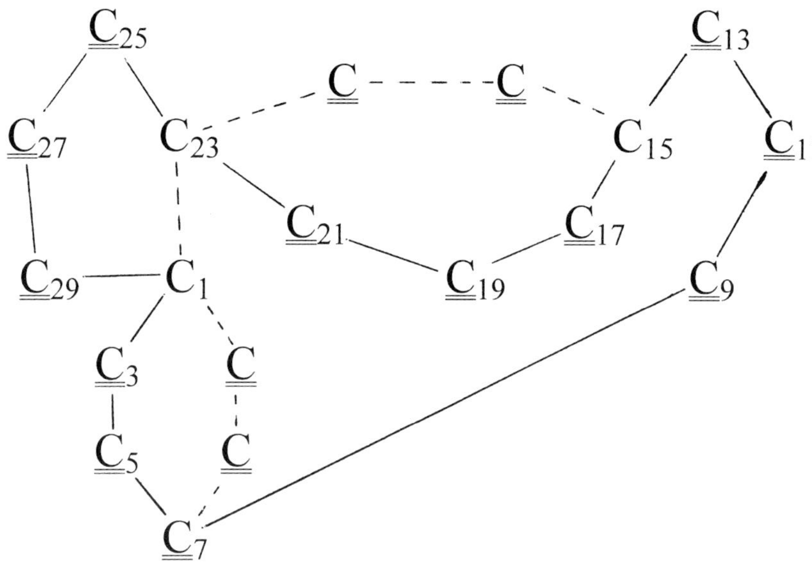

Fig. 28: A reticularly bridged compound

additional bridges and would be classified as reticular, with the systemic name of:

$$\mathrm{C1(\underline{\underline{C}}1)_2(\underline{\underline{C}}1)_2(\underline{\underline{C}}1)_2C1(\underline{\underline{C}}1)_3{:}^{(1\text{-}7,15\text{-}23)}(1(\underline{\underline{C}}1)_2);^{(1\text{-}23,9\text{-}15)}(1)} \qquad (62)$$

As well as the existence of many traditional organic compounds having multiple spiro (carbon) atoms, a comparable scenario is also found in the traditional inorganic domain. For example, attention is directed to an important industrial palladium catalyst* [15], which, by the perspective of considering hydrogen bonding as forming rings, may be considered as a multiple spiro compound. The respective Cartesian (Figure 29) and redundant path (Figure 30) systemic names for this compound are:

$$\mathrm{Pd1P1O\alpha H\alpha O1P1{:}\{^{(1\text{-}1)}(1C\ell 1Pd1C\ell 1){:}^{(5\text{-}5)}(1P1O\alpha H\alpha O1P1){:}} \\ \mathrm{^{(3,3,11,11)}[1C(1\underline{\underline{C}}1H{:})_3]\};^{(3,3,11,11)}[1C(1\underline{\underline{C}}1H{:})_3]} \qquad (63)$$

and

---

* DuPont's POPd1

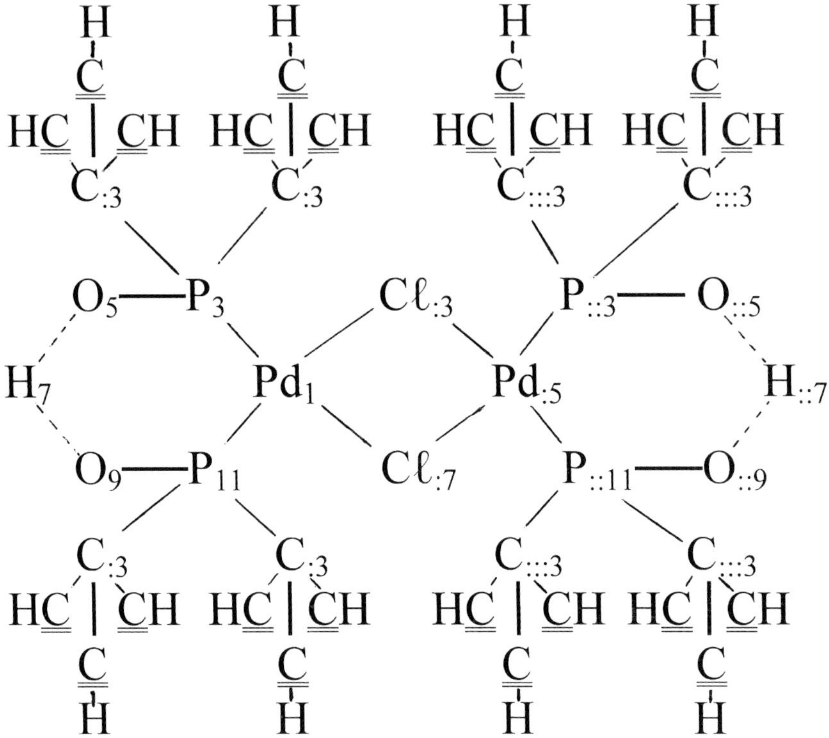

Fig. 29: A multi-spiro organometallic compound - cartesian locant numbering

$$\mathrm{Pd1C\ell1Pd1P1O\alpha H\alpha O1P1Pd^{(17=5)}1C\ell1^{(21=1)}Pd1C\ell1Pd1P1O\alpha H\alpha O1P1}$$
$$:^{(7,7,15,15,23,23,31,31)}\,[\mathrm{1C(1\underline{C}1H:)_3}] \qquad (64)$$

Note that by counting the number of colons in a sequence between semicolons, in formula (63) one can determine whether a given atom in the canonical name is at the secondary vs. tertiary vs. etc. levels.

As mentioned above, a logical extension of the spiro connection is that of a ring assembly, wherein two rings are joined, not at a common vertex, as in spiro compound, but rather by a single edge, or a sequence of edges. In other words, the removal of a bond (an edge in the graph), but NOT the two vertices (atoms) it connects, disconnects the graph. To assign a canonical name to such a ring assembly as well as to multiple ring

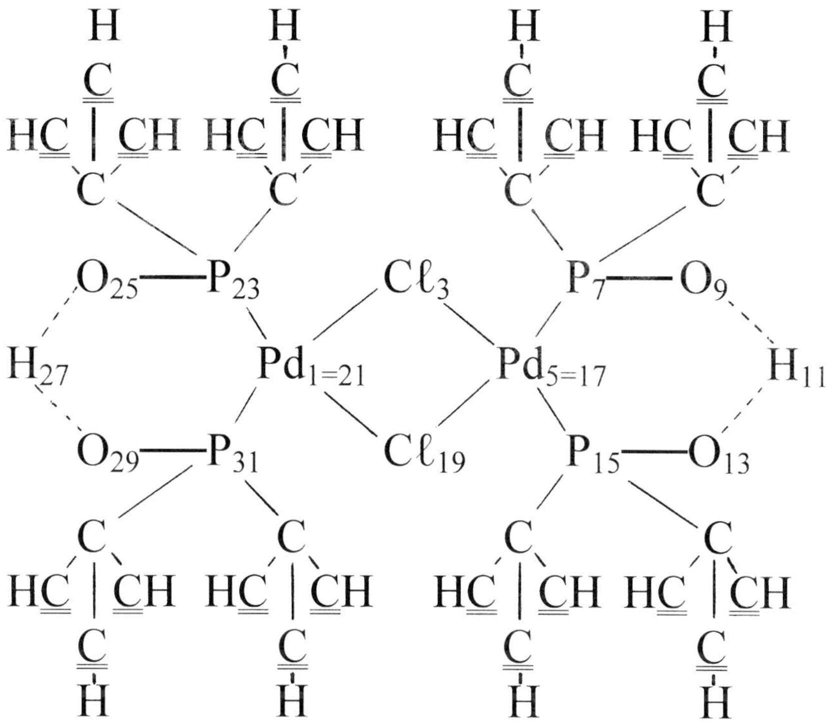

Fig. 30: A multi-spiro organometallic compound - redundant path locant numbering

assemblies, one may use either the bridged or the redundant path methods described above for spiro compounds. This is illustrated in Figure 31 for the double ring assembly alkane combination of five, five and four member rings using the bridged method (65), and the redundant path method (66). Note that the molecule pictured in Figures 31 and 32 was formed by changing the spiro connections in the seven member and six member rings of Figures 26 through 28 into bridges with the formation of two five member rings. The respective systemic names thus became:

$$(\underline{\underline{C}}1)_3(\underline{\underline{C}}1)_2(\underline{\underline{C}}1)_2(\underline{\underline{C}}1)_2(\underline{\underline{C}}1)_3(\underline{\underline{C}}1)_2(\underline{\underline{C}}1)_3\underline{\underline{C}}:^{(9\text{-}15)}(1\underline{\underline{C}}1);^{(1\text{-}7,17\text{-}25,27\text{-}35)}(1) \qquad (65)$$

and

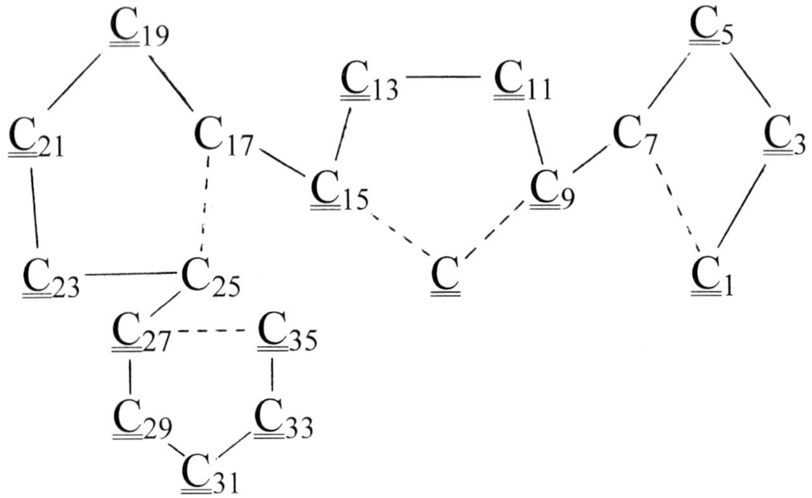

Fig. 31: Nomenclating a multiple ring assembly – open path bridged method

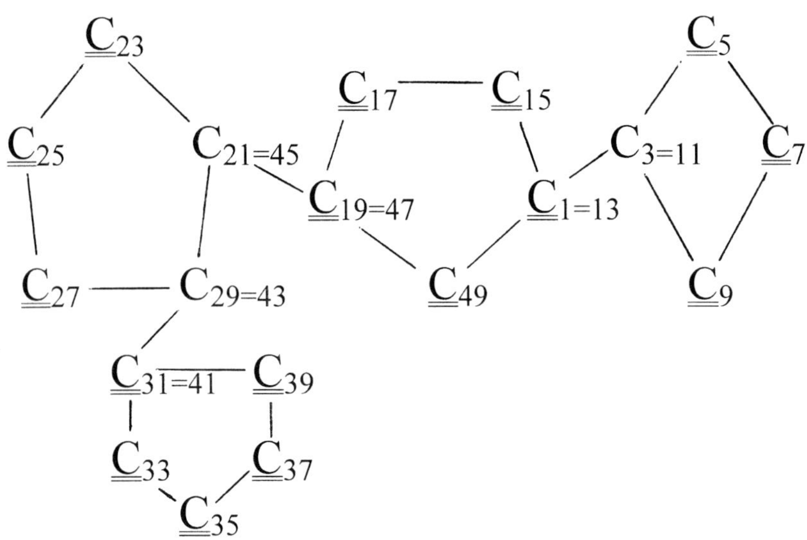

Fig. 32: Nomenclating a multiple ring assembly - redundant path cycle method

$$(\underline{\underline{C}}1)_2(\underline{\underline{C}}1)_3\underline{C}^{(11=3)}1^{(12=2)}\underline{C}^{(13=1)}1(\underline{\underline{C}}1)_2(\underline{\underline{C}}1)_2(\underline{\underline{C}}1)_3C1(\underline{\underline{C}}1)_5C^{(41=31)}1^{(42=30)}C^{(43=29)}1\ \underline{C}^{(45=21)}1^{(46=20)}\underline{C}^{(47=19)}1\underline{\underline{C}}1 \qquad (66)$$

Although a comparable scenario to the "spiro" vs. "reticular" classifications of "organic chemistry" can be found in coordination compounds, in the traditional nomenclature of inorganic chemistry, IUPAC has chosen to handle it in a diametrically opposite manner. Figures 33 and 34 are examples 4 and 5 in Paragraph 7.314 of [16]. Note that here the IUPAC nomenclature focus is on the salicylic acid derivative of the "anion", with the "bridging" aspect of the ethylene being relegated to a minor role, in contrast to its pre-eminence from an organic chemistry perspective. In contrast to these IUPAC names, the systemic names, as shown for Figures 33 and 34 respectively, appear to be:

$$\mathrm{Cu[1O2\underline{C}1C\beta(\underline{C}\beta)_4C1O1:^{(7\text{-}17)}(\beta);^{(13)}(1F)]_2} \qquad (67)$$

and

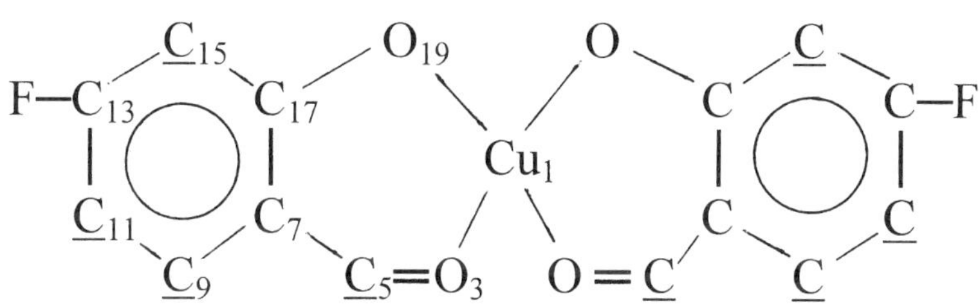

Fig. 33: Molecule having IUPAC inorganic name: bis(4-fluorosalicylaldehydato) copper (II)

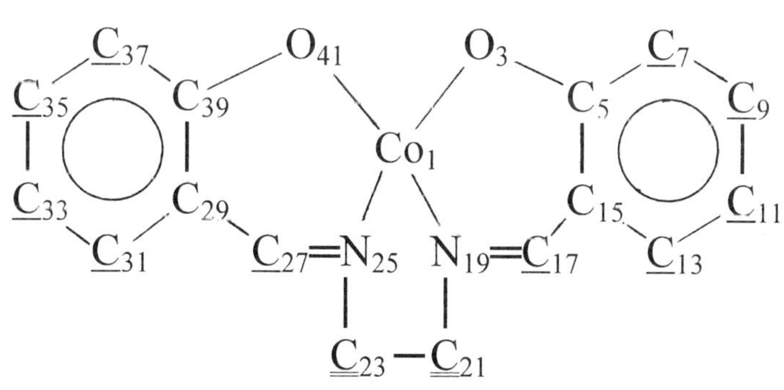

Fig. 34: Molecule having IUPAC inorganic name: N,N'-ethylenebis(salicylidene iminato)cobalt(II)

$$\mathrm{Co1O1C\beta(\underline{C}\beta)_4C1C2N1(\underline{\underline{C}}1)_2N2C1C\beta(\underline{\underline{C}}\beta)_4C1O1:^{(5\text{-}15,29\text{-}39)}(\beta);^{(1\text{-}19,1\text{-}25)}(1)} \tag{68}$$

However, some additional comments about IUPAC's representation in Figures 33 and 34 should be recognized:

(1) The electron pairs on the oxygen and nitrogen atoms, along with the two electrons on the metal atom suggest that the conjugation be extended through four rings, rather than there being conjugation only in the two benzene rings. In the five member ring of Figure 34 there no conjugation; however, since this is a five member ring coplanarity throughout the entire molecule is expected. This is in contrast to Figure 33 wherein the component ring pairs are most likely perpendicular to one another. This idea will be revisited in the discussion of biphenyl vs. biphenylene later in this chapter.

(2) The apparent valence of 3, which was, as expected in Figure 34 for the nitrogen atoms, is explained for the oxygen atoms in Figure 33 by the bonds being an α and a β bond, rather than the illustrated single and double bonds — in other words, the traditional valence of two for oxygen is NOT discarded. This same α and β bond explanation applies in both figures to the other oxygen atom (at the top of the figure) as well as the coordination of 4, but valence of 2 for the "spiro" metal atom. Moreover the unpaired electrons in the fluorine atoms contribute to stabilizing the molecule shown in Figure 33. This is reflected in the nomenclature using beta bonds where they were not traditionally applicable; namely the fluorine-to-carbon bond in the ring. In other words, (67) and (68) become respectively:

$$\mathrm{Cu[\alpha O\beta\underline{C}\beta C\beta(\underline{C}\beta)_4C\beta O\alpha:^{(7\text{-}17)}(\beta);^{(13)}(\beta F)]_2} \tag{69}$$

and

$$\mathrm{Co\alpha O\beta C\beta(\underline{C}\beta)_4C\beta\underline{C}\beta N1(\underline{\underline{C}}1)_2N\beta\underline{C}\beta C\beta(\underline{\underline{C}}\beta)_4C\beta O\alpha:^{(5\text{-}15,29\text{-}39)}(\beta);^{(1\text{-}19,1\text{-}25)}(\alpha)} \tag{70}$$

Returning focus to the traditional organic domain, one notes that the bridged method is especially useful for naming disjoint rings connected by one or more atoms in a linear chain, such as the molecule shown in Fig. 35. Note that IUPAC names this molecule with a nested set of integers as prefix and ligand groupings as suffix: 2-{4- [4-(2-Bromoethyl)phenyl]butyl}cyclo hexanecarboxylic acid [17]. By contrast, in the system being formulated, rather than a confusing sequence of integers followed by a reverse order

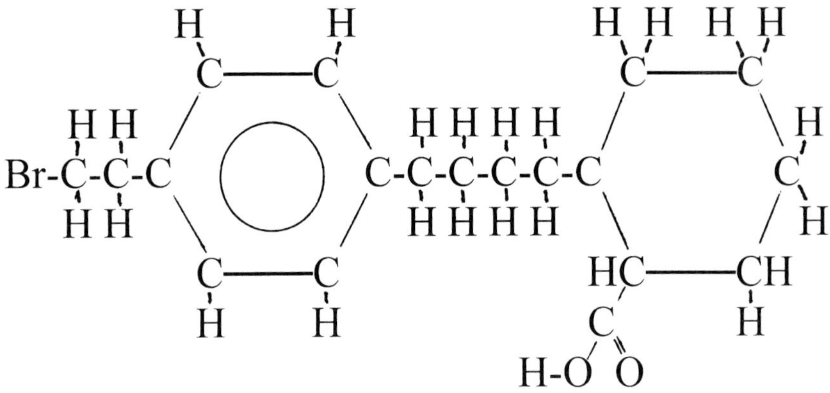

Fig. 35: Molecule IUPAC names using nested sequence of integers and ligands

sequence of ligand groups, interspersed with a jumble of different shaped separators (curved brackets, square brackets, parentheses), no change in protocol is desired. The canonical name is created by, simply, naming the longest continuous chain, augmented with bridges, as needed:

$$\mathrm{Br1(\underline{C}1)_2C\beta(\underline{C}\beta)_2C1(\underline{C}1)_4\underline{C}1(\underline{C}1)_5C1O1H{:}^{(7\text{-}13)}[\beta(\underline{C}\beta)_2];^{(23\text{-}33)}(1);^{(35)}(2O)} \qquad (71)$$

One particular domain in which the alternate method of naming is of significance is that of natural products (perhaps, because of the occurrence of many ring assemblages). Note that this is a domain which, because of the complexity of the IUPAC nomenclature, has opted to formulate its own set of parochial rules of nomenclature, in much the same way as the organic chemistry community formulated IUPAC nomenclature to include 35 "basis" aromatic compounds on which all other "comparable" compounds were to be named [18]. This is in contradistinction to various systematic approach, such as a geometry-based proposals for the fusion of benzene modules [19] and for general arenes [20].

In the now well-entrenched sub-discipline ("fiefdom") of natural products, Buckingham [21] advises: "There is one major departure from recommended IUPAC nomenclature rules that is made by almost all those publishing in natural products... that each name shall have only one principal group." In other words, a major subset of "organic" chemists do not subscribe to the official nomenclature of their own domain, and a uniform nomenclature of the type herein proposed would be no more "foreign" to this group than the supposedly agreed upon one. To illustrate this, consider the compound with the common name of cannogeninic acid, which the natural products community nomenclates as [22]: 3β,14β,-Dihydroxy-5β-card-20-enolid-19-oic acid.

Observe that this name requires one to know a new "parent [23] compound" (cardanolide) as well as its concomitant consensus (i.e., fiat) locant numbering [24]. In the proposed system, on the other hand, although the names are longer, there is no ***yet another table*** (of "parent" compounds) to memorize. Also, all aspects of the proposed name are straight-forward, including the allocation of locant numbers, so that the ability to correlate

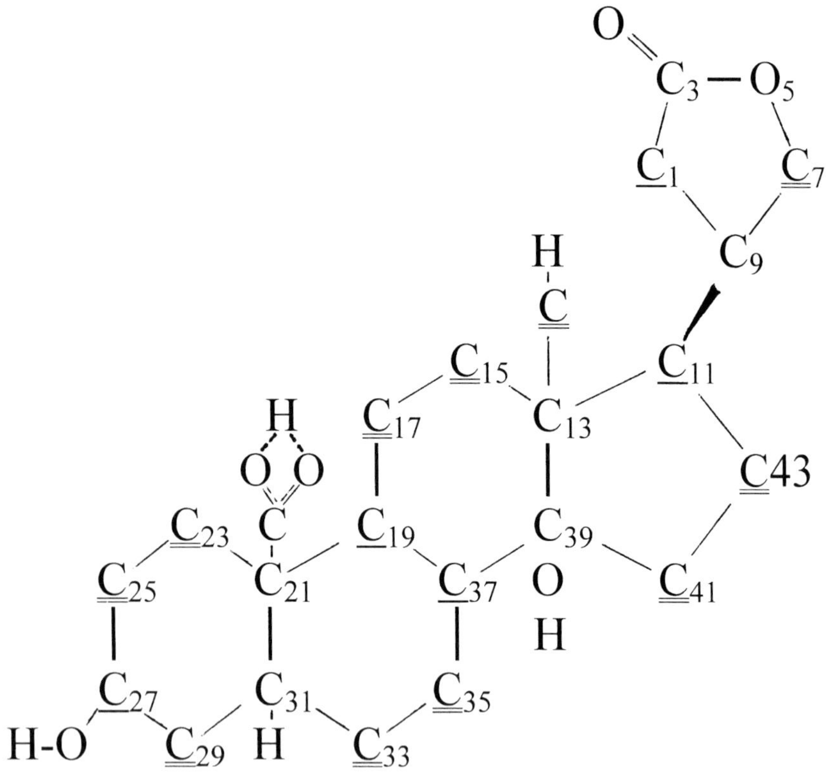

Fig. 36: Nomenclating cannogeninic acid using bridged locant numbering

names and structures is available to all, rather than just a small group of specialists. This is illustrated in Figures 36 (bridged) and 37 (redundant path):

$$\underline{C}1C1O1\underline{\underline{C}}1C1\underline{C}1C1(\underline{\underline{C}}1)_2\underline{C}1C1(\underline{\underline{C}}1)_2\underline{C}1\underline{\underline{C}}1\underline{C}1(\underline{\underline{C}}1)_2\underline{C}1C1\underline{\underline{C}}1\underline{\underline{C}}:^{(1\text{-}9)}(2);{}^{(\overline{11}\text{-}43,13\text{-}39,\overline{19}\text{-}37,21\text{-}\overline{31})}(1);{}^{3}(2O);{}^{13}(1\underline{\underline{C}}1H);{}^{21}[1(C\beta O\alpha H\alpha O\beta)];{}^{(27,39)}(1O1H) \quad (72)$$

and

$$\underline{C}1C2\underline{C}1C1O1\underline{\underline{C}}1\underline{C}^{(13=3)}1^{(14=2)}\underline{C}^{(15=1)}1C1(\underline{\underline{C}}1)_2\underline{C}1C1(\underline{\underline{C}}1)_2\underline{C}1\underline{\underline{C}}1\underline{C}1(\underline{\underline{C}}1)_2\underline{C}1C1\ \underline{\underline{C}}1\underline{\underline{C}}:^{(\overline{3}\text{-}5)}(2);{}^{(1\text{-}4\overline{7},1\overline{7}\text{-}\overline{43},23\text{-}41,25\text{-}35)}(1);{}^{(7)}(2O);{}^{(\overline{17})}(1\underline{\underline{C}}1H);\ {}^{5}[1(C\beta O\alpha H\alpha O\beta)];{}^{(\overline{31},\overline{43})}(1O1H) \quad (73)$$

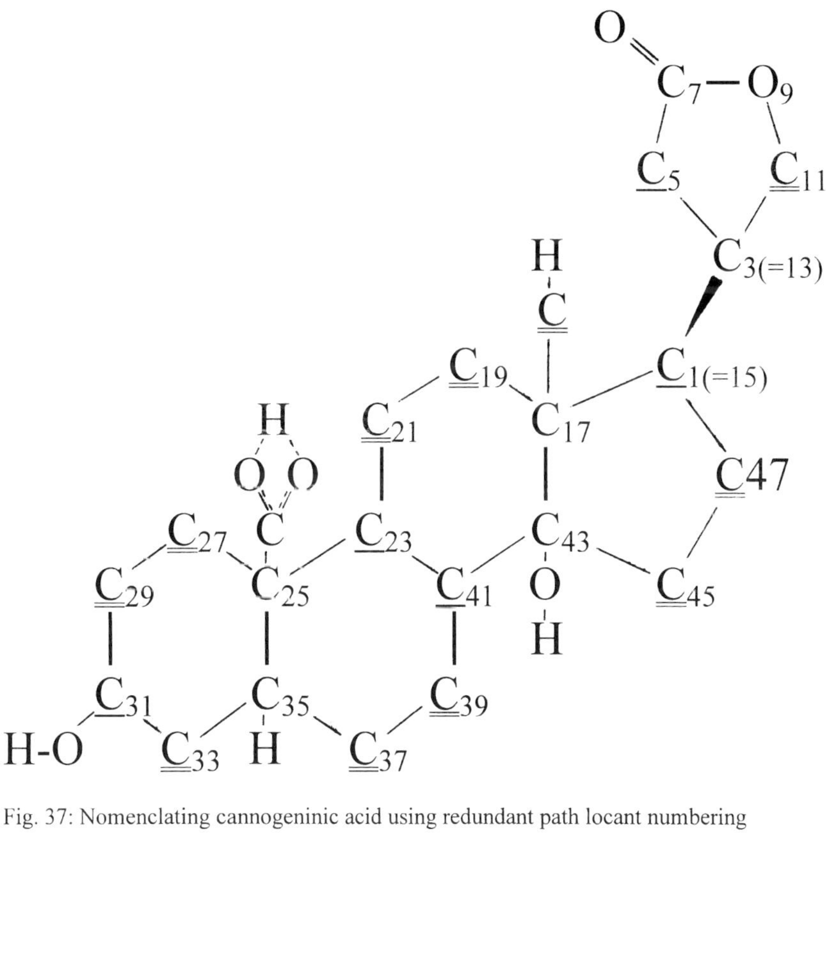

Fig. 37: Nomenclating cannogeninic acid using redundant path locant numbering

One further word is included here with respect to ring assemblies that involve modules that are intrinsically three-dimensional. As well as spiro compounds in which the joining edge of a ring assembly has been replaced by a single atom (limited to inorganic compounds due to the need of a coordination of 6 or higher), the following ideas shall be useful in examining the class of topological isomers (See Chapter 7). For this purpose, attention

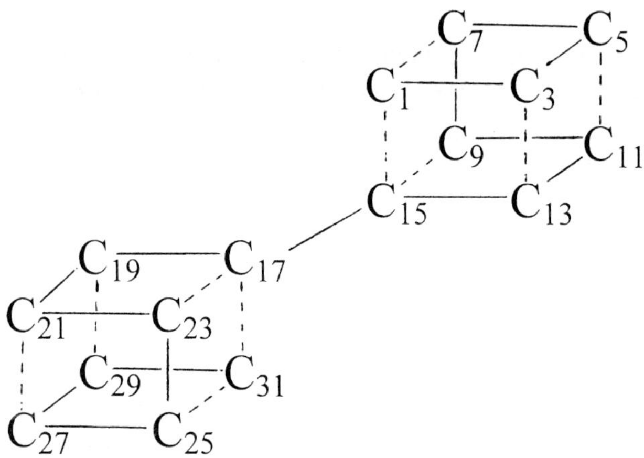

Fig. 38: Cubanylcubane - bridged nomenclature

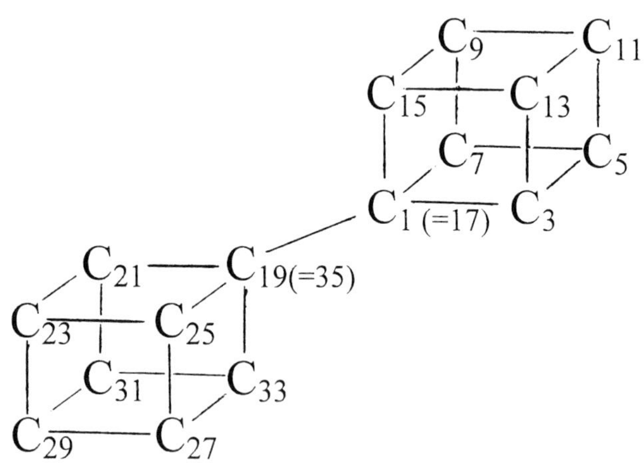

Fig. 39: Cubanylcubane - redundant path nomenclature

is focused on cubanylcubane [25]. Figure 38 and (74) illustrate the locant numbering and canonical name for the bridged nomenclature of this compound; while the redundant path nomenclature is presented in Figure 39 and (75).

$$(\underline{C}1)_7(C1)_2(\underline{C}1)_7:^{(1\text{-}7,1\text{-}15,3\text{-}13,5\text{-}11,17\text{-}23,17\text{-}31,19\text{-}25,21\text{-}27)}(1) \quad (74)$$

$$C1(\underline{C}1)_7C^{(=17)}1C1(\underline{C}1)_7C^{(=19)}1^{(=18)}:^{(1\text{-}7,3\text{-}13,5\text{-}11,9\text{-}15,19\text{-}25,19\text{-}33,21\text{-}31,23\text{-}29)}(1) \quad (75)$$

Having introduced the nomenclature for ring assemblies, the focus is now returned to some attributes associated with the use of fixed single and double bonds vs. beta bonds. Comparing biphenyl (Figure 40) and biphenylene (Figure 41), one notes that the connecting bond in biphenyl is a single bond, with the concomitant result that the pi cloud (aromaticity) of each ring is independent of the other. Furthermore, although each ring by itself is co-planar, these two rings form a non-zero dihedral angle in 3-space.

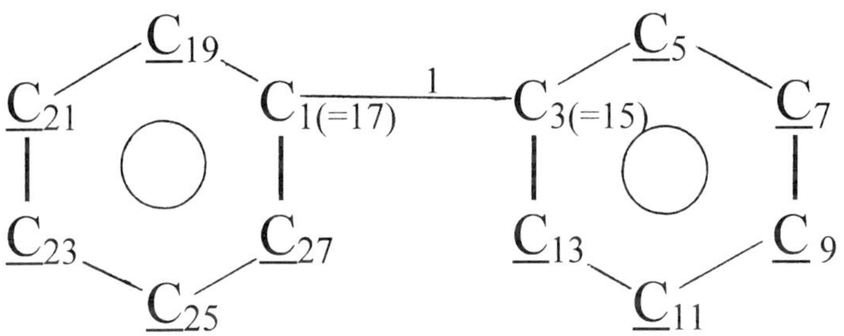

Fig. 40: Biphenyl

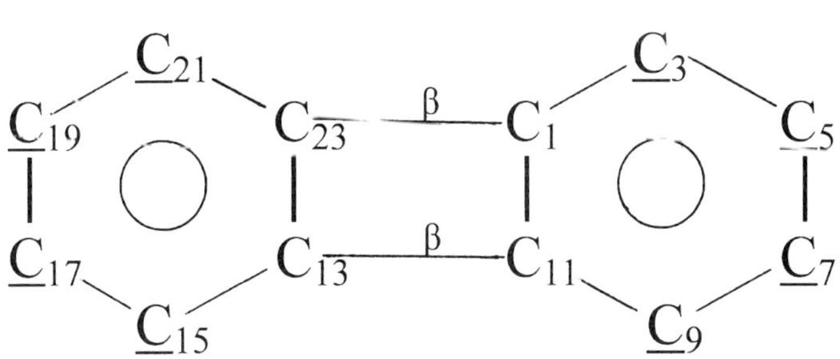

Fig. 41: Biphenylene

Consequently, the name assigned to this molecule is:

$$C1C\beta(\underline{C}\beta)_5C^{(15=3)}1^{(16=2)}C^{(17=1)}\beta(\underline{C}\beta)_5 \quad (76)$$

This is reflected in the nomenclature by the presence of the 1 (rather than a β) for the bond order of the edge between the rings. On the other hand, the presence of the second bond in biphenylene connecting the two rings (Figure 41) propels the two rings into a common plane and extends the aromaticity over the entire molecule. For biphenylene the two connecting bonds are part of an extended pi cloud. Moreover, the nomenclature for biphenylene indicates that the conjugation is NOT broken by the canonical name:

$$[C\beta(\underline{C}\beta)_4C\beta]_2{:}^{(1\text{-}11,13\text{-}23)}(\beta) \quad (77)$$

containing no 1's, only β's, as carbon-carbon bond orders.

Up until this point all examples of symmetry and repetition have concentrated on an atom being the focal center; however, when dealing with spherical nomenclature, the rules can be extended so that the focus is on a bond instead. For example, biphenyl (Figure 40) is symmetric about the connecting bond; consequently, instead of the above Cartesian name (76), which required the specification of additional locant numbers, one *might* assign as the spherical name:

$$1[C\beta(\underline{C}\beta)_5]_2 \quad (78)$$

One *could* similarly extend spherical names to biphenylene (Figure 41) with a name such as:

$$\{\beta[C\beta(\underline{C}\beta)_4C{:}^{(2\text{-}12)}(\beta)]\}_2 \quad (79)$$

however, the simplification over (77) is minimal and the price paid is that the locant numbers for atoms on the principal chain are no longer always the odd integers and the bond the even ones.

Next, just as the synergy between nomenclature and chemical structure was illustrated for biphenyl vs. biphenylene, attention is directed to a newly discovered compound having a bent sp-hybridized skeleton [26]. Due to the postulation of the bet bond (Chapter 2), the proposed systemic nomenclature for the trisilaallene molecule does not convey the false picture of linearity that any IUPAC name would. Instead one would derive a more realistic chemical picture using either the proposed Cartesian name (Figure 42)

$$\mathrm{H1\underline{\underline{C}}1Si1C1(\underline{\underline{C}}1)_2C1Si\overset{=}{\supset}Si\overset{=}{\supset}Si1C1(\underline{\underline{C}}1)_2C1Si1\underline{\underline{C}}1H{:}^{(7\text{-}15,19\text{-}27)}(1);\ ^{(5,5,29,29)}(1\underline{\underline{C}}1H);\ ^{(7,13,13,13,21,21,27)}(1Si1\underline{\underline{C}}1H){:}^{(3,3)}(1\underline{\underline{C}}1H)]} \qquad (80)$$

or the much simpler spherical name (Figure 43):

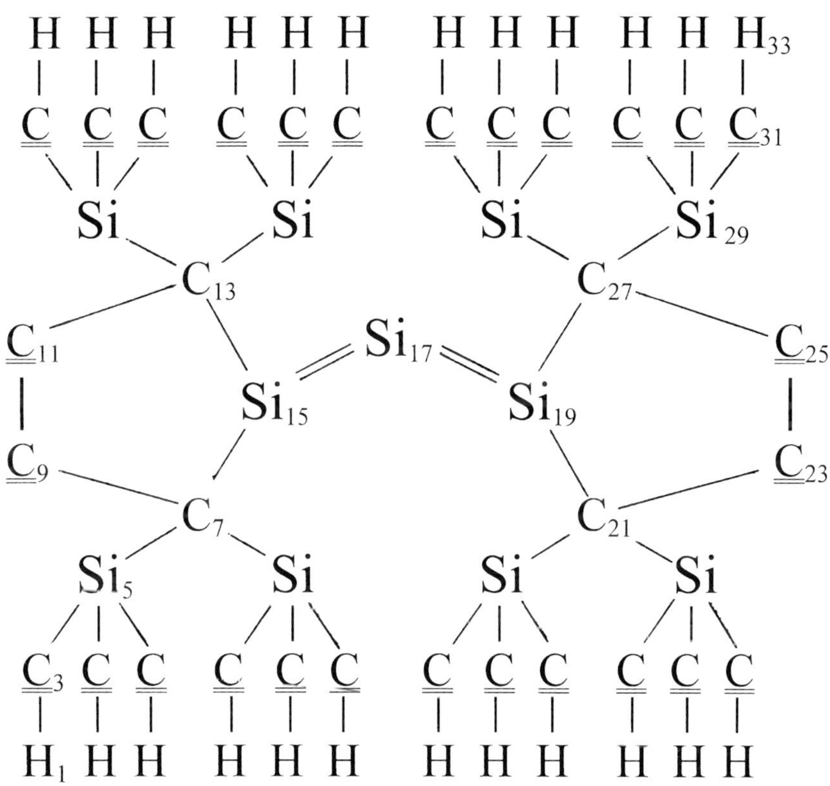

Fig. 42: A trisilaallene with bent sp-hybridization: cartesian locant numbering $\{\beta[C\beta(\underline{\underline{C}}\beta)_4C{:}^{(2\text{-}12)}(\beta)]\}_2$ (79)

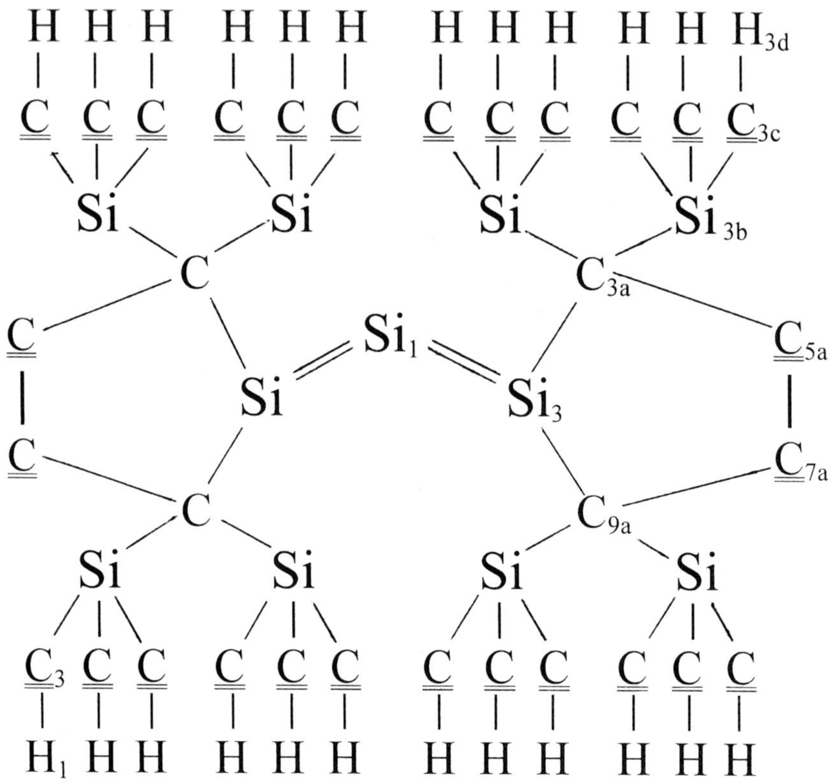

Fig. 43: A trisilaallene with bent sp-hybridization: spherical locant numbering

$$\mathrm{Si\{\sqsupset Si{:}^{(3\text{-}3)}[1C1(\underline{C}1)_2C1{:}^{(3,3,9,9)}(1Si1\underline{C}1H){:}^{(3,3)}(1\underline{C}1H)]\}_2} \quad (81)$$

Returning the focus to assigning canonical names to spiro compounds, attention is directed to multi-spiro compounds that have historically been in the domain of organo-metallic compounds; namely, consider the ion shown in Figure 44. Note that the primary feature emphasized in this representation is an aggregation of eight three-atom cycles (each containing a molybdenum and two sulfur atoms) with the formula given as: $[Mo_2(S_2)_6]^{2-}$. There is, however, the question as to whether this standard grouping of sulfur atoms in pairs is distorting? Perhaps, to the contrary, instead of this being the focal point, which is the perspective one would choose if they were thinking in terms of an SSSR nomenclature, a more accurate picture of this ion might be

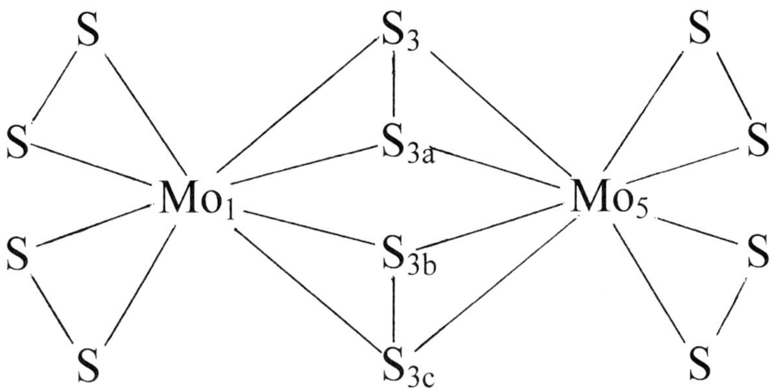

(a) Cylindrical name

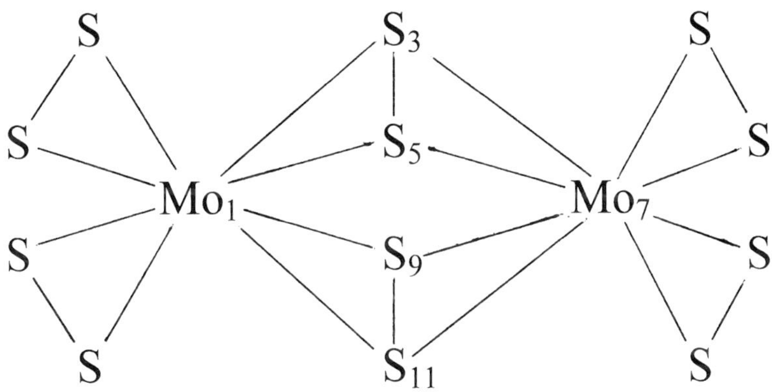

(b) Cartesian name

Fig. 44: A multiple multiply-spiro organo-metallic compound

one of a central octahedron with the molybdenum atoms at two axial vertices of the octahedron and sulfur atoms at the four equatorial vertices. Moreover, the "equatorial" sulfur atoms are represented as being joined in pairs. Such a picture is cylindrically symmetrical and suggests that one nomenclate this ion as:

$$[\mathrm{Mo(1S1)_4Mo:}^{(1\text{-}1,1\text{-}1,5\text{-}5,5\text{-}5)}\mathrm{(1S1S1)};^{(3\text{-}3a,3b\text{-}3c)}(1)]^{2-} \quad (82)$$

Alternately, one could assign a Cartesian name to this ion based on a six atom path with two bridges that is flanked by same four axial three-member spiro rings:

$$[(Mo1S1S1)_2:^{(1\text{-}5,1\text{-}9,3\text{-}7,7\text{-}11)}(1);^{(1\text{-}1,1\text{-}1,5\text{-}5,5\text{-}5)}(1S1S1)]^{2-} \quad (83)$$

Before leaving the subject of spiro combinations and ring assemblies, an intriguing topological idea that has **so far** not produced viable molecules is that of internally and externally tangent "spheres".* Figure 45 illustrates the two different ways in which a pair of different size spheres can be tangent to each other. Although such a spiro combination may be far into the future, the chemically much simpler combinations of ring assemblies of chemical polyhedra (rather than spheres) seems not only viable but actually probable sometime soon. For such an assembly only four ligands need to emanate from a single atom and such is already known for the interior vs. the exterior bonds in a molecule of buckminsterfullerane (See Chapter 3, formula 35 and the discussion there). The theoretical idea is now to attach a cubane or similar small module to a protruding in vs. an out bond of a mixed fullerane/ene. The latter of these syntheses should be relatively straight-forward in much the same manner as had been the production of a bicubyl molecule [27]; i.e., $C_{16}H_{14}$; wherein two cubane molecules are joined at a vertex with the elimination of a hydrogen atom from each of them. The laboratory synthesis problems of building a fullerene/ane superstructure around a cubane module will be more difficult but is theoretically within reason. Moreover, if this bridge were to be removed, a molecule in which a

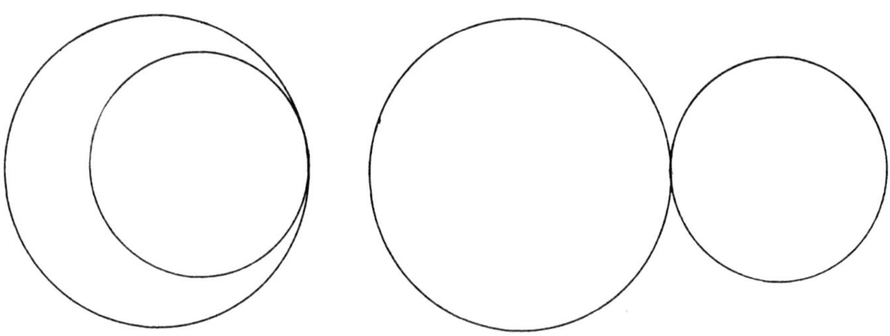

a. Internally tangent b. Externally tangent

Fig. 45: Tangent spheres

*See footnote #17 in [26].

cubane module was free to rattle around inside a fullerene cage (See Chapter 7) would be formulated. To-date only single atom, linear chains and small planar cycles have been so encapsulated. Although the above described internally combined modules to form a single "molecule" have only extended to simple spiro and catenane forms, external combinations of quite great complexity have been created.

REFERENCES

[1] F. Harary, Graph Theory, Addison-Wesley, Reading, Ma. 1969, 26.
[2] International Union of Pure and Applied Chemistry, Nomenclature of Organic Chemistry, Section A, Pergamon Press, Oxford, U.K., 1979, 38.
[3] H. Margenau and G.M. Murphy, The Mathematics of Chemistry and Physics; Van Nostrand Co.; Toronto, Canada, 1943, 173-187.
[4] E.O. Fischer, W. Pfab, Z.Naturforsch 76 (1952) 377.
[5] G. Wilkinson, M.C. Rosenblum and R.B.; Woodward, J.Am.Chem.Soc., 76 (1952) 2123.
[6] R.S. Cahn and O.C. Dermer, "Introduction to Chemical Nomenclature", 5-th Ed. 1979, Butterworths, London, 30.
[7] T.W.G. Solomons, Organic Chemistry, 4-th Ed.; Wiley, New York, 1988, 716-7.
[8] S.B. Elk, J.Chem.Inf.Comput.Sci., 25 (1985) 17.
[9] S.B. Elk, J.Chem.Inf.Comput.Sci., 24 (1984) 203.
[10] M. Tsunoda and F.P. Gabbai, J.Am.Chem.Soc., 122 (2002) 8335.
[11] P. Margl, K. Schwarz, P.E. Bloechl, J.Am.Chem.Soc. 116 (1994) 11177.
[12] G. Suss-Fink, Angew.Chem.Int.Ed., 41 (2002) 99.
[13] B.R. James, Chem.&Eng.News, 3/11/02, Letter to Editor, 8.
[14] H. Piotrowski, G. Hilt, A. Schulz, P. Mayer, K. Polborn, K.Severin, Chem.Eur.J. 7 (2001) 3197.
[15] P.S. Zurer, Chem.&Eng.News, 7/8/02, 27.
[16] International Union of Pure and Applied Chemistry, Nomenclature of Inorganic Chemistry, 2-nd Ed., Definitive Rules, 1970, Butterworths, London, p.44.
[17] "Chemical Nomenclature"; Editor: K. J. Thurlow; Kluwer Academic Publishers, Dordrecht, The Netherlands, 1998, 113.
[18] International Union of Pure and Applied Chemistry, Nomenclature of Organic Chemistry: Section A, Pergamon Press: Oxford. U. K., 1957, 23.
[19] S.B. Elk, MATCH, 8 (1980) 121.
[20] S.B. Elk, MATCH, 13 (1982) 239.
[21] Ibid #17, 164.
[22] Ibid.
[23] S.B. Elk, MATCH, 36 (1997) 157.
[24] Ibid #17, 176.
[25] S. Ishida, T Iwamoto, C. Kabuto and M. Kira, M., Nature, 421 (2003) 725.
[26] Ibid #8, 22.

Chapter 7

Topologically restrained compounds

CHAPTER ABSTRACT:
Whereas neither I.U.P.A.C. nor nodal nomenclature bother to extend their system to what is, at present, a very small group of topologically restrained compounds, the proposed system allows for a simple extension to the set of catenanes and rotaxanes. Also, one may readily adapt this system to canonically naming endothelial compounds, such as endothelial fullerenes, etc. Moreover, extension to "chemical" knots having multiple interconnections and windings, without having to resort to the tedium employed by Schill to extend I.U.P.A.C. nomenclature to these compounds, is straight-forward.

Another use for the zero integer superscript occurs in the domain of catenanes and rotaxanes. Although the same examples as the ones cited in Schill's monograph [1] for the development of a suitable nomenclature have been selected for discussion, his organizational scheme -- which is merely an extension of I.U.P.A.C.'s organic chemistry nomenclature -- has not been. Instead, the nomenclature developed earlier for more traditional compounds is easily expandable to include these "compounds". Toward this end, note that in the formulation described in Chapter 1, the semi-colon was selected as a co-ordinate separator and the colon as a sub-ordinate separator of strings of nomenclature code. One now needs only to introduce the use of the zero superscript* to indicate that the choice of locant numbering is irrelevant. Note that there is no coordination between the atoms in different rings of a catenane, or the atoms in a ring and a rod of a rotaxane. Consequently, an alkane catenane with ring sizes m and n would be named as:

* This is just a small extension of the scenario in Chapter 3 in which the use of an in-line zero bond to emphasize that in a given molecule two specific atoms were not connected — either covalently or with hydrogen bridges — was introduced.

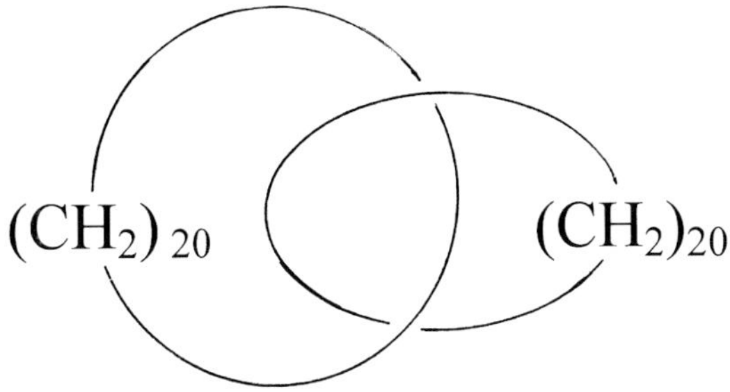

Fig. 1: A typical catenane

$$(\underline{\underline{C}}1)_m;^0(\underline{\underline{C}}1)_n. \qquad (1)$$

This is illustrated in Figure 1 with m = n = 20.

In a similar manner, a simple rotaxane with m atoms in the ring and n atoms in the rod could be named either as:

$$(\underline{\underline{C}}1)_m;^0[X1(\underline{\underline{C}}1)_nX] \quad \text{or} \quad X1(\underline{\underline{C}}1)_nX;^0(\underline{\underline{C}}1)_m \qquad (2)$$

where X represents some end group that is large enough so that the two components are physically constrained; i.e., that the "cyclic" part can not be physically separated from the "straight" part. Notice that, unlike for a ring structure with a side chain, in which the ring is **always** given priority in assigning a canonical name, for rotaxanes, it is of pragmatic advantage that the two components be deemed as equal in importance, so that one could list either of them first. Consequently, multiple rings about a single rod will be nomenclated with the rod first, while multiple rods inside a single ring will

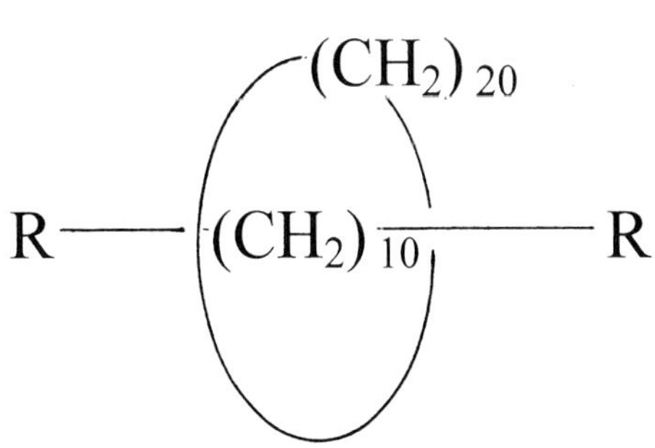

Fig. 2: A typical rotaxane

name the ring first. Note that this last category, while mathematically possible, has at present no known members.

Meanwhile, the directness and simplicity of the proposed names is evident when comparing the names that Schill has been forced to rely upon in conformity with IUPAC practice; namely: for that catenane having both m and n = 20: [2]-[cycloeicosane]-[cycloeicosane]-catenane (Figure 1) and for that rotaxane having m=20, n=10: [2]-[1,10-diaryldecane]-[cycloeicosane]- rotaxane (Figure 2). Furthermore, for these two "simple" topologically restrained molecules, whether to call the components co-ordinate (to use a semi-colon) or sub-ordinate (to use a colon) is immaterial. However, for more complex joining of rings and chains, such will not always be the case. In Figure 3, for example, the linear string of rings, which IUPAC names as: [3]- [cycloeicosane]-[cyclohexacosane]-[cycloeicosane]-catenane is co-ordinate, and in the proposed system would be named:

$$(\underline{\underline{C}}1)_{20};^{0}(\underline{\underline{C}}1)_{26};^{0}(\underline{\underline{C}}1)_{20} \qquad (3)$$

As indicated in an article in Chemical and Engineering News [2], the synthesis challenges of making catenanes, once considered "freaks of the laboratory"[3], has been reduced to almost routine. This is exemplified by a five ring catenane which models the five ring symbol of the Olympic Games, dubbed olympiadane, that was produced by Amabilino et al [4]. Although drawing this molecule is tedious, to describe and canonical name such a catenane is simple; namely:

$$A;^{0}B;^{0}C:^{0}B;^{0}A \qquad (4)$$

where the modules A, B and C are respectively the monocycles:

$$[N\beta(\underline{C}\beta)_2C1C\beta\ (\underline{C}\beta)_2N1\underline{\underline{C}}1C\beta(\underline{C}\beta)_2C1\underline{\underline{C}}1]_2:^{(1\text{-}7,9\text{-}15,19\text{-}25)}(\beta\underline{\underline{C}}\beta\underline{\underline{C}}\beta) \qquad (5)$$

$$[O1C\beta(\underline{C}\beta)_3(C\beta)_2(\underline{C}\beta)_3C1(O1\underline{\underline{C}}1\underline{\underline{C}}1)_4]_3:^{(3\text{-}13,11\text{-}21)}(\beta) \qquad (6)$$

$$\{N\beta(\underline{C}\beta)_2C1C\beta(\underline{C}\beta)_2N1\underline{\underline{C}}1[C\beta(\underline{C}\beta)_2C1]_2\underline{\underline{C}}1:^{(1\text{-}7,9\text{-}15,19\text{-}25,27\text{-}33)}(\beta\underline{\underline{C}}\beta\underline{\underline{C}}\beta)\}_2 \qquad (7)$$

As well as a linear string of ring modules, one can also create a branched string of rings, as illustrated in Figure 4. IUPAC names this catenane as: [4]- [cycloeicosane]-2-[cyclodotriacontane-2_1-cycloeicosane]-

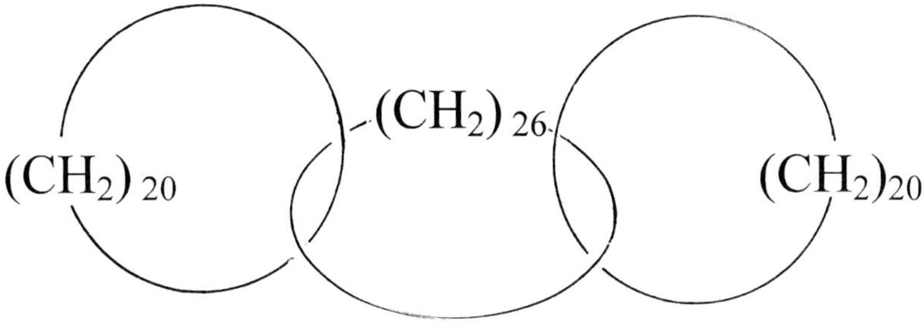

Fig. 3: A typical linear catenane

3-[cycloeicosane]-catenane. In the canonical system being developed there is the much simpler Cartesian name of:

$$(\underline{\underline{C}}1)_{20};{}^{0}[(\underline{\underline{C}}1)_{32}:{}^{0}(\underline{\underline{C}}1)_{20}];{}^{0}(\underline{\underline{C}}1)_{20} \qquad (8)$$

Note that there was no need for an additional augmenting set of subscripted locants to indicate the branching of the rings. Moreover, because of the symmetry, there exists an even simpler spherical name:

$$(\underline{\underline{C}}1)_{32}:{}^{0}[(\underline{\underline{C}}1)_{20}]_{3} \qquad (9)$$

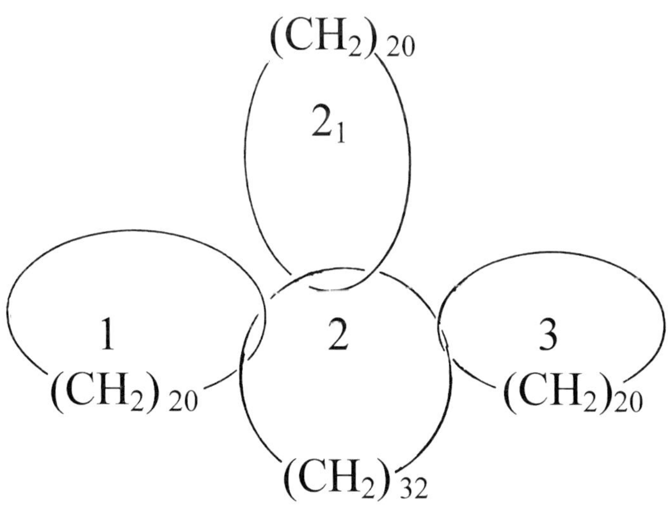

Fig. 4: A typical branched catenane

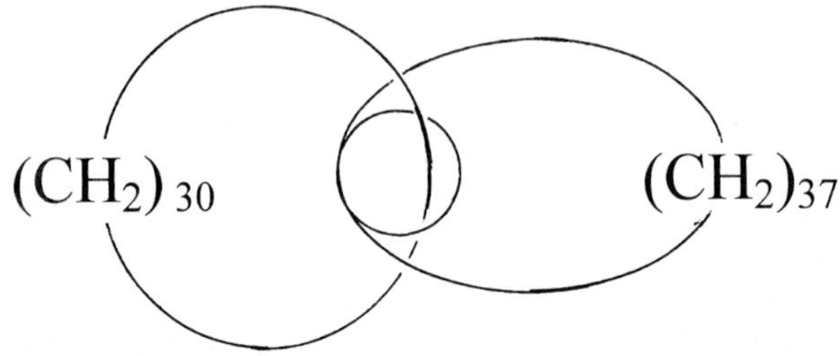

Fig. 5: A multiply-wound catenane which IUPAC designates with α = 2

Next, in the proposed system, multiple winding of a chain (Figure 5) is nomenclated using fractional subscripts:

$$(\underline{\underline{C}}1)_{30};^{0}(\underline{\underline{C}}1)_{37/2} \qquad (10)$$

This is in contradistinction to requiring the appending of yet another parameter, the "winding number" α, as is done for the IUPAC name: [2]-[cyclotricontane]-[cycloheptatricontane]-catenane (α = 2). Additionally, even this added parameter is an ad hoc one. In the IUPAC name it is not unreasonable to assume that the α = 2 might be applied to either the 37- or the 30-member ring. By contrast, in the proposed system, there is no ambiguity as to which of the two rings is the one that is multiply wound. In other words, instead of having to assume that the α = 2 was intended to be applied to the last named ring, the proposed system specifies precisely which ring is being multiply wound. Moreover, one is not limited to only one of the rings having multiple windings. Although no such combinations such as:

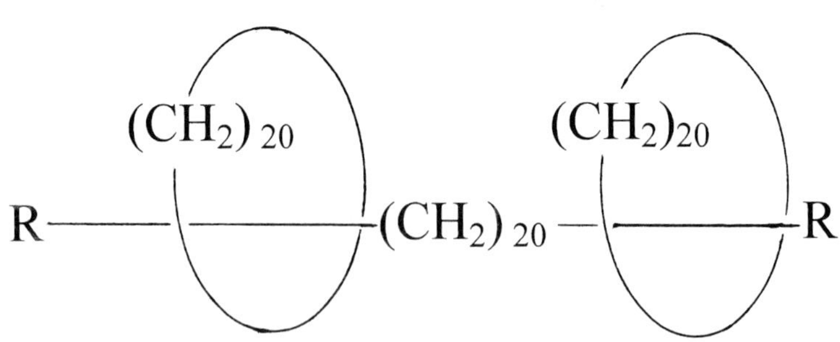

Fig. 6: A multiple ring rotaxane

$$(\underline{\underline{C}}1)_{60/3};{}^{0}(\underline{\underline{C}}1)_{40/2} \qquad (11)$$

where the larger ring threads through the common "circle" three times and the smaller ring twice, etc. are known, there is no mathematical limitation to such windings and thus there should not be one in the nomenclature.

Continuing to other combinations, one could expand upon Schill's work and note that multiple strands in a rotaxane can be of two types: The first of these, which Schill included, is to have multiple rings surround a single axis, as is illustrated in Figure 6. IUPAC names this aggregation: [3]-[1,20-diaryl-eicosane]-[cycloeicosane]-[cycloeicosane]-rotaxane in contrast to the much simpler systemic name of:

$$[R1(\underline{\underline{C}}1)_{20}R]:{}^{0}(\underline{\underline{C}}1)_{20};{}^{0}(\underline{\underline{C}}1)_{20} \qquad (12)$$

Additionally, although not treated by Schill, there is the mathematical possibility of having multiple axes through a single ring. The systemic name for such a combination is:

$$(\underline{\underline{C}}1)_{m}:{}^{0}[X_{1}1(\underline{\underline{C}}1)_{n}X_{2}];{}^{0}[X_{3}1(\underline{\underline{C}}1)_{p}X_{4}] \qquad (13)$$

where the X_i's are any large enough group to prevent the rods from disengaging from inside the ring. Such an aggregation could also arise if a ring was big enough to allow one of the rods to slip out but because there were two (or more) intertwined rods, all are constrained to stay within the ring.

One further aggregation, which Schill illustrates, involves rings interdigiting to form a "super-ring", as in Figure 7. IUPAC names this ring of rings: [4]-1-[cyclohexacosane]-2-[cyclohexacosane]-3-[cyclotriacontane]-4 [cyclohexacosane]- cyclocatenane. We, on the other hand, need only:

$$(\underline{\underline{C}}1)_{30};{}^{0}[(\underline{\underline{C}}1)_{26};]_{3} \qquad (14)$$

Note that the use of the colon in (9) indicated the subordination of the following rings to the central one, whereas the semicolon in (13) indicates the independence of these rings. Also, the final semicolon in (13) represents the continuation (back to the beginning) of the super-cycle.

As well as the ability of atoms to be formed into cyclic compounds and for two or more such cycles to form simultaneously and to interdigit into a catenane, there is also the ability of these same chains of atoms when

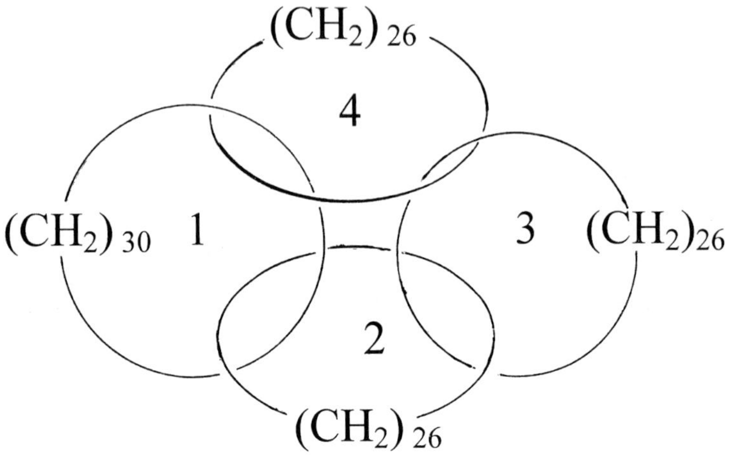

Fig. 7: A cyclic catenane

combining to create a knot.* The topologically simplest knot, called a trefoil (Figure 8), requires that three oriented modules (A-B) be appropriately joined. Pragmatically, it should be noted that the smallest catenane has inner diameter "rings" of over twenty atoms each; otherwise, there would not be enough room for these independent aggregations to pass through one another. An even larger number of atoms in the path is required in order to form a knot of any sort. A molecule in the shape of a trefoil, was created by Voegtle [4] by combining a diamine and a dicarboxylic acid dichloride. The diamine module that he chose (Figure 9) had five benzene rings, two

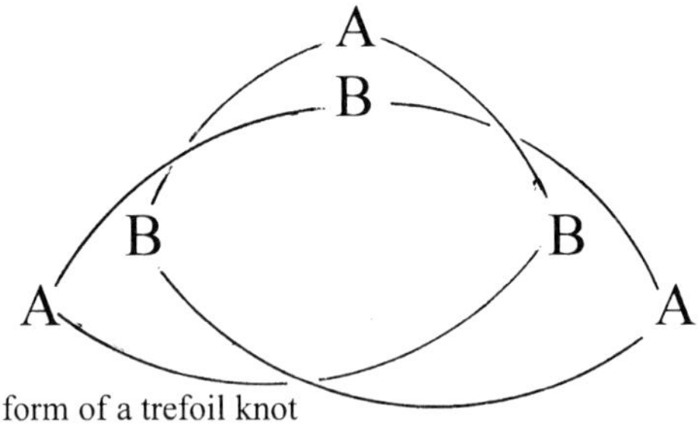

Fig. 8: Modularized form of a trefoil knot

*Because of space constraints this formation of knots occurs far less frequently than the simple addition of an atom onto a previously formed strand; however, such restrained molecules, while rare, are not unheard of.

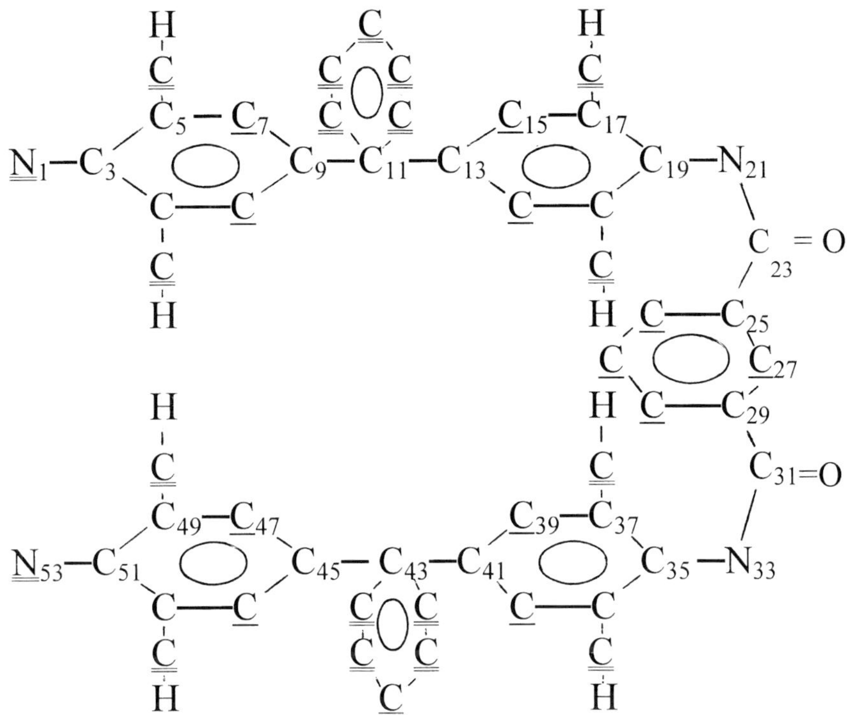

Fig. 9: Diamine used by Voegtle in formulating chemical trefoil knot

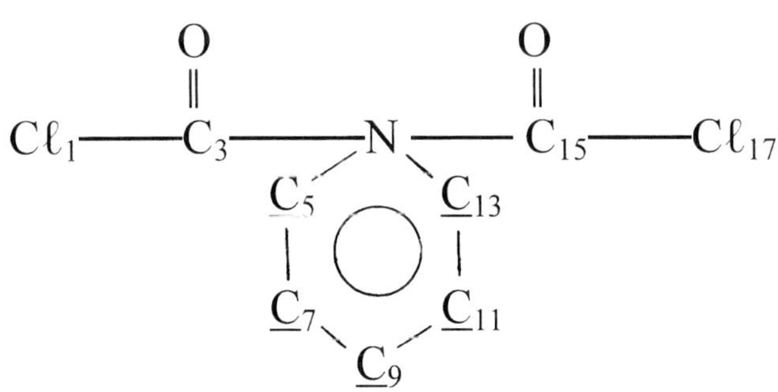

Fig. 10: Dicarboxylic acid dichloride used by Voegtle in formulating chemical trefoil knot

cyclohexyl groups, eight methyl groups, two amide groups and two amine groups. The dicarboxylic acid dichloride chosen (Figure 10) was much simpler; namely a pyridine molecule with acylchloride ligands in the 2 and 6 position. The systemic names for these two modules, using the pictures given, appear to be:

$$\underline{\underline{\mathrm{N}}}1(\mathrm{C}\beta)_2\underline{\mathrm{C}}\beta(\mathrm{C}1)_2\mathrm{C}\beta\underline{\mathrm{C}}\beta\mathrm{C}\beta\mathrm{C}1\underline{\mathrm{N}}1\mathrm{C}1\mathrm{C}\beta\underline{\mathrm{C}}\beta\mathrm{C}\beta(\mathrm{C}1)_2\underline{\mathrm{N}}1(\mathrm{C}\beta)_2\underline{\mathrm{C}}\beta(\mathrm{C}1)_2\mathrm{C}\beta\underline{\mathrm{C}}\beta\mathrm{C}\beta\mathrm{C}1\underline{\underline{\mathrm{N}}}:^{(3\text{-}9,35\text{-}41)}[\beta\mathrm{C}\beta\underline{\mathrm{C}}\beta:^{(3)}(1\underline{\underline{\mathrm{C}}}1\mathrm{H})];^{(11,43)}[1(\underline{\underline{\mathrm{C}}}1)_5];^{(13\text{-}19,45\text{-}51)}[\beta\underline{\mathrm{C}}\beta\mathrm{C}\beta:^{(5)}(1\underline{\underline{\mathrm{C}}}1\mathrm{H})];^{(23,31)}(2\mathrm{O}) \qquad (15)$$

and

$$\mathrm{C}\ell1\mathrm{C}1\mathrm{C}\beta(\underline{\mathrm{C}}\beta)_3\mathrm{C}1\mathrm{C}1\mathrm{C}\ell:^{(5,13)}(1\mathrm{N}1);^{(3,15)}(2\mathrm{O}) \qquad (16)$$

respectively. In actuality, the bonds involved in the amide groups should indicate the extension of aromaticity and be beta bonds, rather than the indicated single and double bonds of formulas (10) and (11); i.e., a better representation of these modular names is:

$$\underline{\underline{\mathrm{N}}}\beta(\mathrm{C}\beta)_2\underline{\mathrm{C}}\beta(\mathrm{C}1)_2\mathrm{C}\beta\underline{\mathrm{C}}\beta(\mathrm{C}\beta)_2\underline{\mathrm{N}}\beta(\mathrm{C}\beta)_2(\underline{\mathrm{C}}\beta)_3(\mathrm{C}\beta)_2\underline{\mathrm{N}}\beta(\mathrm{C}\beta)_2\underline{\mathrm{C}}\beta(\mathrm{C}1)_2\mathrm{C}\beta\underline{\mathrm{C}}\beta(\mathrm{C}\beta)_2\underline{\underline{\mathrm{N}}}:^{(3\text{-}9,35\text{-}41)}[\beta\mathrm{C}\beta\underline{\mathrm{C}}\beta:^{(3)}(1\underline{\underline{\mathrm{C}}}1\mathrm{H})];^{(11,43)}[1(\underline{\underline{\mathrm{C}}}1)_5];^{(13\text{-}19,45\text{-}51)}[\beta\underline{\mathrm{C}}\beta\mathrm{C}\beta:^{(5)}(1\underline{\underline{\mathrm{C}}}1\mathrm{H})];^{(23,31)}(\beta\mathrm{O}) \qquad (17)$$

and

$$\mathrm{C}\ell1(\mathrm{C}\beta)_2(\underline{\mathrm{C}}\beta)_3(\mathrm{C}\beta)_2\mathrm{C}\beta\mathrm{C}1\mathrm{C}\ell:^{(5,13)}(\beta\mathrm{N}\beta);^{(3,15)}(\beta\mathrm{O}) \qquad (18)$$

Whether formula (18) has gone far enough in incorporation of the aromaticity is left for further study; however, it is not unlikely as was seen in Chapter 5 with the discussion of the aluminum chlorine bonding that perhaps the aromaticity extended to the chlorine atoms as well and thus an even better name would be:

$$\mathrm{C}\ell\beta(\mathrm{C}\beta)_2(\underline{\mathrm{C}}\beta)_3(\mathrm{C}\beta)_2\mathrm{C}\beta\mathrm{C}\beta\mathrm{C}\ell:^{(5,13)}(\beta\mathrm{N}\beta);^{(3,15)}(\beta\mathrm{O}) \qquad (19)$$

Meanwhile note that when a chlorine atom of the dichloride is brought into contact with a hydrogen atom on an amine groups of the diamine a covalent bond is formed between the neighbors of these atoms with the concurrent release of a molecule of hydrogen chloride gas. As well as the expected catenane formed when three of each module combined, in 20% of the cases, there was also the surprising result that these modules combined into a trefoil. Moreover, because of the reflective symmetry of both components, the combining of three pairs of modules results in a molecule having now a six fold rotational symmetry; thus the graph theoretical part of the systemic name of this chemical trefoil is:

$$\{\underline{N}\beta(C\beta)_2\underline{C}\beta(C1)_2C\beta\underline{C}\beta(C\beta)_2N\beta(C\beta)_2(\underline{C}\beta)_3(C\beta)_{\underline{2:}}{}^{(3\text{-}9)}[1C1\underline{C}1:{}^{(3)}(1\underline{\underline{C}}1H)];\ {}^{(11)}[1(\underline{\underline{C}}1)_5];{}^{(13\text{-}19)}[1\underline{C}1C1:{}^{(5)}(1\underline{\underline{C}}1H)];\ {}^{(23,35)}(2O);{}^{(25\text{-}33)}(\beta\underline{C}\beta)\}_6 \tag{20}$$

Note that this same part of the canonical name will be valid whether or not this 108 atom chain has a topologically restraining loop or is straight. Graph theory can not distinguish between such topological isomers. Also note that there are no α bonds between one part of the chain and any other part of it. How the nomenclature is able to deal with such ideas will be addressed in a follow-on article.

Another arena in which topologically restrained molecules are of importance involves endothelial compounds; for example, a small molecule or ion inside a fisular cage, such as a fullerene. The first of these to be discovered occurred when Diedrich et al. [5] doped C_{60} with potassium to encapsulate 3 potassium atoms inside the cage. The systemic name for this compound would now be formed by augmenting the name for fullerene that had been developed earlier in Formulas (42) or (43) of Chapter 3 with a semicolon, followed by a zero numbered locant branch; e.g., ${}^{0}(K,K,K)$ -- assuming that the three potassium atoms are independent of one another; i.e.,

$$C[\beta(C\beta)_8]_3C:{}^{(3\text{-}7)}(\beta(C\beta)_3);{}^{(9\text{-}59,13\text{-}61,15\text{-}65,23\text{-}37,29\text{-}81,31\text{-}83,35\text{-}89,43\text{-}99,45\text{-}101,51\text{-}11,55\text{-}7)}(\beta(C\beta)_2);{}^{(5\text{-}21,11\text{-}57,17\text{-}67,25\text{-}63,27\text{-}77,33\text{-}85,\ 39\text{-}53,41\text{-}79,47\text{-}103)}(\beta C\beta);{}^{(49\text{-}95,69\text{-}87,71\text{-}113,75\text{-}93,91\text{-}109,97\text{-}107,105\text{-}117,115\text{-}119)}(\beta);{}^{0}(K,K,K) \tag{21}$$

Similarly, if these three potassium atoms were joined in a three member ring, then the augmenting part of the name would be: ${}^{0}[(K1)_3]$, etc.

REFERENCES:

[1] G. Schill, "Catenanes, Rotaxanes, and Knots", Academic Press, New York, 1971, 7-10.

[2] R. Dagani, Chem & Eng News, 8/29/94, 28.

[3] Ibid

[4] D.B. Amabilino, Angew.Chem.Int.Ed.Engl. 33 (1994)1286.
[5] F. Voegtle, Angew.Chem.Int.Ed.,39 (2000) 1616.
[6] K. Holczer, O. Klein, S-M. Huang, R.B. Kanes, K-J. Fu, R.L. Whitten and F. Diederich, Science (1991) 252, 1154.

Chapter 8

Polymers

CHAPTER ABSTRACT:
Denotations and connotations of the term "polymer" and its associated building block, termed "monomer", are probed. The nomenclature previously developed in order to canonically name finite length molecules is extended so as to apply to unlimited repeats of the monomer. A system of taxonomy based on dimension underlies the choice of canonical ordering of "polymers", as well as that aggregation of atoms which lacks the "regularity" to meet the proposed limitation to the definition of the term "polymer" (herein called "multimer") is introduced. The extension from Cartesian nomenclature to spherical nomenclature introduced in Chapter 6 is further developed for "dendritic" molecules.

Delineation as to what precisely characterizes a "polymer" [1] has a great deal of ambiguity. The heuristics of an unending concatenation of congruent modules, referred to as "monomers", is what is implied by the term. So long as one restricts the focus to such an aggregation, the mathematical ideal embodied by the term "polymer" **may** be nomenclated. This, however, is chemically unrealistic. The number of atoms in the universe is finite; consequently, no molecule **could** have an infinite number of atoms in it.* Nevertheless, molecules with thousands, or even millions, of atoms in a single aggregation are viable. It is these that have been the pragmatic focus of laboratory chemists, even though such a perspective is at

---

***It can not be emphasized too strongly** that expressions (often seen in non scientific publications, and unfortunately sometimes in scientific ones as well) such as "almost infinite", "nearly infinite", etc. are oxymorons — and the users of such statements lose all credibility. A number can be very large, such as a google: 10 raised to the 100-th power, or a googleplex: a google raised to the google power, etc. but such numbers are still finite. Adding 1 to such a number produces another number that IS larger — NOT EQUAL, but larger. This is in contradistinction to the term infinity, for which many of the rules of elementary arithmetic are NOT applicable.

odds with what is necessary in formulating a canonical (mathematical) system that will be expandable as new examples of this set are discovered.

Because of these different approaches to "polymers" two conflicting systems of nomenclature have historically evolved. The first, a "source-based" system, like the synthetic nomenclature of the polybenzenes [2], suffers from consistency limitations and is of negligible interest in establishing a canonical system of nomenclature that will be applicable to all of chemistry. The second, a "structure-based" system examines the final molecule in terms of its geometrical connectivity. Because this latter perspective focuses on the ultimate arrangement of atoms in the "forming" monomer, the beginning and ending atoms have not been pre-selected by any production process, but rather by atomic number. In other words, there may be different criteria to be applied if one wants to canonically name the monomer of a mathematically abstract (namely infinite) "polymer" vs. the repeating sequence found in a finite molecule.

Meanwhile note that Chemical Abstract Services (CAS), and also IUPAC, regard not only the monomer, but also the "end groups" as part of the name of the moiety. This is done by assigning a name to each of the two end groups, which are designated with **α** as the first end group and **ώ** as the last end group; as well as assigning a third name to the repeating "monomer" – which is that aggregation of atoms that has been repeated some arbitrarily large* number of times. The number of repeats is designated by the letter **n** as a subscript.† Figure 1 is a reproduction of one such example [3]. Both CAS and IUPAC [3] classify this macromolecule as a polymer and name it:

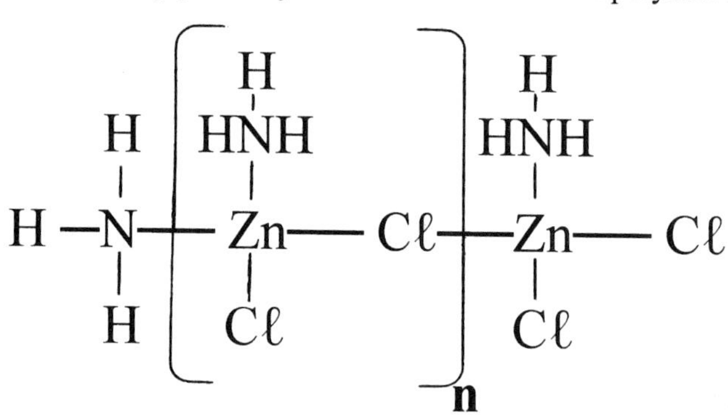

Fig. 1: Finite "non-polymer" with end groups

* Large, but still finite.

† This is in contradistinction to the mathematical ideal in which the number of repeats is indicated by the infinity symbol, ∞.

α-ammine-ω-(amminedichlorozinc)-*catena*-poly[(amminechlorozinc)-μ-chloro]. This is in contradistinction to the perspective of the proposed nomenclature, which, because the end groups are known and can be included, views this aggregation as merely a very large molecule, rather than being the mathematical ideal of a polymer.

At this point note that in order to assign a canonical name to such a macromolecule, one lists, in sequence, for the principal chain:

(1) The higher priority end group,
(2) The bond connecting end group 1 to the monomer,
(3) The monomer -- inside a set of parentheses or brackets (written with the highest priority atom designated as locant #2. Locant #1 is the bond connecting the monomer to the end group)
(4) The subscript n (when the only difference between successive molecules is the number of repeats of the monomer) or a definite large integer (when all of the molecules have the same known number of repeats of the monomer),
(5) The bond connecting the other end group to the monomer,
and (6) The lower priority end group.

Next, it is important to observe that where one chooses to draw the lines which separate the aggregation to be known as the "monomer" from the remaining atoms, referred to as the "end groups" is ambiguous for many large repeating macromolecules, as well as for "polymers". For example, for the macromolecule pictured as Figure 1, an alternate way to partition this molecule is illustrated as Figure 2; additionally, as when nomenclating a cycle, one can traverse the path in either the clockwise or counterclockwise direction. Consequently, analogous to the description of cycles, one: (a)

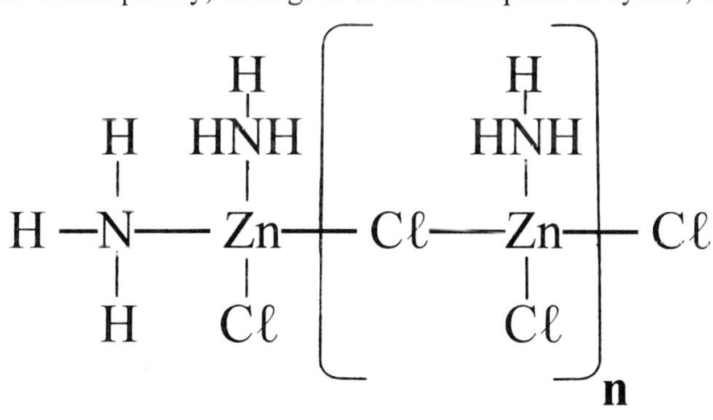

Fig. 2: Alternate grouping of atoms in Fig. 1

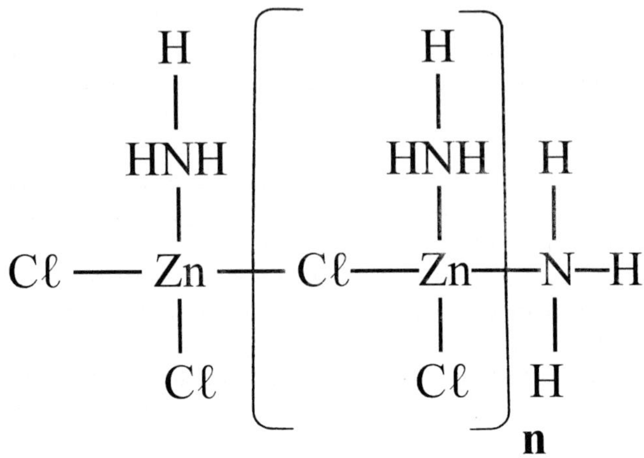

Fig. 3: Reorientation of atoms in Fig. 1

considers a repeating sequence of the atoms, (b) selects the largest atom as locant #2 and (c) selects the direction of transit that one which has the highest priority bond or atom next. By this choice one-half of the bond that precedes locant number 2 starts the canonical designation of the monomer and one-half of the bond preceding the repeat of atom designated as locant #2 ends the monomer. Specification of what is to be called the initial and terminal end groups is now whatever atoms and bonds have been left over and can not be included as another monomer in the repeating chain. This is illustrated for the macromolecule of Figures 1 and 2 as Figure 3.

Next, reiterating that this molecule does not meet the above stated mathematics requirements for being classified as a "polymer", one observes that the use of the α and ώ affixes circumvents the need to know the exact number of repeats; however, use of such affixes create other inconsistencies. For example, one is advised: "The α end group is that attached to the left side of the senior radical in the repeating unit." [4]. This is notwithstanding that all usage of drawings (in particular to indicate orientation, such as left vs. right or up vs. down, etc.) in formulating geometry, and thus also in formulating a nomenclature based on geometry, is passe. The use of "visual evidence" in modern geometry is fraught with difficulties, which when logically applied results in the ability to "prove" ridiculous statements, such as "All triangles are isosceles!" [5]

Additionally, a different criterion (than what has been used up to this time) for seniority of the heteroatoms would be prescribed by [4]; namely, that one should start from the upper right hand corner of the periodic table and progress down each column before moving leftward to the next column. By such a criterion one prioritizes the chlorine atom ahead of the zinc atom, and thus give a different name than the one listed. The problem becomes even more confusing when, as in Figures 1 through 3, the two immediate neighbors of a "bivalent" atom (chlorine in this case) are identical, but the neighbors of these neighbors (i.e., atoms at graph theoretical distance equals two) are NOT. One is thus confronted with the choice as to whether this molecule should be regarded as drawn in Figure 1 or with the monomer as indicated in Figure 2. In either case, in the proposed nomenclature, the zinc atom has the desired priority over the chlorine and thus the monomer is the same.

It might next appear that, because of the order in which the atoms of this molecule has been written, there is a different specification as to which group of atoms is to be designated as **α** vs. **ώ**. Such a question, however, is moot for a "true" polymer (which has no end groups). Similarly, for a "macromolecule" the priority rules established in Chapter 1 designates one of the terminal chlorine atoms of the end group as locant #1, rather than one of the hydrogen atoms. Consequently, Figure 3 (which is merely Figure 1 written in the reverse order) is the canonical representation*.

Like so much of the rest of chemistry, the problem of assigning . canonical names to polymers is further complicated by the historical development of the field. In the evolution of chemical knowledge, the concept now categorized as "polymers" was initially confused with isomers and polymorphs. Recognition of the differences between these ideas was first described by Herman Staudinger [6], who assigned as the delineation of the term *polymer*: "a macromolecule having repeating units". However, because all of the early polymers were created either by adding monomers to an existent chain by means of a free radical mechanism (addition polymers) or else by joining two shorter strands of atoms by eliminating a small molecule (usually water; i.e., condensation polymers), the initial nomenclature for polymers was the above mentioned "source-based" one.

Before progressing further, it should be noted that the development of a system of source-based nomenclature inevitably leads to problems of consistency and expandability. For instance, consider the polymer formed

---

*The assertion that the end $Zn(NH_3)_2Cl$ group is just another copy of the monomer is considered to be in error. Rather, inasmuch as each internal chlorine atom has a coordination of 2 vs. a coordination of only 1 for a terminal chlorine atom, these are viewed as different; namely, the free electron needed for bonding to another monomer is missing in the end group.

by the condensation of CH_2N_2 in BF_3. This condensation liberates nitrogen gas and forms $-CH_2-$ modules, which combine to form what would be called "polymethylene". On the other hand, as described in [1], the source of the polymer with common name "polyethylene" is the diradical of ethene ($\cdot CH_2\text{-}CH_2 \cdot$). However, the resulting polymers are identical and have, in both cases, the same repeating unit: $\cdot CH_2 \cdot$.

This was just one of the reasons why a more mathematically based system was developed. Unlike source-based nomenclature which merely attached the prefix 'poly' to the name of some real or assumed monomer (the source), "structure-based" nomenclature analyzes the structure of a repeating unit in a molecule and selects as the desired "monomer" the smallest unit that consistently repeated in the aggregation.

Adopting the convention that a monomer begins and ends with one half of a bond, the structure-based canonical name assigned to the monomer of "polyethylene" is:

$$({}^{1}/_{2})\ \underline{\underline{C}}\ ({}^{1}/_{2}) \qquad (1)$$

and the full macromolecule ("polymer") name is:

$$H({}^{1}/_{2})\ [({}^{1}/_{2})\ \underline{\underline{C}}\ ({}^{1}/_{2})]_{n}\ ({}^{1}/_{2})H \qquad (2)$$

When locant numbers are required in the process of nomenclating an end group, the algorithm prescribed, for both infinite polymers and for finite macromolecules that approximate polymers, uses single prime superscripts to indicate locant numbering on the initial end group and double prime superscripts for the terminal end group. Also, one uses the subscript ∞ for a polymer vs. **n** for a finite aggregation.

Applying this algorithm to the molecule illustrated in Figures 1 through 3, the repeating unit (i.e., the monomer) is nomenclated as:

$$½Zn1C\ell½{:}^{(2)}(1C\ell);^{(2)}(1\underline{\underline{N}}1H) \qquad (3)$$

and the canonical name of the whole molecule is:

$$C\ell½[½Zn1C\ell½{:}^{(2)}(1C\ell);^{(2)}(1\underline{\underline{N}}1H)]_{n}½Zn1\underline{\underline{N}}1H{:}^{(2'')}(1C\ell);^{(2'')}(1\underline{\underline{N}}1H) \qquad (4)$$

Observe that while the number of atoms that comprise the monomer can range from a single atom to a large aggregation of similarly arranged

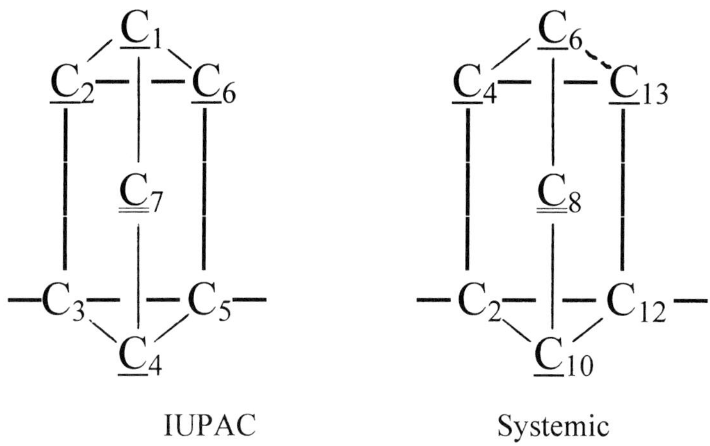

Fig. 4: IUPAC and systemic locant numbering used to assign names to a linear polymer having three dimensional monomers

atoms, nomenclature-wise this is of minor consequence. The important parameter is the length of the graph theoretical path before the repeat of the sequence — the two atoms on the principal path in (4) is the parameter of significance rather than the seven atoms which comprise the monomer.

A slightly more complicated example in which the monomer is essentially three dimensional even though the connection remains linear is illustrated in Figure 4 [7]. Using IUPAC nomenclature this polymer (whose monomer contains two pentagons, one rectangle and two triangles) is named: Poly(tricyclo$[2.2.1.0^{2,6}]$hept-3,5-ylene). Its systemic name is:

$$[{}^{1}/_{2}C1(\underline{C}1)_{2}\underline{\underline{C}}1\underline{C}1C{}^{1}/_{2}:^{(6\text{-}12)}(1\underline{C}^{(=13)}1);\ ^{(2\text{-}10,2\text{-}12,4\text{-}13)}(1)]_{n} \qquad (5)$$

When assigning canonical names to moieties which fit the above limitation of a polymer, one especially focuses on the repeating unit in a structure-based nomenclature, and biases the organization according to dimension. For the completely general cases, one has:

(1) A monomer having a pair of half bonds emanating from atoms A and B (B may be the same as A) such that successive monomers are joined* by

* Note that in the nomenclature system being developed, it would be unusual, but NOT wrong, to have the semantics of the word "joined" include hydrogen bonding, as is the case of a concatenation of hydrogen bonded HF modules. However, for pragmatic purposes, the ease with which such bonds are broken (a property referred to as "labile")

similar half bonds*; thereby forming an unending chain that can be construed as "tessellating"† a one-dimensional‡ space. For such one-dimensional polymers the ideal is an infinite (in both directions) chain of atoms. In other words, as described above, exactly the same criterion for formulating the canonical name of one-dimensional polymers as had earlier been created for cycles is employed; namely, because atom chain ...ABCDABCD... can start with either A or B or C or D and can progress both forward and backward, select the highest atomic number atom as the first atom of the canonical name. The second character of the name is then determined by the larger bond or atom that is first encountered by traversing in either direction, etc.

As the next example, attention is directed to a polymer in which the end groups are not significant, namely the polymer which is commonly named as polyethylene terephthalate, abbreviated PET (Figure 5). IUPAC

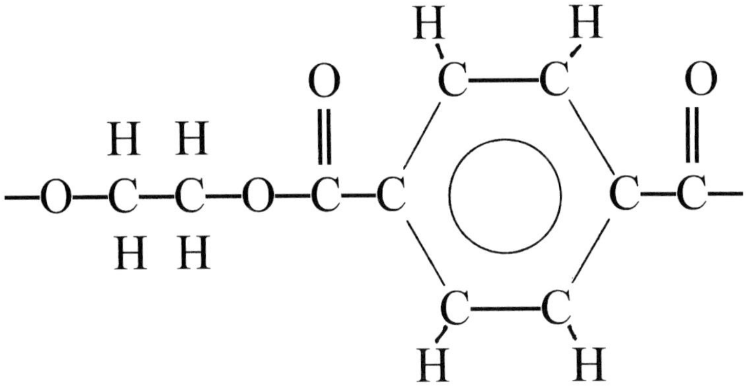

Fig. 5: IUPAC's representation of the monomer of polyethylene terephthalate

---

precludes such a perspective. The question as to whether the alpha bonding of bond order = 0.5 in a boron concatenation is of sufficient stability for there to form a "polymer" is unresolved at present. Perhaps the smaller size of the boron atom will allow for enough added stability that at least, what is called an "oligimer" (the prefix "olig-" denotes "few", with the connotation of repeating congruent units), if not a full "polymer" could exist.

* Because of the demand for concatenation of congruent modules, one can not have a monomer that starts with half a single bond and ends with half a double bond, etc.

† For further discussion of the term "tessellation", which was defined in [8] as "any arrangement of polygons fitting together so as to cover the whole plane without overlapping", and its expansion and application to chemistry, see [9].

‡ No consideration of straightness is implied, just that a uniquely ordered (single) strand can be formulated..

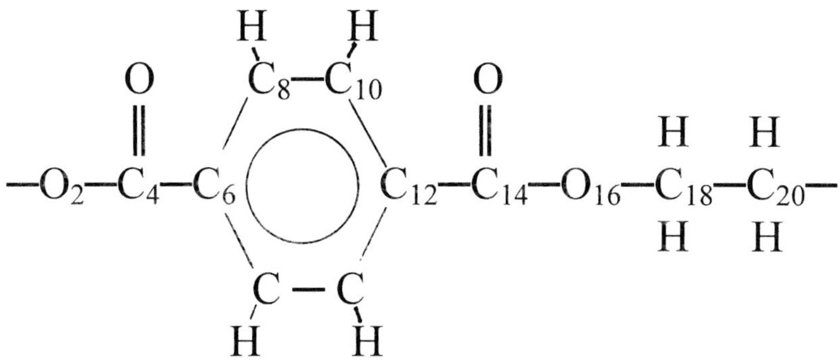

Fig. 6: Systemic representation of the monomer of PET

names this polymer: poly(oxyethyleneoxyterephthaloyl).

In determining the systemic name for this polymer, mathematical consistency dictates starting the locant numbering at one of the oxygen atoms of the principal chain and then progressing (either clockwise or counterclockwise) so that the path goes through the benzene ring before it goes through the ethane chain (Figure 6); i.e., the sequence from an oxygen atom which goes through part of the benzene ring first (1-C-1-C-β) has priority over the sequence that passes through the ethane first (1-C-1-C-1). Consequently, because the first bond or atom difference occurs at position 6, the canonical name assigned to this polymer is:

$$[{}^{1}/_{2}O(1C)_{2}(\beta\underline{C})_{2}\beta(C1)_{2}O(1\underline{\underline{C}})_{2}{}^{1}/_{2}:^{(6\text{-}12)}(\beta\underline{C})_{2}\beta;^{(4,14)}(2O)]_{\infty} \quad (6)$$

(1a) Although most familiar polymers have single-bonded connected monomers, this is not a necessity. Figure 7, which IUPAC calls *catena-*

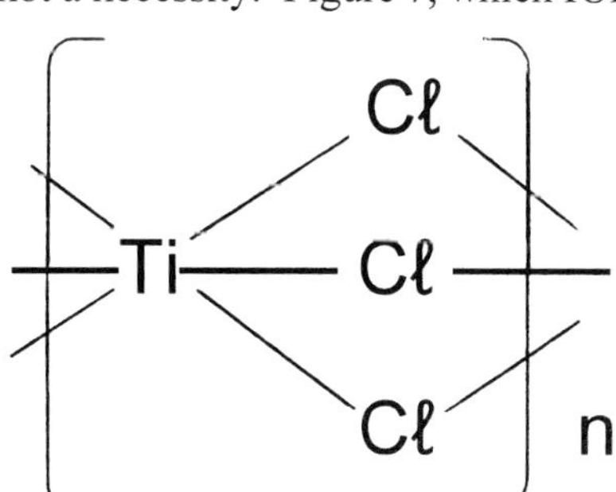

Fig. 7: A quasi-linear monomer

poly[titanium-tri-μ-chloro] has three single bonds as both the starting and ending part of the monomer. One can accommodate such "quasi-linear" aggregations with a multiplicity and an asterisk followed by a fractional bond, etc.; e.g., one *could* name the monomer in Figure 7 as:

$$[3{*}{}^{1}\!/_{2}\mathrm{Ti}1\mathrm{C}\ell^{1}\!/_{2}{:}^{(2\text{-}6,2\text{-}6)}(1\mathrm{C}\ell^{1}\!/_{2})] \qquad (7)$$

However, a better form of this relation, especially for the mathematical ideal of a polymer as an infinitely repeating moiety, is expressed using that congruence property of number theory referred to as "modulus"* [10]. In this form, a prime symbol† is affixed to a repeating locant number in the next module, rather than including more numbers than exist in the monomer. (7) thus becomes:

$$[3{*}{}^{1}\!/_{2}\mathrm{Ti}1\mathrm{C}\ell^{1}\!/_{2}{:}^{(2\text{-}2',2\text{-}2')}(1\mathrm{C}\ell^{1}\!/_{2})]_{n} \qquad (8)$$

(1b) Another familiar quasi-linear polymer, this one represented by a bipartite graph, is formed when two different monomers (call them A and B) are joined so that each has three neighbors of the other monomer. Despite the geometric heuristic of viewing this polymer as being isomorphic with a linear polymer having two entry and two exit points, this "ladder" polymer

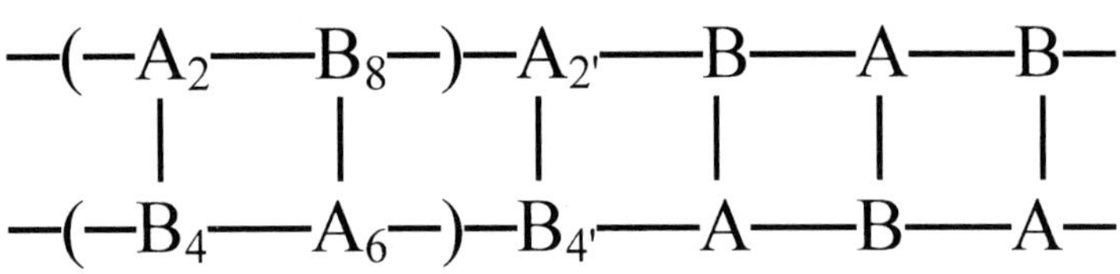

Fig. 8: A Ladder "Pseudo-Linear" Polymer

---

* The term "modulo" in number theory designates only the remainder of a division; e.g., 13 = 1 (mod 4) since 13/4 = 3 + 1/4. The whole number (3 in this case) is ignored and only the numerator of the fraction (whose denominator is the modulo) is important. Such a usage in slightly modified form is familiar in everyday activities wherein modulo 12 describes how one reads a clock. The number that follows 12 o'clock is 1 o'clock (except in military time which uses a modulo 24 system).

†Although there have been two different uses of the prime notation when dealing with polymers, because this usage is limited to the monomer part of the name there should not be confusion with the occasional use of the prime symbol as part of the initial end group

(Figure 8) may be viewed as having been formed by tracing a path starting from one of the modules (A) and traversing in both directions through two adjacent, non-co-linear modules (B). There is now exactly one adjacent but non-co-linear module A to which this just named B module may be joined. This process* may be continued as more modules are addended producing the canonical name for the polymer:

$$[{}^{1}/_{2}A1B1A1B{}^{1}/_{2}{:}^{(2\text{-}8)}(1);{}^{(6\text{-}4')}({}^{1}/_{2})]_{\infty} \quad (9)$$

A more complicated example of a polymer, which IUPAC would view as a ladder with two entering and exiting ports and thus would name as: poly(perimidine-6,7:1,2-tetrayl-1-carbonyl) [11], may be envisioned as a

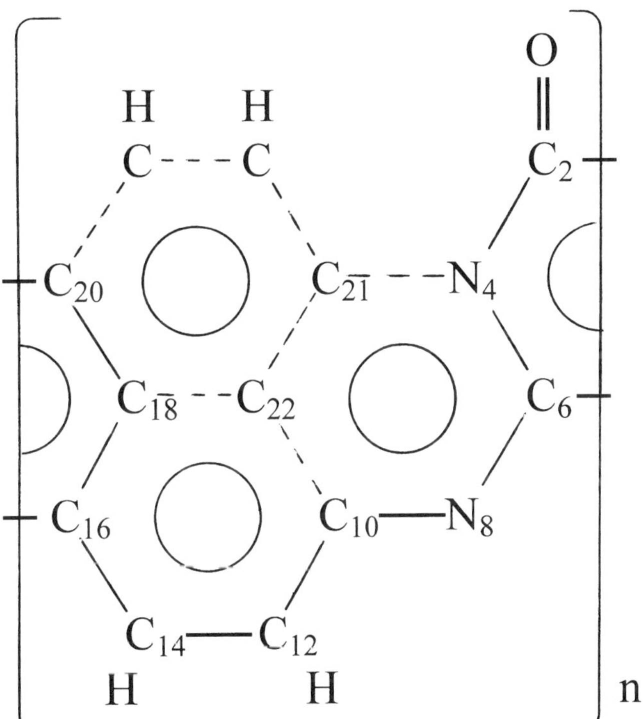

Fig. 9: A polymer which I.U.P.A.C. considers to have two entering and exiting ports

* A familiar example of this structural relationship, which might be worth nomenclating by this scheme, is an RNA molecule, wherein the A's are the pentose-base combinations and the B's the phosphate ion.

single bridged strand with a longest path length of 10. (Figure 9 is a reproduction of the monomer as drawn in [11] — which IUPAC assumed to be appropriate.) Two points of contention with such a depiction are:

(1) Rather than the conjugation being restricted to the three indicated rings, there exists a fourth ring C_2-N_4-C_6-C_{16}-C_{18}-C_{20}, where the oxygen connected at locant 2 and the extra electron on the nitrogen complete the conjugation.

(2) There is no logical reason to start the locant numbering as indicated. To the contrary, one of the nitrogen atoms should be in the premier locant position; i.e., locant #2.

Consequently, rather than the systemic canonical name for this polymer being:

$$[^{\beta}/_{2}(C\beta N\beta)_2(C\beta)_5C^{\beta}/_{2}]:^{(4\text{-}10)}[\beta C^{(=21)}\beta \underline{C}\beta];^{(20\text{-}21)}(\beta C^{(=23)}\beta C^{(=24)}\beta);\ ^{16\text{-}6',18\text{-}22)}(\beta);\ ^{(2)}(2O);^{(12,14,23,24)}(1H) \qquad (10)$$

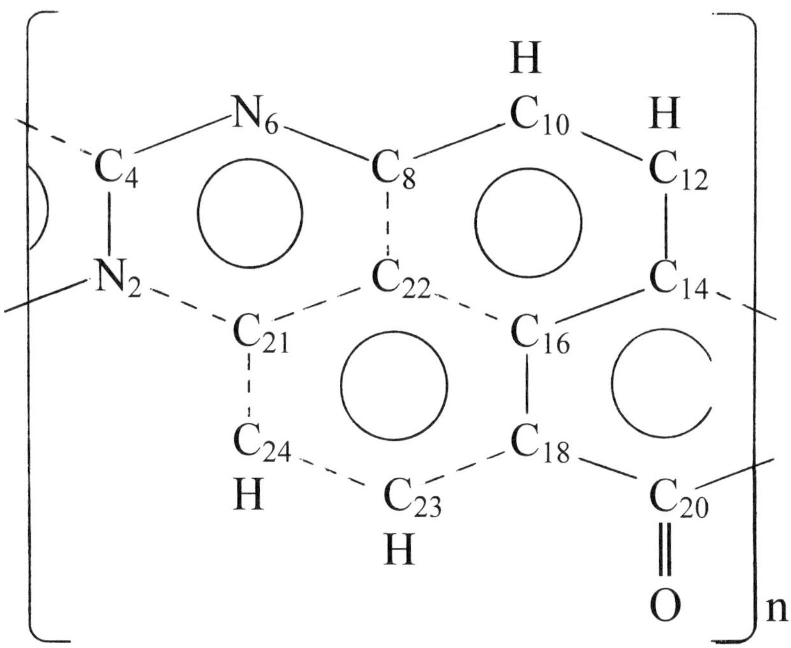

Fig. 10: Re-allocation of canonical locant numbers for polymer depicted in Figure 9

as Figure 9 would imply, the appropriate locant numbering is as shown in Figure 10 — which produces as the desired canonical name:

$$[{}^{\beta}/_{2}(N\beta C\beta)_{2}(C\beta)_{5}C^{\beta}/_{2}]:^{(2\text{-}8)}[\beta C^{(=21)}\beta\underline{C}\beta];^{(18\text{-}21)}(\,\beta C^{(=23)}\beta C^{(=24)}\beta);^{(\;)};^{(14\text{-}4',16\text{-}22)}(\beta);\;^{(20)}(2O);^{(10,12,23,24)}(1H) \qquad (11)$$

Note in the above sections (1a) and (1b) the graph of the polymer has used a single, congruent aggregation of atoms as the repeating parameter. This is consistent with the line of demarcation prescribed as the definition of "monomer". This parameter, which is repeated n (= ∞) times, is the basis for IUPAC's terminology of calling the above-named polymer a "regular polymer" [12], in contradistinction to its use of the more familiar term

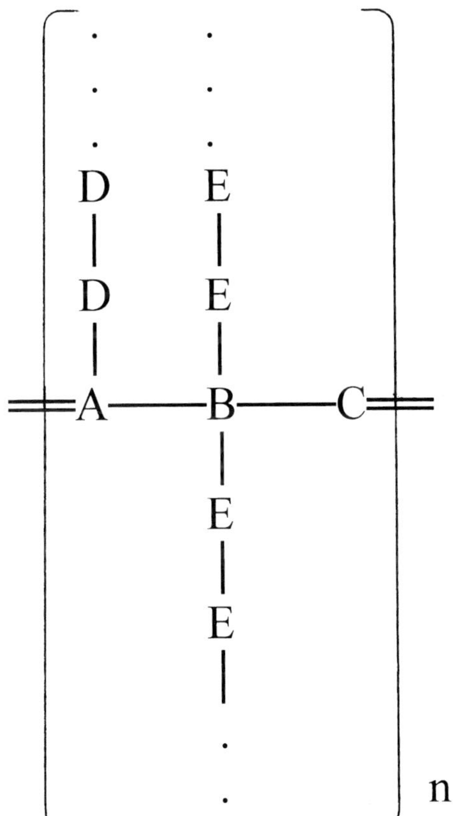

Fig. 11: A multiply-one-dimensional polymer

"homopolymer" [13], which is process-based. At this point, it should be reiterated that, by the definitions chosen, the monomer of this polymer is the entire repeating aggregation and there is no use for the term "co-polymer".
(1c) Before turning to polymers that are "intrinsically" higher dimensional, observe that there exist a large number of known polymers that, even though they require the use of the second and sometimes even the third dimension because of bridges and/or finite chains emanating from the principal chain, are classified as "multiply one-dimensional". A mathematical model of such a polymer is one containing two or more independently repeating units. Even for multiple combinations of unending sequences such as indicated in Figure 11, such an aggregation is easily named in the proposed system as:

$$({}^{2}/{}_{2}A1B1C^{2}/{}_{2})_{\infty}{:}\,{}^{2}(1D)_{\infty};{}^{(4,4)}(1E)_{\infty}\text{, etc.} \tag{12}$$

(1d) One logical extension of multiply one-dimensional polymers occurs when the chains emanating from a single atom (A) are alike (B) but with module B ending with atom A, which now becomes the center of a new radial ("polymeric") expansion. But this is precisely what had been described as a dendrite in Chapter 6, and as such may be viewed in terms of "polar coordinates". The formula for a dendrite with m "spokes" in the graph theoretic "wheel" is:

$$A\{BA{:}[BA{:}(BA{:}\ \ldots)_{m}]_{m}\}_{m} \tag{13}$$

where the ellipses indicates an unending continuation of the process. To indicate an ending process one uses precisely the desired number of parentheses/brackets.
(2) The next level of complexity for "polymers" occurs when there is an infinite cross-linking between the various modules so that the mathematics model requires a higher dimension for its description. The simplest examples of this, even though they traditionally are not considered as "polymers", are two of the allotropes of carbon — graphite and diamond. Before naming these, however, the focus is returned to the above section (1) and a third and fourth allotrope of carbon are examined. The simplest theoretically-possible example of a one-dimensional polymer is the repetitive cumulene. The systemic name for this allotrope is:

$$({}^{2}/{}_{2}C^{2}/{}_{2})_{\infty} \tag{14}$$

Also, in this category of a mathematical ideal carbon compound that tessellates the one dimensional space is the infinite acetylene, which might be named as either*:

$$({}^{3}/_{2}\mathrm{C}1\mathrm{C}{}^{3}/_{2})_{\infty} \qquad \text{or} \qquad ({}^{1}/_{2}\mathrm{C}3\mathrm{C}{}^{1}/_{2})_{\infty} \tag{15}$$

However, such is not physical reality. To the contrary, Lagow et al. [14] showed in the laboratory that long repetitive acetylenes and cumulenes tend to curl and form rings. Consequently, unless future studies reestablishes linearity or some other "mathematically-heuristic" form, such as an infinitely long helix formation, the systemic names (14) and (15) will have to be discarded, when better chemistry is available.

Next, in the development of the concept of a mathematically ideal polymer, one examines graphite. Here each carbon atom is surrounded by three coplanar β bonds — with bond order exactly 4/3, which would have systemic name:

$$[({}^{\beta}/_{2}\mathrm{C}{}^{\beta}/_{2}):^{(2)}({}^{\beta}/_{2})]_{\infty} \tag{16}$$

An interesting feature of this formulation, in particular the use of the infinity subscript, is here noted; namely, the unending concatenation of half of two beta bonds to form a single beta bond in the principal chain is supplemented by half of a beta bond at locant #2 – the carbon atom. This half beta bond is now joined by the beginning half bond of the next module thereby extending the infinite bonding pattern into the second dimension. Because the heuristic concept of straightness is neither wanted nor implied, the traditional 120° angle is readily accommodated, along with the tessellation of the plane. Likewise, diamond has each carbon surrounded by four single bonds in a three dimensional embedding space, with systemic name:

$$[({}^{1}/_{2}\mathrm{C}{}^{1}/_{2}):^{(2,2)}({}^{1}/_{2})]_{\infty} \tag{17}$$

At this point a major omission in (16) is noted; namely, such a formula implies that each of the layers of graphite is completely independent of its neighbors. To the contrary, one finds that there is a correlation

---

* The first of these is the canonical name when this allotrope is considered alone; however, because an infinite number of carbon allotropes can be formed by replacing congruently placed single bonds in graphite or diamond with acetylenic linkages of the second form, this name is also important.

between layers, which are separated from one another by from 335 to 344 nm [15] — which is precisely what gives graphite its traditional feel; namely the ability of layers to slide one over the next. This *could* be accommodated in the nomenclature by the inclusion of alpha bonds between carbon atoms in different layers; namely, if one were to have such an alpha bond at every one of the carbon atoms, the systemic name for graphite would then be:

$$[({}^{\beta}/_{2}C^{\beta}/_{2}):^{(2)}({}^{\beta}/_{2});^{(2)}({}^{\alpha}/_{2})]_{\infty} \qquad (18)$$

However, because of the lability of these bonds and also the fact that the length of this alpha bond is so much longer than even the weak alpha bonds (290 nm) in tetracyanoethene (see Figure 6 in Chapter 2), it is probably overkill to include a alpha bond for every one of the carbon atoms. To the contrary, use of (16) with an accompanying note (when extreme precision is desired) is the best that nomenclature can supply without micromanaging and thus making the nomenclature unusable. On the other hand, recent studies of graphite under 17 gigapascals of pressure [16] reveals a molecular arrangement in which half of the carbon-carbon pi bonds in a layer break and the resulting half-bonds align with carbon atoms directly above and below them forming full long sigma bonds between carbon atoms in adjacent layers. If one were to now assume a uniform alternation above and below the plane of a single graphite layer, akin to an isotacticity*, this would require a repeat of four modules in order to have the repeating units properly aligned and thus would be nomenclated as:

$$[({}^{\beta}/_{2}C^{\beta}/_{2})_{4}:^{(2,4,6,8)}({}^{\beta}/_{2});^{(2,6)}({}^{\alpha}/_{2});^{(4,8)}(0)]_{\infty} \qquad (19)$$

Whether this is an accurate assumption, or if the randomness of atacticity is a more realistic model, will require further studies.

Returning focus to the more traditional use of the term "polymer", Fox [18] examines what, in common parlance, is referred to as "regularly crosslinked" vs. "randomly crosslinked" polymers. As an example of a general two dimensional regularly crosslinked polymer, consider the polymer depicted in Figure 12. Although this polymer has fourteen

---

*An isotactic polymer is defined as: "A regular polymer whose molecules can be described by only one species of configurational base units in a single sequential arrangement."[17]

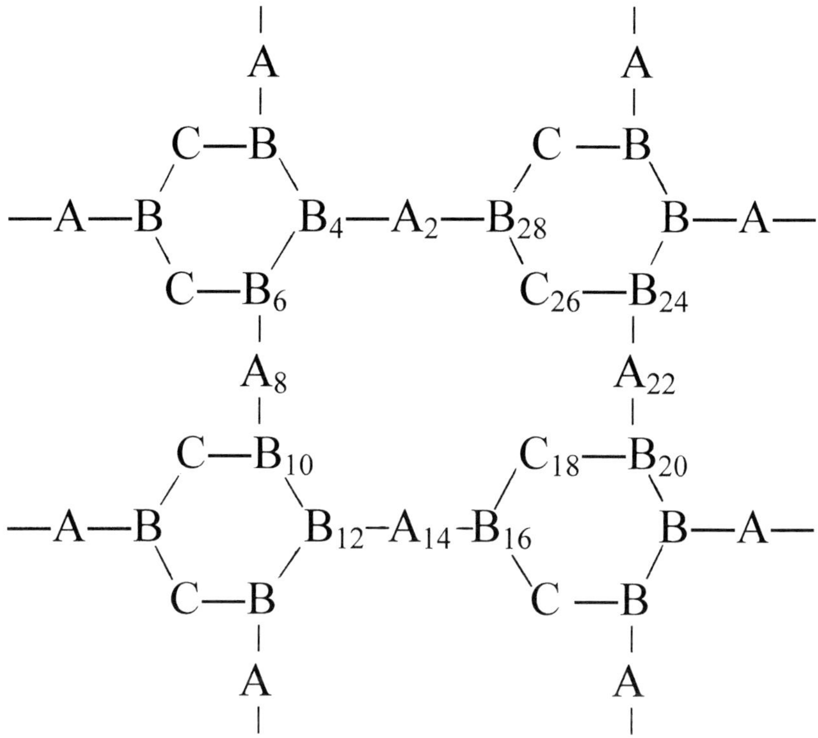

Fig. 12: A regularly crosslinked polymer and its largest cycle

members* as its largest cycle, such a cycle is not relevant in selecting the monomer, or the systemic canonical name. Instead, one can readily find a smaller congruent repeating module that tessellates the polymer. In fact, two different eight member aggregates are evident: Upon scrutinizing Figure 12, one notes that the combination of members B-A-B had been repeated twice (at both the top and bottom and at the left and right side of the picture); however, in the emergent polymer they should have been counted only once. Consequently, the fourteen member cycle is irrelevant, and one might delete the repeated members leaving an eight member linear repeating sequence as the monomer. Moreover, surrounding this monomer are eight congruent monomers which are now designated as **a** through **h** starting horizontally

---

*In this context the term "member" refers to congruent combinations of an atom followed by a bond. These may be designated alphabetically with the order of priority for canonically nomenclating the polymer being A > B > C, etc.

right and progressing counterclockwise, exactly as one assigns angle values in trigonometry. Meanwhile observe that Figure 13 is a deliberately distorted picture that has the nine blocks improperly aligned in order that each block contain a single complete monomer, rather than parts from two different monomers. Using such a picture the name of this polymer would be:

$$[{}^{1}/_{2}C1B1A1B1B1A1B1C1{}^{1}/_{2}{:}^{(2\text{-}14f,4\text{-}10g,8\text{-}16e,10\text{-}4c,14\text{-}2b,16\text{-}8a)}({}^{1}/_{2})]_{n} \qquad (19)$$

However, because there is another monomer which also covers all of the bonds and edges as well as tessellating the plane; namely, the six-member repeating ring with two additional members illustrated in Figure 14, both of these names must be examined before selecting which one is to be

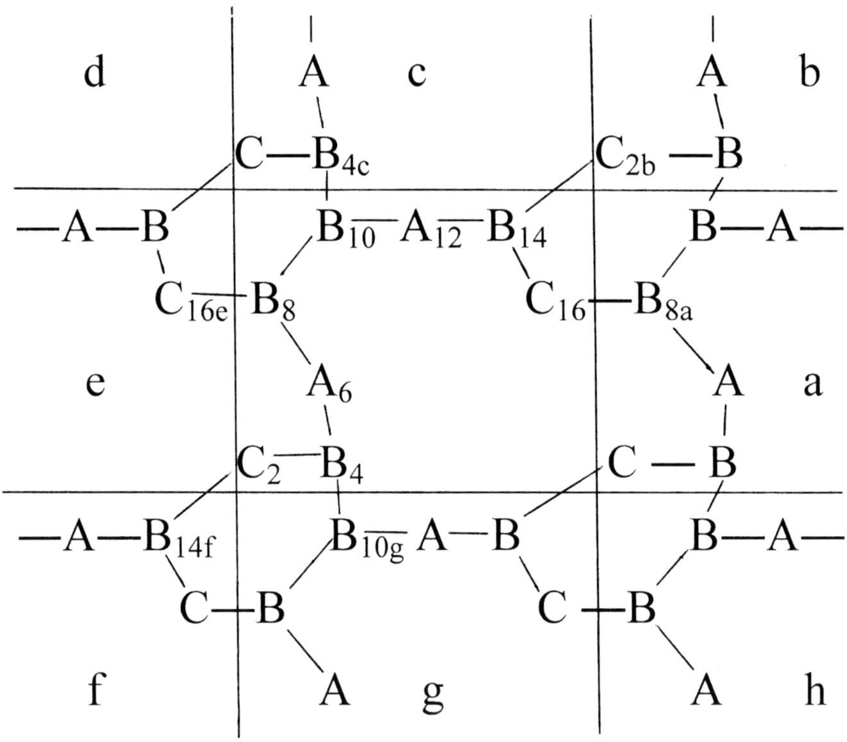

Fig. 13: A picture which emphasizes the designated linear monomer and the divisions of the tessellated plane for the previous polymer

designated as the canonical name. Using the model of Figure 14 and again surrounding the monomer with congruent sectors, one now obtains a cycle, which by the convention established in Chapter 1 has priority over a path. Consequently, the canonical name is:

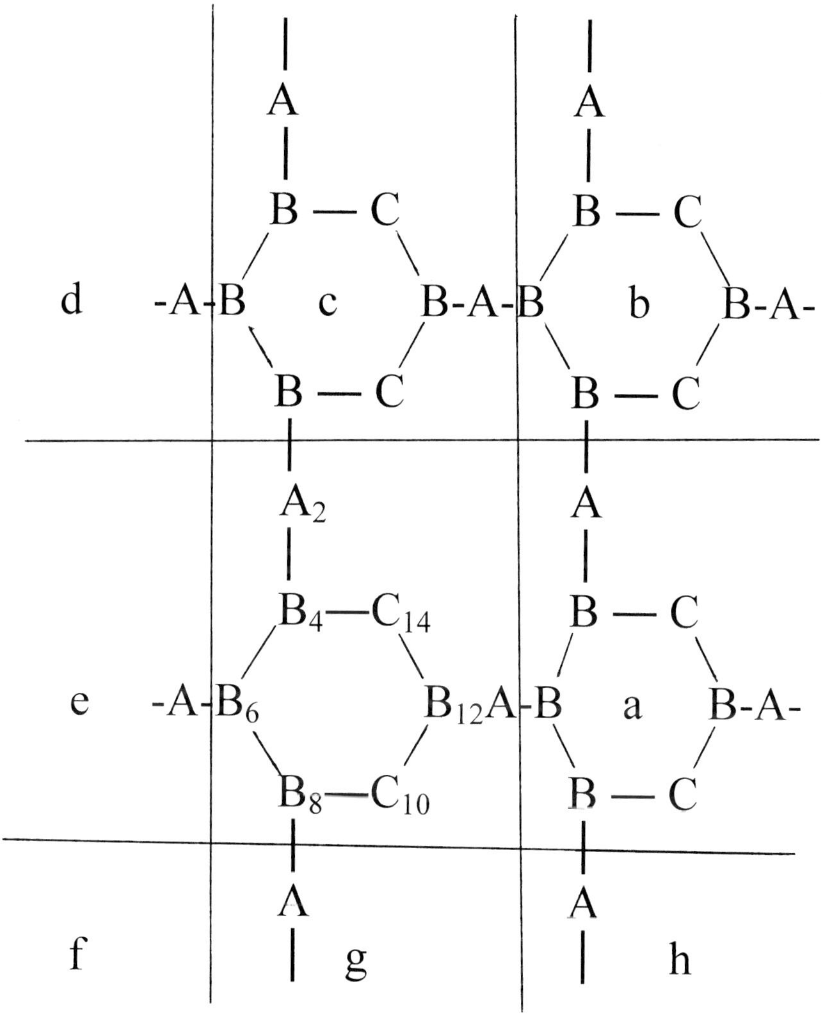

Fig. 14: A picture of the previous regularly crosslinked polymer which emphasizes the designated cyclic monomer, while maintaining the divisions of the tessellated plane. This model produces the canonical systemic name

$$\{{}^{1}/_{2}\mathrm{A1}[(\mathrm{B1})_{3}\mathrm{C1\ B1C1}:^{(6\text{-}12e)}(1\mathrm{A}^{1}/_{2});^{(2\text{-}8c,8\text{-}2g,12\text{-}2a)}({}^{1}/_{2})]\}_{n} \qquad (20)$$

Note that the class which Fox designates as "randomly crosslinked polymers" (see Figure 15) belong to what has been designated in [1] as "multimers". For multimers, there is no generalizable combination that is amenable to a standardized nomenclature.

Another item of importance in the nomenclating of polymers is that not only can polymers be formed by the concatenation of distinct monomers, but also these monomers may be joined in different orientations. This results in an expansion of the concepts inherent in DuPont's "structural repeating unit" (SRU) [19], as well as IUPAC's "constitutional repeating unit (CRU) [20], to one involving orientation that has been assigned the name

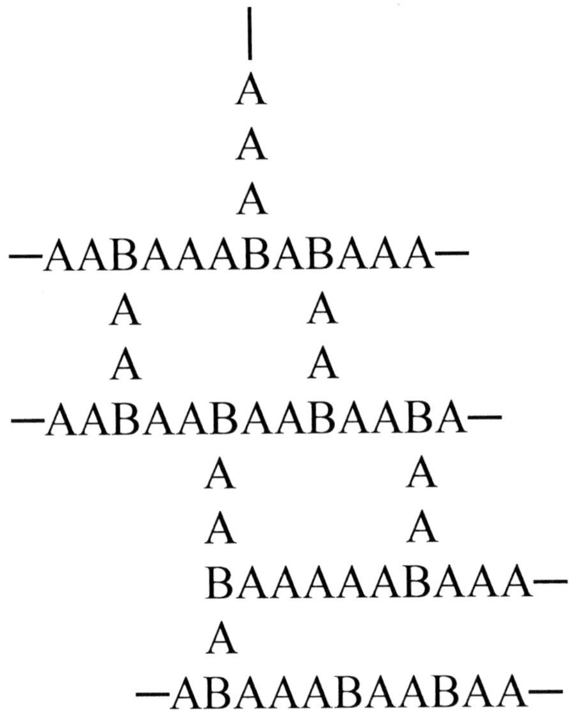

Fig. 15: A randomly crosslinked polymer

ORU (Oriented Repeating Unit).*

Before leaving the subject of nomenclating polymers, some additional comments on the geometry and, thus, the nomenclating of dendimers is in order. To begin, remember that a dendimer is an oligimer that expands radially, rather than in the linear manner that polymers do. Consequently,

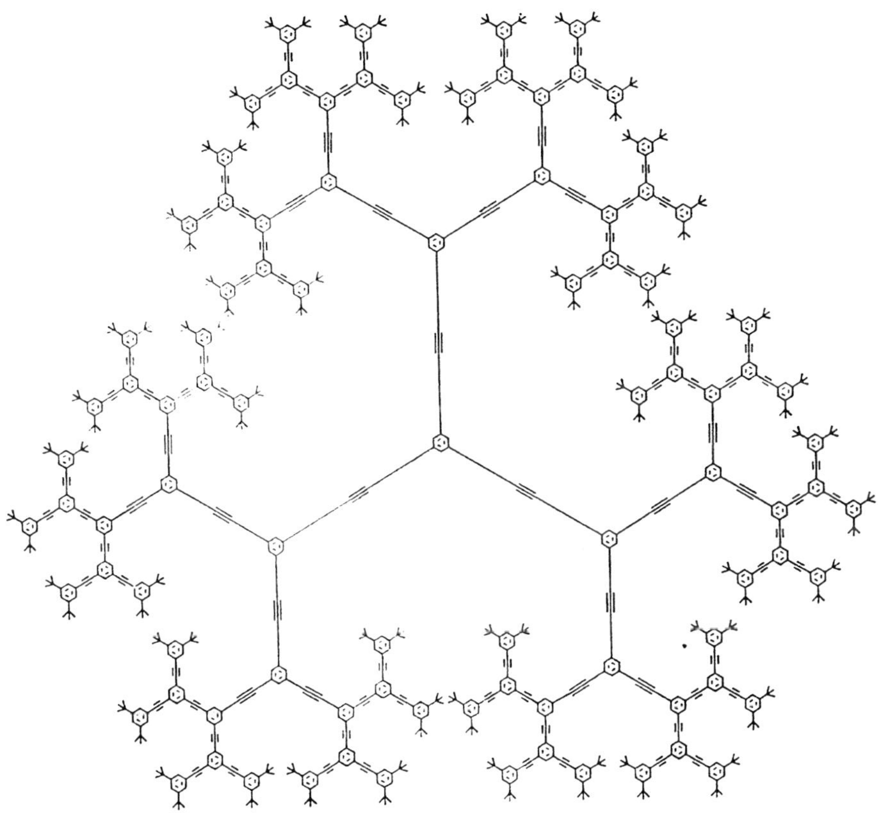

Fig. 16: The phenylacetylene 94-mer prepared by Moore

* The term ORU [21] was created in order that the repeating unit differentiates between a string of congruent vs. symmetric modules. In this way the term "syndiotactic" is no longer of importance, having been replaced by an isotacticity with double the length of the repeating unit. Similarly, "atactic" units will be relegated to the class of "multimers" - with the inherent added complexity of naming due to this lack of "regularity", rather than being described by the mathematically simpler class of "polymers".

the same advantages that accrue to describing a geometric figure using polar, in contrast to Cartesian, coordinates will accrue. Unlike polymers, in which a reaction seemed to continue unabated creating very long chains, often by a continuing free radical reaction, dendrimers are formed by often tedious planned mechanisms in which adding each successive "layer" becomes more difficult than the previous one. Both hydrocarbon and hetero-atom dendrimers have been formed. The first of these [22], with molecular formula $C_{1134}H_{1146}$, consists of 94 phenylacetylene monomers is illustrated in Figure 16. Its common name is 94-mer. The Cartesian systemic canonical name for the monomer of this dendrimer is:

$$(^1/_2)\mathrm{Ph}{:}^{(5,9)}1\mathrm{C}3\mathrm{C}(^1/_2) \tag{21}$$

The polar form of the name of this monomer in this particular case is nearly the same as the rectangular form; however, it has the advantage that it emphasizes the radial symmetry associated with dendimers; viz.:

$$\mathrm{Ph}{:}^{(1,5,9)}1\mathrm{C}3\mathrm{C}(^1/_2) \tag{22}$$

A "second stage" oligimer, which has a similar monomer at each of the three bonds indicated with a $(^1/_2)$ in (22), in common parlance is referred to as a "4-mer". Its systemic polar canonical name is:

$$\mathrm{Ph}{:}[^{(1,5,9)}1\mathrm{C}3\mathrm{C}(^1/_2){:}(^1/_2)\mathrm{Ph}{:}^{(1,5,9)}1\mathrm{C}3\mathrm{C}(^1/_2)] \tag{23}$$

Similarly, the third stage oligimer, referred to as a 10-mer, again repeats the monomer following a colon; namely:

$$\mathrm{Ph}{:}\{^{(1,5,9)}1\mathrm{C}3\mathrm{C}(^1/_2){:}[(^1/_2)\mathrm{Ph}{:}^{(1,5,9)}1\mathrm{C}3\mathrm{C}(^1/_2){:}\ (^1/_2)\mathrm{Ph}{:}^{(1,5,9)}1\mathrm{C}3\mathrm{C}(^1/_2)]\} \tag{24}$$

At this point an abbreviation is introduced; namely:

$$\mathrm{Ph}{:}\{^{(1,5,9)}1\mathrm{C}3\mathrm{C}(^1/_2){:}\ \ldots n \ldots{:}\ (^1/_2)\mathrm{Ph}{:}^{(1,5,9)}1\mathrm{C}3\mathrm{C}(^1/_2)\ \} \tag{25}$$

Here the n indicates that this radial repetition occurs n times. Consequently the fourth stage dendrimer is a 31-mer, while the molecule pictured in Figure 16 with n=5 is a 94-mer.

A similar early example of another fifth generation dendrimer that is formed from the reaction of 1,4-diaminobutane, acrylonitrile, a Raney cobalt catalyst and hydrogen is illustrated in Figure 17. Its systemic canonical name is:

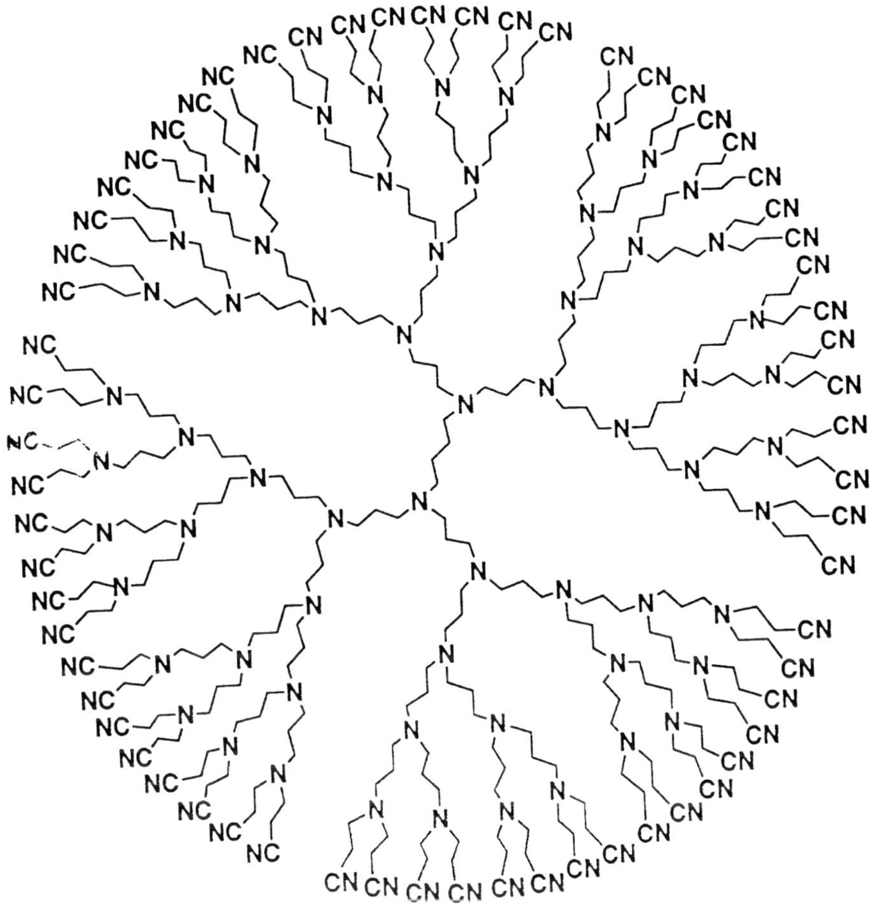

Fig. 17: A fifth generation poly(propylenimine) dendrimer

$$\mathrm{N}{:}\{^{(1,5,9)}1(\underline{\underline{\mathrm{C}}}1)_4(^1/_2){:}\ \ldots\ 5\ \ldots\ (^1/_2)\ \mathrm{N}{:}\{^{(1,5,9)}1(\underline{\underline{\mathrm{C}}}1)_4(^1/_2) \qquad (26)$$

Many additional dendrimers have been introduced in the past decade; however, no additional nomenclature problems are, at present, evident.

REFERENCES:

[1] S.B. Elk, THEOCHEM 589 (2002) 27.

[2] S.B. Elk, MATCH, 8 (1980) 121.
[3] W.V. Metanomski, International Union of Pure and Applied Chemistry: 1990 Compendium of Macromolecular Nomenclature, Blackwell Scientific Publisher, Oxford, U.K., (Introduction to Macromolecular Nomenclature, 1974), p.6.
[3] Ibid.
[4] Macromolecules, Vol. 1, Issue 3, 1968, American Chemical Society, p.193.
[5] W. Prenowitz and M. Jordan, Basic Concepts of Geometry, Blaidell Publishing Co., Waltham, Mass., 1965, p.3.
[6] H. Staudinger, Chem.Ber. 57 (1924) 1203.
[7] Ibid #5, p.194.
[8] H.S.M. Coxeter, Introduction to Geometry, Wiley, New York, 1961, p. 52.
[9] Ibid # 2.
[10] O. Ore, Number Theory and Its History, McGraw-Hill, New York, 1948, pp.209-233.
[11] Ibid #5, p.197.
[12] Ibid #3, p.15.
[13] Ibid, p.18.
[14] R.J. Lagow, J.J Kampa, H-C Wei, S.L. Battle, J.W. Genge, D.A. Laude, C.A. Harper, R. Bau, R.C. Stevens, J.F. Haw and E. Munson, Science, 267 (1995) 362.
[15] A.F. Wells, Structural Inorganic Chemistry, 3. Ed., Oxford Univ. Press, London, 1962, 709.
[16] W.L. Mao, Science 302 (2003) 425.
[17] Ibid #3, p. 16.
[18] Fox, R.B., J.Chem.Doc., 7 (1967) 74.
[19] J.A. Patterson, J.L. Schultz and E.S. Wilks, J.Chem.Inf.Comput.Sci., 35 (1995) 8.
[20] Ibid #3, p.13.
[21] Ibid #1.
[22] J.S. Moore and Z. Xu, Angew.Chem.Int.Ed.Engl. 32 (1993) 246.
[23] E.M. deBrabander and E.W. Meijer, Angew.Chem.Int.Ed.Engl. 32 (1993) 1308

Chapter 9

Molecular Rearrangement

CHAPTER ABSTRACT:
Transition states may be viewed as possessing alpha bonds to both an "entering" and a "leaving" group simultaneously. Such a state may now be described by a unique canonical name. Similarly, tautomeric equilibria are amenable to description in terms of a "ring" possessing alpha and beta bonds. This ring may be regarded as a single moiety to be canonically named, rather than separating "enol" from "keto" (as well as the "imine" from "enamine" and "oxime" from "nitroso") forms. The logical extension of those tautomerisms which describe the movement of an atom, usually hydrogen, "sliding" over a long distance in lieu of a diradical is also readily treated in this nomenclature by the addition of phantom locant descriptors.

As indicated in Chapter 1, the constitution of a given molecule, monomer, or ion is the foundation on which the entire concept of chemical nomenclature has been based. With this in mind, there is a convenient representation of such moieties using a connectivity matrix*. As an example, compare the antiquated perspective of the benzene molecule as 1,3,5-cyclohexatriene (Table 1) with the chemically more accurate matrix (Table 2) that has β bonds between adjacent carbon atoms. Table 3 is an abbreviation of Table 2 using the $\underline{C}$ designation.

Because of the introduction of α bonds, one may consider a transition state as possessing such bonds and thus assign nomenclature that reflects this temporary state. For example, consider the S_N2 reaction of a hydroxide ion attacking bromomethane: $OH^- + CH_3Br \rightarrow Br^- + CH_3OH$
In the transition state, the carbon atom has five bonds — three traditional single bonds to each of the hydrogen atoms as well as alpha bonds to the leaving Br atom and the income O of the hydroxyl group. The connectivity matrix (Table 4), although requiring the segregation of the constitutionally

* By a "connectivity matrix" is meant a square array in which the n individual atoms are listed across the top row and down the left side forming an n x n table. The entries in this symmetric table are the bond orders between the indicated pairs of atoms.

Table 1
Bond incidence matrix for benzene using traditional single and double bonds

	C_1	H_1	C_2	H_2	C_3	H_3	C_4	H_4	C_5	H_5	C_6	H_6
C_1	-	1	2	0	0	0	0	0	0	0	1	0
H_1	1	-	0	0	0	0	0	0	0	0	0	0
C_2	2	0	-	1	1	0	0	0	0	0	0	0
H_2	0	0	1	-	0	0	0	0	0	0	0	0
C_3	0	0	1	0	-	1	2	0	0	0	0	0
H_3	0	0	0	0	1	-	0	0	0	0	0	0
C_4	0	0	0	0	2	0	-	1	1	0	0	0
H_4	0	0	0	0	0	0	1	-	0	0	0	0
C_5	0	0	0	0	0	0	1	0	-	1	2	0
H_5	0	0	0	0	0	0	0	0	1	-	0	0
C_6	1	0	0	0	0	0	0	0	2	0	-	1
H_6	0	0	0	0	0	0	0	0	0	0	1	-

Table 2
Bond incidence matrix for benzene using beta bonds

	C_1	H_1	C_2	H_2	C_3	H_3	C_4	H_4	C_5	H_5	C_6	H_6
C_1	-	1	β	0	0	0	0	0	0	0	β	0
H_1	1	-	0	0	0	0	0	0	0	0	0	0
C_2	β	0	-	1	β	0	0	0	0	0	0	0
H_2	0	0	1	-	0	0	0	0	0	0	0	0
C_3	0	0	β	0	-	1	β	0	0	0	0	0
H_3	0	0	0	0	1	-	0	0	0	0	0	0
C_4	0	0	0	0	β	0	-	1	β	0	0	0
H_4	0	0	0	0	0	0	1	-	0	0	0	0
C_5	0	0	0	0	0	0	β	0	-	1	β	0
H_5	0	0	0	0	0	0	0	0	1	-	0	0
C_6	β	0	0	0	0	0	0	0	β	0	-	1
H_6	0	0	0	0	0	0	0	0	0	0	1	-

different hydrogen atoms needs only one entire for all constitutional equivalent atoms (see Figure 1). The systemic name* for this transition state is:

$$Br\alpha\underline{\underline{C}}\alpha O1H_a{:}^{3}1H_b \qquad (1)$$

* The use of subscripted letters was selected to distinguish constitutionally different hydrogen atoms in (1). This is different from the earlier use of subscripted numbers, which indicates the repeat of the preceding atom the number of times stated by the subscript.

Table 3:
Modularized bond incidence matrix for benzene using beta bonds

	$\underline{C}_1$	$\underline{C}_2$	$\underline{C}_3$	$\underline{C}_4$	$\underline{C}_5$	$\underline{C}_6$
$\underline{C}_1$	-	β	0	0	0	β
$\underline{C}_2$	β	-	β	0	0	0
$\underline{C}_3$	0	β	-	β	0	0
$\underline{C}_4$	0	0	β	-	β	0
$\underline{C}_5$	0	0	0	β	-	β
$\underline{C}_6$	β	0	0	0	β	-

Table 4:
Connectivity matrix for transition state of hydroxide ion attacking bromomethane

	Br	C	O	H_a	H_b
Br	-	α	0	0	0
C	α	-	α	0	1
O	0	α	-	1	0
H_a	0	0	1	-	0
H_b	0	1	0	0	-

At this point it is important to reiterate that the term "transition state" usually implies the existence of two distinct molecules in a "before" vs. "after" configuration, rather than mixed percentages of each over a period of time; consequently, although the nomenclature protocol for the transition state is exactly the same as for any stable molecule, when formulating a system of nomenclature, attention is normally restricted to stable aggregations (molecules, ions, or monomers). This is equivalent to asserting that, because there are no new principles involved, from a pragmatic perspective, the naming of transition states is of minimal importance. In fact, naming of a transition state shall be of significance **only** when the use of more traditional names are ambiguous, or even wrong. For instance, to name benzene as 1,3,5-cyclohexatriene would imply a wrong chemistry; however, to imply that the two main resonance structures of benzene were in a state of transition between each other, while inferior to other descriptions, has a modicum of merit, despite that it ignores other resonance structures,

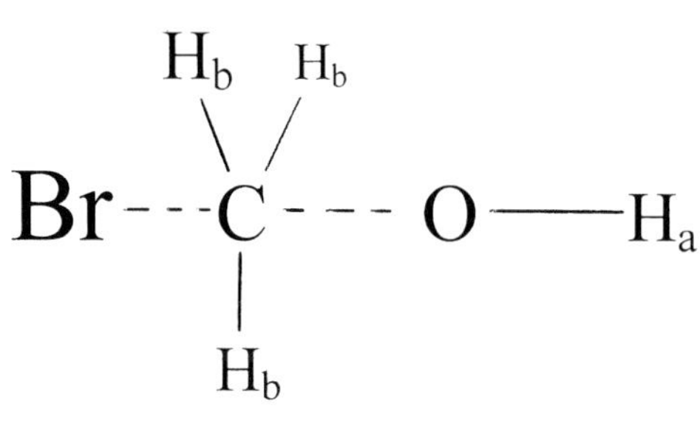

Fig. 1: A typical transition state

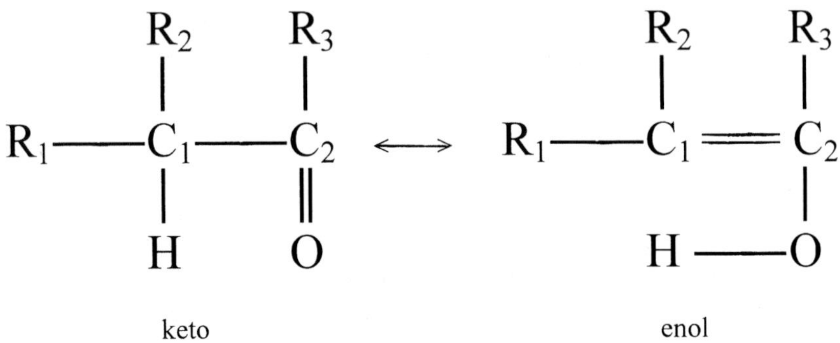

Fig. 2: Traditional representation of a keto-enol tautomerism

such as the Dewar benzenes. Note that because of the introduction of non-integer bonds such problems are not encountered.

As well as the ability to manipulate the non-zero terms in the connectivity table to indicate what has traditionally been described by resonance structures, the use of partial bonds and their place in the table also gives a convenient means of describing tautomerisms; namely, one or more of the selected non-zero terms in the matrix becomes zero while previously zero term(s) take on positive value(s). In keto-enol taumerism, for example, rather than considering that the principal path for nomenclating is the longest chain in either form of this acyclic aggregation of atoms, namely $R_1C_1C_2R_3$ in Figure 2, the locus of action is a four atom virtual ring (Figure 3) containing an oxygen, two carbons and a hydrogen atom, along with two

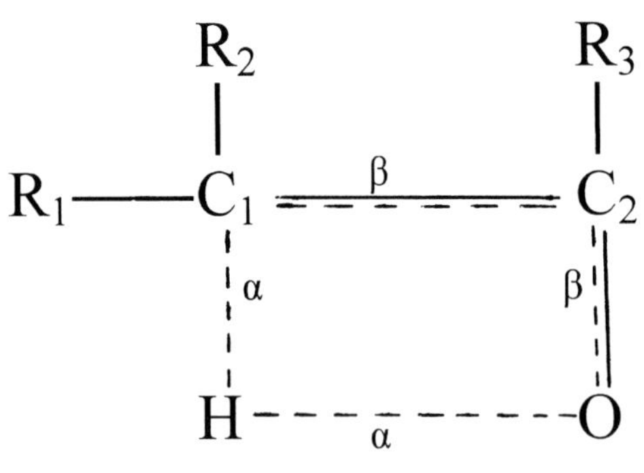

Fig. 3: Systemic form for nomenclating a keto-enol tautomerism

beta and two alpha bonds. Because rings of any size take precedence over longer chains in the proposed nomenclature, the name for this "compound" is:

$$O\beta C_2\beta C_1\alpha H\alpha{:}^{3}1R_3;^{5}1R_1;^{5}1R_2 \tag{2}$$

The only difference between this formulation and that of benzene is the historical view of benzene as a single molecule with two "main" overlapping resonance structures that were sufficiently similar that they could be coalesced into a single mental construct. On the other hand, the chemical difference between an alcohol and a ketone make such a viewpoint seem strange, if not completely untenable. Only because of the introduction of an extended bond set does such a perspective become viable.

The virtue of (2) is that, by considering these four atoms as a module, in exactly the same manner as had been done for the six bonds between neighboring carbon atoms in benzene, one may describe the chemistry of this aggregation as one which functions simultaneously in both keto and enol forms. Such a perspective differs from the traditional picture of a transition state in that no unusual coordination is indicated and that the height of the energy hill in a free-energy diagram is very small. It is, therefore, pragmatic to use a **BOTH**, rather than an *either/or*, form of the molecule to be named whenever considering "tautomeric" molecules.

Similar to the tautomers of the keto-enol form, one also encounters imine-enamine tautomerisms wherein a nitrogen atom replaces the oxygen atom (Figure 4) and also oxime-nitroso tautomers where a nitrogen atom replaces a carbon atom (Figure 5). Each of these is analogous to the keto-enol tautomerism in that a four member ring containing two alpha and two beta bonds is the action pathway. The only difference in the nomenclature is

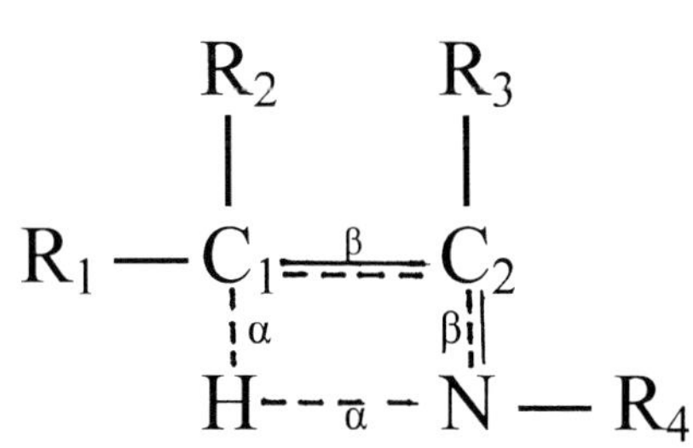

Fig. 4: Systemic formula for imine-enamine tautomerism

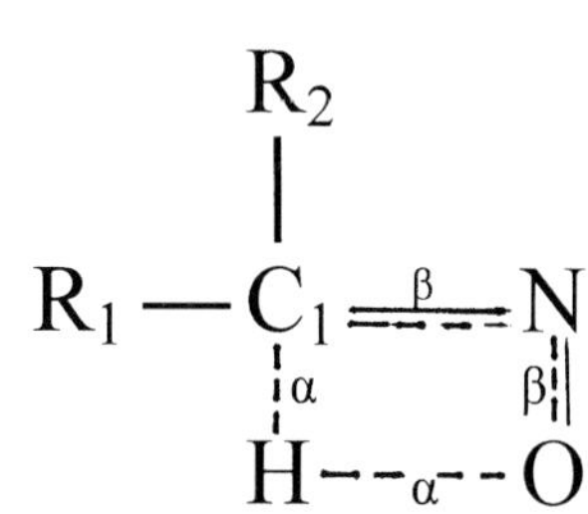

Fig. 5: Systemic formula for oxime-nitroso tautomerism

the number of R groups involved: For imine-enamines, the name is:

$$N\beta C_2\beta C_1\alpha H\alpha{:}^1R_4{}^31R_3;{}^51R_1;{}^51R_2 \quad (3)$$

while for oxime-nitroso tautomers, it is:

$$O\beta N\beta C\alpha H\alpha{:}^51R_1;{}^51R_2 \quad (4)$$

With this focus on virtual rings, one may employ the protocol endemic to IUPAC nomenclature and represent each such aggregation by a special symbol, such as δ. Moreover, because of the different types of tautomerisms, this symbol may be viewed as representing a function, rather than being just a single "scalar" symbol; namely, let the symbol δ describe any tautomeric virtual ring, which contains as its arguments the "anchor" atoms of the ring. In particular, let the symbol δ(C,O) denote $O\beta C_2\beta C_1\alpha H\alpha$, with the added specification of locant numbers for the R groups as shown in Figure 3; namely: ${}^31R_3;{}^51R_1;{}^51R_2$. Consequently, these locant numbers are implied and (2) simplifies to:

$$\delta(C,O){:}1R_1,1R_2,1R_3 \quad (5)$$

In a similar manner, (3) and (4) become, respectively:

$$\delta(C,N){:}1R_1,1R_2,1R_3,1R_4 \quad (6)$$

and

$$\delta(N,O){:}1R_1,1R_2 \quad (7)$$

By this designation the four atom-four bond (three actual and one virtual) system should be regarded in much the same manner as the abbreviation PhH for the benzene ring (C_6H_6) in the metallocenes in Chapter 6. One difference, however, is that this functional symbol indicates not only that the pseudo-ring has two distinct forms (keto and enol) but also it canonically assigns locant numbers to the atoms, bonds and substituent R groups.

Furthermore, the nomenclature being developed is readily amenable to most of the various rearrangements that are familiar to chemists. Unlike tautomers, however, the two forms of such a rearrangement are usually in a "before" vs. "after" environment -- each of which is normally stable at its respective temperature (and pressure). Furthermore, they are not, in large measure, rapidly interchanging. For example, in the Claisen rearrangement (Figure 6), at low temperature the molecule is in the form of allyl phenyl ether [part (a) of Figure 6], which has as its systemic name*:

$$\mathrm{Ph1O1}\underline{\underline{\mathrm{C}}}\mathrm{1}\underline{\mathrm{C}}\mathrm{2}\underline{\mathrm{C}}\mathrm{1H} \qquad (8)$$

On the other hand, when this substance is heated above 200°C, the only product is o-allylphenol [part (c) of Figure 6], which has systemic name†:

$$[(\mathrm{C}\beta)_2\underline{\mathrm{C}}\beta]_4\mathrm{:}^{(1)}\mathrm{1O1H;}^{(3)}\mathrm{1}\underline{\underline{\mathrm{C}}}\mathrm{1}\underline{\mathrm{C}}\mathrm{2}\underline{\mathrm{C}}\mathrm{1H} \qquad (9)$$

Meanwhile note that part b is a typical unstable intermediate that is traditionally given in textbooks [1-2]. The systemic name for this intermediate‡ is:

$$(\underline{\mathrm{C}}\mathrm{2}\underline{\mathrm{C}}\mathrm{1})_2\mathrm{C1}\underline{\mathrm{C}}\mathrm{1:}^{(9)}\mathrm{(2O);}^{(11)}\mathrm{(1}\underline{\underline{\mathrm{C}}}\mathrm{1}\underline{\mathrm{C}}\mathrm{2}\underline{\mathrm{C}}\mathrm{1H)} \qquad (11)$$

Meanwhile, one could form a different intermediary in which the conjugation of the benzene ring is not disrupted (and thus which should be expected to be energetically more favorable); namely, consider the formation of four single electron (alpha) bonds in a tri-cyclic ring system, as well as momentarily extending the conjugation through the creation of beta bonds. This is illustrated in Fig. 7 and has as its systemic name:

$$\mathrm{O1C}\beta(\underline{\mathrm{C}}\beta)_4\mathrm{C}\alpha\underline{\underline{\mathrm{C}}}\beta\underline{\mathrm{C}}\beta\underline{\underline{\mathrm{C}}}\alpha\mathrm{:}^{(1\text{-}13)}(\alpha\mathrm{H}\alpha)\mathrm{;}^{(3\text{-}13)}(\beta) \qquad (12)$$

Whether such a transition is viable has not been investigated.

These ideas are next extended to that logical extension of tautomerism created by an atom, usually hydrogen, "sliding" over a long distance in a

---

* Note that the use of the abbreviation Ph obviated the separation by a colon of the ring from the rest of the nomenclature for the side chain.

† The logical extension of systemic name from the oPh, mPh and pPh abbreviations introduced in Chapter 2 to a more general functional abbreviation such as Ph(1,3) might be undertaken at this point; however, care must be taken to indicate that the arguments of this function follow the prescribed alternating atom bond sequence, rather than the traditional atom only sequence; i.e., Ph(1,3) is ortho, not meta. When such is done, the colon is a necessary part of the name and (9) can be written as:

$$\mathrm{Ph(1,3):}^{(1)}\mathrm{1O1H;}\ ^{(3)}\mathrm{1}\underline{\underline{\mathrm{C}}}\mathrm{1}\underline{\mathrm{C}}\mathrm{2}\underline{\mathrm{C}}\mathrm{1H} \qquad (10)$$

‡ Because of the break in conjugation, the Ph(a,b) abbreviation is not relevant.

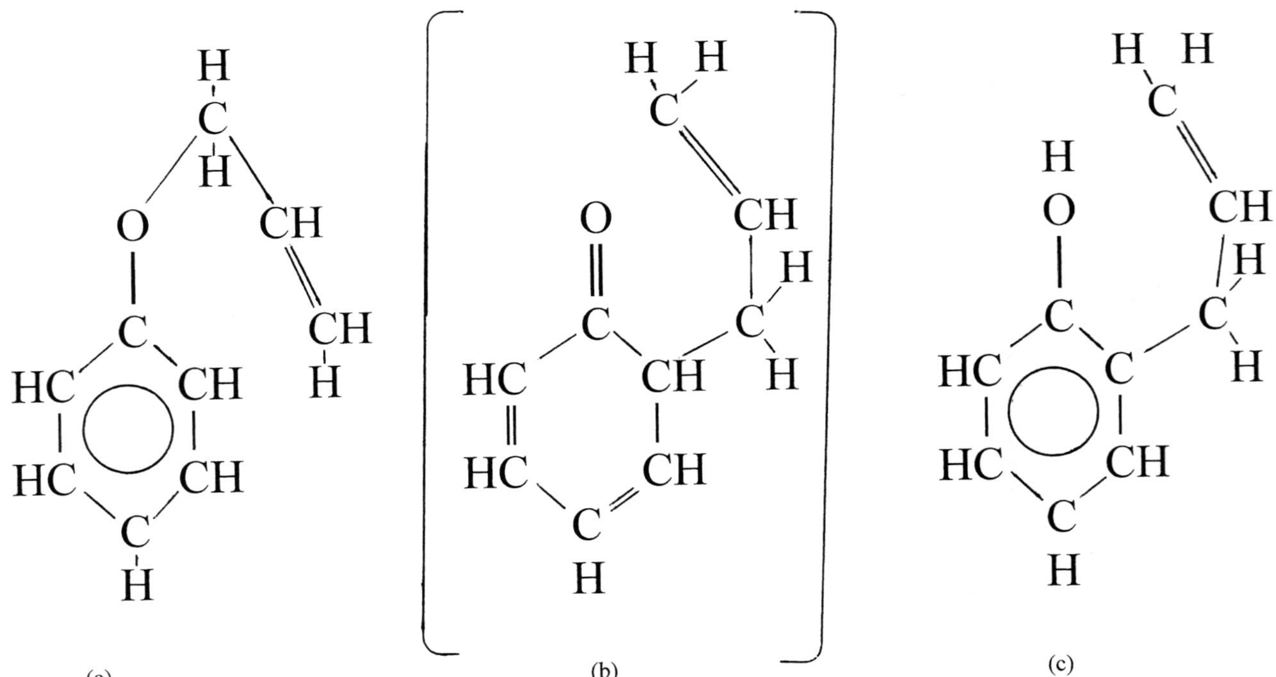

(a) (b) (c)

Fig. 6: A Claisen rearrangement

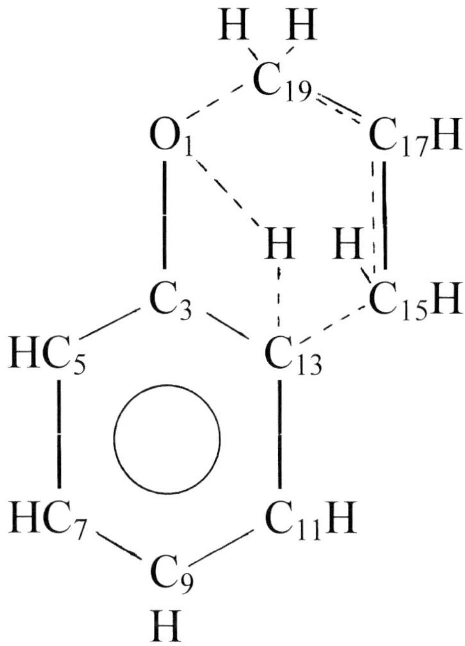

Fig. 7: Systemic intermediary for the above Claisen rearrangement

molecule. For this purpose let the focus be redirected to the twisted zethrene molecule illustrated in Figures 35 and 36 of Chapter 3. For this aggregation of atoms there is no way to allocate double and single bonds so that each carbon atom has a valence of four. One partial Kekule structure (Figure 8) has a conjugated perimeter; however, this leaves the two interior carbon atoms having three single bonds and an extra electron emanating from each; i.e., a diradical. Alternately, one can view this molecule as having 22 beta bonds and six single bonds. Now, in order to have a neutral molecule, consider that there is an additional atom, usually hydrogen, at each of these two vertices (Figure 9)*. Moreover, it should be noted that these various diradicals (and the neutral molecules having the two addended atoms) are always an even graph theoretical distance (GTD) apart from each other. For example, this perimeter conjugated molecule has GTD = 6 between the two addended hydrogen atoms, and thus would be named in the proposed

* The partial circle along with the normal edge indicates a β bond at the respective edges; namely, 2,4,....

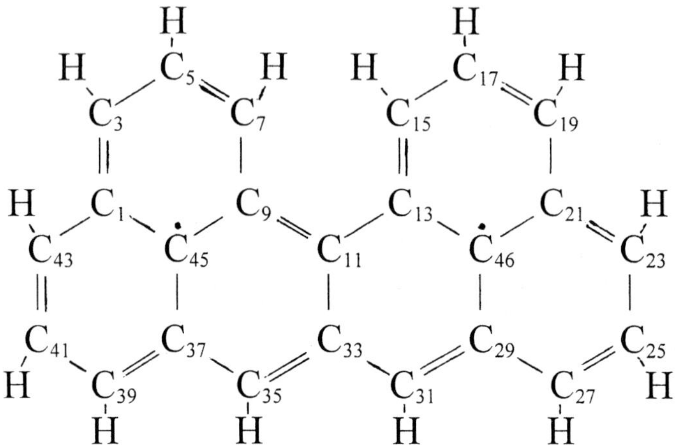

Fig. 8: A partial Kekule structure for "twisted zetherene"

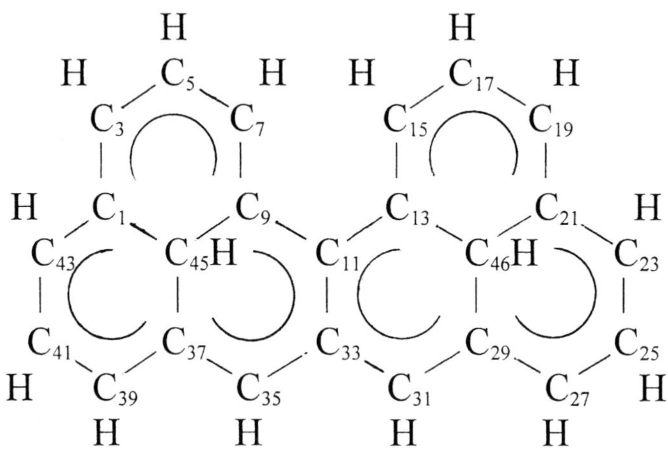

Fig. 9: The beta-bonded structure with added hydrogen atoms for "twisted zetherene"

system as*:

---

* This is one of the few names in which the underscored hydrogen abbreviation produces heuristic inferiority; namely:

$$C\beta(\underline{C}\beta)_3(C\beta)_3(\underline{C}\beta)_3C\beta(\underline{C}\beta)_3(C\beta\underline{C}\beta)_2C\beta(\underline{C}\beta)_3{:}^{(1\text{-}9)}(1C^{(=45)}1;^{(13\text{-}21)}(1C^{(=46)}1;^{(11\text{-}33)}\beta;\ _{(29\text{-}46,37\text{-}45)}(1);^{(45,46)}(1H) \qquad (14)$$

$$(C\beta)_{22}{:}^{(1\text{-}9)}(1C^{(=45)}1);^{(13\text{-}21)}(1C^{(=46)}1);^{(11\text{-}33)}(\beta);^{(29\text{-}46,37\text{-}45)}(1);\;^{(3,5,7,15,17,19,23,25,27,31,35,39,41,43,45,46)}(1H) \quad (13)$$

In a similar manner, diradicals, and hydrogen addended molecules with GTD=4, 8 and 10 are easily formed — each having a name similar to (13). Additionally, no diradicals or hydrogen addended molecules will have the GTD equal to any odd integer.* This property may now be used to find a general formula for such an extension of tautomerism. Next, observe that the lowest locant numbers for the bridges is achieved using the locant numbering illustrated in Figure 9 with the extra hydrogen atoms being attached to carbon atoms having locant numbers 3 and 15, instead of 45 and 46. This produces as the systemic name:

$$(C\beta)_{22}{:}^{(1\text{-}9)}(1C^{(=45)}1);^{(13\text{-}21)}(1C^{(=46)}1);^{(11\text{-}33)}(\beta);^{(29\text{-}46,37\text{-}45)}(1);\;^{(3,3,5,7,15,15,17,19,23,25,27,31,35,39,41,43)}(1H) \quad (14)$$

a name that differs from (13) only in the location of the floating hydrogen atoms. Consequently, if one names the phantom locants as **a** and **b**, a general formula will be:

$$(C\beta)_{22}{:}^{(1\text{-}9)}(1C^{(=45)}1);^{(13\text{-}21)}(1C^{(=46)}1);^{(11\text{-}33)}(\beta);^{(29\text{-}46,37\text{-}45)}(1);\;^{(3,5,7,15,17,19,23,25,27,31,35,39,41,43,a,b)}(1H) \quad (15)$$

where a will be a repeat of one of the integers between 3 and 23 and b a repeat between 15 and 39.

REFERENCES:

[1] T.W.G. Solomons, Organic Chemistry, 5-th Ed., New York 1992, 949.

[2] A. Streitwieser and C.H. Heathcock, Introduction to Organic Chemistry, 2-nd Ed., Macmillan, New York, 1981, 1012.

[3] E. Clar, The Aromatic Sextet, Wm. Clowes & Sons, Ltd., London, 1972, 103.

Additionally, one is unable to use the functional Ph symbol to advantage due to the fusing of benzene rings, rather than having them connected in ring assembly fashion.

* If this GTD were to be odd, the bonds may be rearranged so that the two free electrons were adjacent and together would form a single bond.

INDEX* (bold face font page numbers indicate systemic definitions)

abbreviations 108
aceanthrylene 77,87
acenaphthylene 75,87
acephenanthrylene 77,87
acetylenic chain extenders 232
acetylides 174
acyclic polyenynes 40
adamantane 126-128
adamanzane 128
addenda **14**
addition polymer 273
adjacent integer bond orders 51
affix 44
alchemy 2
aleph (א) bond **53**,219-225
aleph subscript 219
aleph subscript superscript 220
algebraic language 2
algorithm **9**
aliphatic compounds 117
alkali oxides 168-169
alkanecarboxylic acid 57
allotrope 282
"almost infinite" 269
alpha (α) bond **52**
alphabet **16**
alphabetical ordering 31,125
aluminum dimers/trimers 199-204,214-215
analytic nomenclature 32,118
anchor atoms 135,156
anthracene 23-24,26,76
anti-aromatic 62
anti-bonding orbitals 50
arachno 184
arene **82**,118
aroma 8
aromatic 8, 21,62
aromaticity 53,54,79
atactic **284**,289
atomic bridge **22**
average 15
azide ion 170
azulene 73,82,86
balancing chemical equations 167
benz[a]anthracene 83
benz[d]anthracene 83
benzene 55,102
benzenoids **67**,160
benzo[c]phenanthrene 84
benzyne 98-99
beryllocene 216,223-225
Berzelius 2
bet (ב) bond 52,252-254
beta (β) bond **52**
bicolorable graph **17**
bicyclodecane 63-65
bigraph **17**
binary carbon compounds 174-177
biochemistry 98
biparametric 5
bipartite graph **17**,27,67
biphenyl 251-252
biphenylene 74,251-252
bis-iron complexes 103-105,213-214,236
blank space 41
block **207**
bond 50
bond angle 50
bond length **50**
bond incidence matrix 294
bond order 50,52
bond strength **50**
bonding orbitals 50
boranes 180-200
borides 176-177
boron bridge **181-183**,187,191
boron fluorides 175-176
boron hydrides 180-200
boundary-defined definition 117
boundary 150
bridge 22,**40**,118
bridgehead atoms 138
bridging groups 107
buckminsterfullerene 147-150,159
butadiene 56
butane conformers 15

* For terms that are used frequently, only a few selected page references are included

butyl cyclopropyl malonate 41-42
butyl cyclopropylmalonate 42
butylcyclopropyl malonate 42-43
butyllithium 18
cage 145
calixarenes 232-235
cannogeninic acid 248-249
canonical 9
carbon oxides 171-174
carboxylic acid 57-58
cardanolide 248
Cartesian coordinates 135
Cartesian nomenclature 135
CAS = Chemical Abstracts Services
CAS Registry 28
cata-condensed **67**
catena 54
catenanes 258-263
$\underline{C}_6$-compounds‡ 70-72
chain 29
chelation **4**,124-125
chemical bond 50
chrysene 78
Claisen rearrangement 299-301
closo 184
cluster name 118
colon 31
complete bigraph **17**
complete graph **16-17**
complexity **5**,282
condensation polymer 273
configuration **10**
conformation **10-11**
conformers 15
congruent modules 276
conjugated 13,51-**52**,56
connectivity matrix 67,293-295
connotation **7**
constitution **9**
constitutional isomer 11
constitutional repeating unit 288
constitutionally different atoms 294
content-defined definition 117
continuous variable 50
coordination number 124
Copernicus 123
co-polymer 282
corona-condensed **67**,159
cross-linking 282
crown ether 231
cryptands 135-138
CRU = constitutional repeating unit
C_6H_6 compounds† 70
cubane 26-28,159,199
cubanylcubane 250-251
cumulenes 32-33,56
cumulenic 13
cycle 4,**11**
cyclic catenane 264
cyclo 54
cyclobutane 150,199
cyclohexatriene 51
cyclooctatetraene-tricarbonyliron 104-106,235-237
cyclopentadienyl module 87,108,215-218
cyclopent[fg]acenaphthylene 84
cyclophanes 101-102,232
cyclopropane 19
cylindrical coordinate system 136
cylindrical molecules 210
cylindrical nomenclature 135,145
cylindrical symmetry 137,145,190
Dagwood sandwich 221-222
decimal subscripts 177-178
delta (Δ) prefix 111
dendritic molecules 211-213,282,288-290
denotation **7**
dentation 125
Dewar anthracene 23-24
Dewar benzene 23
diamantane 130
diamond 126-127,281,282
diazabenzenes 91
diborane 55,182-183
dictionary 16
diisobutylaluminum hydride 201,215

† alphabetized as "C six H six"
‡ alphabetized as "C bar six"

dimer **58**
dimethylaluminum chloride 201,214
dimethylbenzene 55-56
dimethyl ether 10
discrete variable 50
dodecahedrane 140,143,144
dodecahedrene 145
drawing techniques 25
Dyson nomenclature 37-38
effective ionic charge 170
electron-deficient bond 181
Elk-Matula numbers 121
ellipsis 213,282
embedding space 117
end groups 270
endothelial compounds 267
equilibration points 193
Escher 109
Escher-like lutetium compound 107-109
ESER=essential set of essential rings
ESER 123
eta (η) prefix 102
ethanol 10
ethyl alcohol 10
Eulerian **11**,117,119
Euler's Polyhedron Formula 28
Ewens-Bassett system 169
extended pi bond cloud 252
extremely long single bond 60
ferrocene 215-220
fiefdoms 6
fisular **22**,159
fixed single/double bonds 73,75,251
flu **7**
fluoranthene 76
fluorene 75,87,88
fluoropentane 21
fluxional corrections 186-190
font 31-32
formic acid 58
formulae vs. formulas 177
4-center-2-electron bond 60
fractional oxidation numbers 170,177
free radical mechanism 272
fullerenes 147-151
functional group **12**
fused 22
gamma (γ) bond 52-**54**
garnish 221,226
gear mechanism 224
General Rule of Orismology 8
geometrical isomers 63-65
geometrical picture 183
Gillespie 189
gimel (ג) bond 54
global perspective 131
Goedelian impasse 6,41
"good" nomenclature 6-7
google 269
graph theoretic descriptors 65
graph theoretical distance **11**,148
graph theory 3,16,30,40
graphite 282-284
Grignard compounds 23
GTD = graph theoretical distance
half sandwich 220-221
Hamiltonian **11**
Hamiltonian cycle 119,153,156
Hamiltonian path 19,153
Hantzsch-Widman system 104,111
hapto 102
Harary 40
-hedrane 145
-hedrene 145
-hedron 145
helicanes 25
helicenes 25
heliocentric system 123
helix 283
heme 100-101
heptalene 74,82
heuristic **9**
hexagonal grid 118
hexahedrane 140,142,144
homopolymer 282
hydrogen bridge **181**
hydrogen fluoride 62
hydrogen sodide 129-130
icosahedrane 140,143,144
icosahedron 157-158
ill-defined word 40
imine-enamine tautomerism 296-297

indacene 74,82
indene 73,87
infinite acetylene 283
infinite number of atoms 269
infinity 269
infinity subscript 274
influenza **7**
"inside" atoms - fulleranes 150-151
intercalation compounds 175,177
International Convention
of 1892 4
iron oxides 171
isobutane 19
isomer **9**,11,**58**
isomeric pseudoconversion 118
isotactic **283**
IUPAC = Int'l Union of Pure
& Applied Chemistry
IUPAC 4
IUPAC reference compounds 72-81,83-86
IUPAC trivial names 95-97
JEOL Ltd. 70
jump rope effect 135,138
$K_{3,3}$ 17,28
K_5 16-17,28
kappa (κ) affix 109
kappa convention 109-110
Kekule structure 301
keto-enol tautomerism 296-297
knots 264-267
Kuratowski 28
labile 275
ladder polymer 278-280
lambda (λ) affix 44
languages 16
lanthanide carbides 175
Lavosier 2
leBel 24
Lewis 50
Lewis structure 61
ligand **12**
linear Moebiane 131-133
linear molecules 210
local geometry 121
local perspective 131
locant **12**
locant number 12,116
"longest chain" 29-30
machine-readable number 40
macromolecule 271
magnesium boride 176
marked carbon atoms 162-163
Matula nomenclature 38-39,121
meta-connected groups 103,108
metallabenzyne 99-100
meta-phosphate ion 61-62
MD = metric distance
metallocarbohedrenes 151-154
metallocarboranes 153,237
metcars 151
methides 174
methylene group 126
metric descriptors 65
metric distance **11**
minimum set of bridges 118,121
minimum spanning set 28,159
modulo 278-279
Moebiane 132,158
Moebius strip 131
moiety **8**
molecular formula 11
molecular inversion mechanism 225
molecular rearrangement 293-303
molybdenum - half sandwich 106,225-226
molybdenum-sulfur compound 254-256
monocyclic compound 19
monodentate chelation 125
monomer 8,**58**,269
morpheme 9,13,29,39
mu (μ) affix 102-103
multimer 285,287
multiple ring assembly 243-244
multiple winding catenane 262
multiply-connected 159
multiply spiro 207
name **9**
naphthacene 78
naphthalene 73
natural products 247-249
n-butane 14,18
"nearly equal" 20
"nearly infinite" 269

"nearly similar" 20,209
negative bonding 146
negatively-defined **22**
nesting intervals 213,246
nido 184
nitroalkane 56-57
nomenclate **9**
nodal nomenclature 5,35,116,121
-nodane 40
node 30
non-adjacent integer bond orders 51
nonsense words **16**
nonterminal hydrogen atoms 34
"normal" acids 124
octahedrane 140,142,144
octahedron 196
oligimer 276,290-291
olympiadane 260
1-dimensional space 19
open path bridge method 243-244
optical isomerism 25
organoiridium lithium ion 229-231
organometallic chemistry 18,23
orientation 271
oriented repeating unit 288
orismology **8**
ortho-connected groups 103,108
ortho-fused 118
ortho-phosphate ion 61
orthogonal coordinate systems 208
ORU = oriented repeating unit
overlap compound 134-135
oxidation **168**
oxidation numbers 30,167-178
oxime-nitroso tautomerism 296-297
oxy acids 124
oxymoron 269
paddlane 134-136
palladium catalyst 241-243
para-connected groups 108
parameter 15
parent 12,**13**,248
Parent Compound Handbook 14,20
parent heteroatom compounds 92-95
path **11**
pathologic molecules 134
Patterson's rules 23-25,85,118
Pauling 50
pentacene 81
pentalene 73,82,85,87
pentane module fusions 25
pentaphene 80
peri-fused 118
peroxide 57
perylene 80,118-120
PET = polyethylene terephthalate
phantom bonds 186
phenalene 75.87
phenanthrene 76,88
phenyl 55
phosphate ion 61
phosphorus fluorides 196
phosphorus oxides 128-129
picene 79
planar Moebiane 158-159
Platonic solids 140-141,145
pleiadene 79,88
polar coordinates 282,290
polyacetylene 33
polybenzenes 118
polydentate chelate complexes 109
polyenynes 118
polyethylene 274
polyethylene terephthalate 276-277
polymer 8,**58**,269-291
polymethylene 274
positively-defined **22**
priming convention 109-110
primitive word 40
principal chain **31**
priority 61,83,93,271
propellane 133-134
pseudo-cylindrical 134
pseudo-organic name 102
pseudo-rotation 189,197
Ptolemaic geocentric system 123
Ptolemaic principle 178
pyrene 78
pyridazine 91,102
QSAR = quantitative structure
activity relationships 131
quadruple bond 12

quasi-linear 278
radial symmetry 290
randomly cross-linked polymer 284
rectangular nomenclature 135
reduction **168**
redundant path nomenclature 215,238-239,244
regular polyhedron 199
regular polymer 281,287
regularly cross-linked polymer 284
repulsion 145-146
resonance structure 294
reticular **22**,159
ring **4**,117
ring assembly 242-244,250
Ring Index 14
RNA 279
Robinson ring 23,51
rooted trees 38
rotaxanes 258-263
ruthenium carborane complex 237
ruthenium cluster 226-228
saline carbides 174
sandwich compounds 220-223
Schlegel projection 157
self-assembled dodecahedron 154-156
semicolon 31,258
signature **13**,56
"similar" bond orders 52
similar names 41-43,209-211
similarity 10
"simple" 117
simplex 126
simply-connected 159
singly spiro 207
skew hexagons 128
skew polygons 117
skew quadrilaterals 186
slip sandwich 223
smallest set of smallest rings 26,117,123
sorcerers 2
source based polymer 270
spanning cycle 117
spherical molecules 211
spherical nomenclature 60,208
spiro 22,**207**
square pyramid 188
SRU = structural repeating unit
SSSR = smallest set of smallest rings
star **16**
Staudinger 273
stem name 17
stereochemistry 25
stereoisomer **11**
stiochiometric 169
Stock system 169
straight 18,25,276
strained rings 22
structural formula 11
structural isomers 11,36,71-72
structural repeating unit 288
structure-based polymer 270
subscripts 34
sulfate ion 60-61
sulfur oxygen acids 175
supermolecular dodecahedron 154-156
superoxide ion 170
superscripts 31,34
symmetry 136,157,190
syndiotactic 289
synthetic nomenclature 32,118
tangent spheres 256
target polyhedra 200
tautomers 296-298
term **7**
terminal hydrogen atoms 34,116
terminology **8**
tertiary butyl 212
tessellation 118,276
tetrabenzenes 66-70
tetracarbonyl(η-1,5-cyclo-octadiene)molybdenum 106
tetracyanoethene dimer 59-60,282
tetrahedral carbon atom 24
tetrahedrane 140,141,144
tetraphenylene 81
thiosulfate ion 61
3-center-2-electron bonds 55,181
topological picture 183
topologically restrained compounds 258-268
trail **11**
transition state 293,295

trees 38
trefoil knot 264-267
triangular Moebiane 28
triangular prismane 28
triangulene 163-164
trigonal bipyramid 45,139,189-191
tri-iodide ion 170
trimesitylaluminum 201-202,214
trimethylaluminum 200,214
trimethylphenylaluminum 202
tri-μ-carbonyl-bis(tricarbonyliron) 103-104
triphenylaluminum 203,214
triphenylene 77
triple points 90,160
trisilaallene 252-254
twisted zethrene 161-162
typewriter 20
underscoring convention 34
uniparametric 5
unmarked carbon atoms 162-163
unnecessary affixes 111
untwisted Moebiane isomer 131-133,231-232
user-friendly nomenclature 39,40
valence 187
valence bridge **22**
van't Hoff 24
virtual ring 296-298
vocabulary 7
walk **11**
Walba 132
Wells 24,51
wheel 193,195,282
winding number 262
Wiswesser Line Notation 20
word **7**
word stem 40
zero bond order 133-134,258
0-dimensional space 18
zero superscript 258-263
zethrene 90